Springer Proceedings in Physics 10

Springer Proceedings in Physics

Managing Editor: H.K.V. Lotsch

Volume 1 *Fluctuations and Sensitivity in Nonequilibrium Systems*
Editors: W. Horsthemke and D.K. Kondepudi

Volume 2 *EXAFS and Near Edge Structure III*
Editors: K.O. Hodgson, B. Hedman, and J.E. Penner-Hahn

Volume 3 *Nonlinear Phenomena in Physics*
Editor: F. Claro

Volume 4 *Time-Resolved Vibrational Spectroscopy*
Editors: A. Laubereau and M. Stockburger

Volume 5 *Physics of Finely Divided Matter*
Editors: N. Boccara and M. Daoud

Volume 6 *Aerogels*
Editor: J. Fricke

Volume 7 *Nonlinear Optics: Materials and Devices*
Editors: C. Flytzanis and J.L. Oudar

Volume 8 *Optical Bistability III*
Editors: H.M. Gibbs, P. Mandel, N. Peyghambarian, and S.D. Smith

Volume 9 *Ion Formation from Organic Solids (IFOS III)*
Editor: A. Benninghoven

Volume 10 *Atomic Transport and Defects in Metals by Neutron Scattering*
Editors: C. Janot, W. Petry, D. Richter, and T. Springer

Springer Proceedings in Physics is a new series dedicated to the publication of conference proceedings. Each volume is produced on the basis of camera-ready manuscripts prepared by conference contributors. In this way, publication can be achieved very soon after the conference and costs are kept low; the quality of visual presentation is, nevertheless, very high. We believe that such a series is preferable to the method of publishing conference proceedings in journals, where the typesetting requires time and considerable expense, and results in a longer publication period. Springer Proceedings in Physics can be considered as a journal in every other way: it should be cited in publications of research papers as *Springer Proc. Phys.*, followed by the respective volume number, page number and year.

Atomic Transport and Defects in Metals by Neutron Scattering

Proceedings of an IFF-ILL Workshop
Jülich, Fed. Rep. of Germany, October 2–4, 1985

Editors: C. Janot, W. Petry, D. Richter,
and T. Springer

With 171 Figures

Springer-Verlag
Berlin Heidelberg New York Tokyo

Professor Christian Janot
Dr. Winfried Petry
Dr. Dieter Richter
Institut Laue-Langevin, Avenue des Martyrs, Centre de Tri 156 X,
F-38042 Grenoble Cedex, France

Professor Dr. Tasso Springer
Institut für Festkörperforschung, Kernforschungsanlage Jülich GmbH, Postfach 1913,
D-5170 Jülich, Fed. Rep. of Germany

ISBN-13: 978-3-642-71009-4 e-ISBN-13: 978-3-642-71007-0
DOI: 10.1007/978-3-642-71007-0

Library of Congress Cataloging-in-Publication Data. Atomic transport and defects in metals by neutron scattering. (Springer proceedings in physics; v. 10) "Jointly organised by the Institut Laue-Langevin and the Institut für Festkörperforschung of the KFA-Jülich" – CIP pref. Includes Index. 1. Physical metallurgy – Congresses. 2. Transport theory – Congresses. 3. Neutrons – Scattering. I. Janot, C. (Christian), 1936–. II. Institut Laue-Langevin. III. Kernforschungsanlage Jülich. Institut für Festkörperforschung. IV. Series. TN690.A8116 1986 669'.9 86-10146

Softcover reprint of the hardcover 1st edition 1986

2153/3150-543210

Preface

The Institut Max-von-Laue-Paul Langevin (ILL) in Grenoble regularly organises workshops that deal with the various applications of neutrons in physics, chemistry, biology and also in nuclear physics. The workshop "Atomic Transport and Defects in Metals by Neutron Scattering", jointly organised by the Institut Laue-Langevin and the Institut für Festkörperforschung of the KFA-Jülich, was held in October 1985 in Jülich.

The study of problems in metal physics and in physical metallurgy is a traditional field of neutron scattering. The most commonly used methods are diffuse elastic, small-angle and inelastic scattering of neutrons. A number of problems can be identified where neutrons yield information that is supplementary to that from other methods such as x-ray diffraction, synchrotron radiation or electron microscopy. In certain fields, for example spectroscopy for the investigation of atomic motions or for the investigation of magnetic properties, neutron scattering is a unique method.

The facilities at the High Flux Reactor of the ILL, and also at the Jülich and at other medium flux research reactors, have contributed numerous results in these fields. It was the aim of this workshop to give a survey of the present state of neutron scattering in metal physics.

The study of *short-range order* by diffuse scattering, presented in the first session, has achieved important progress thanks to the isotope substitution method. The measured spectra are sufficiently precise that by means of Monte Carlo simulations with very large computers, parts of the phase diagrams of alloys can be reproduced and as-yet-unknown details of the phase diagrams can be predicted. Neutron diffractometry is so far unique in determining the partial structure functions of *amorphous alloys* by means of isotope substitution. By immersing amorphous ribbons in partly deuterated solutions in order to match the scattering density, it is possible to overcome the very intense surface scattering and to access pure bulk properties.

In the session on *precipitation, growth and time-dependent experiments*, high quality experimental results were presented, which in some cases were able to distinguish between spinodal decomposition and nucleation growth mechanisms. In this field particularly, the combination of neutron scattering with complementary methods, e.g. transmission and field emission electron microscopy, plays an important role. On the other hand, theoretical understanding seems to be still at an early stage.

Hydrogen in metals is a traditional field of neutron spectroscopy and diffraction: Quasi-elastic scattering is applied to investigate hydrogen diffusion in crystalline and, recently, amorphous samples. Inelastic experiments reveal the hydrogen vibrational state. In binary hydrides, higher-order excitations have been found using pulsed spallation sources. Diffuse scattering on hydrogen-caused strain fields yields memory effects for fast diffusing hydrogen. High-resolution quasi-elastic neutron scattering has been applied to study *self and impurity diffusion in metals.* Thanks to the new technique, a further level of insight into the microscopic diffusion mechanism in metals has been reached.

It is worth mentioning the pioneering work in the field of *point defects, radiation damage, voids and bubbles.* This field, close to materials science, goes along with efforts of researchers at Harwell to apply small-angle neutron scattering or neutron radiography to problems in practical metallurgy.

The editors hope that this survey of neutron scattering in metal physics may stimulate the further use of this method, especially by those scientists who are not yet familiar with it. Future perspectives are linked to new developments in instrumentation. Scattering of polarised neutrons with spin analysis will greatly enhance the possibilities of separating coherent and incoherent scattering and of studying magnetic scattering. Very high resolution phonon spectroscopy employing the neutron spin-echo technique may allow us to measure phonon widths and therefore phonon-electron relaxation times down to 10^{-10} s. Highly efficient instruments will allow the investigation of kinetic phenomena during phase separation.

The editors would like to thank the speakers for promptly preparing their manuscripts; they hope that readers will tolerate small "imperfections" in the manuscripts which the editors have accepted for the benefit of fast publication. We express our gratitude to the Institut Laue-Langevin and to the KFA-Jülich for their technical and financial help, and to the KFA staff who helped to organise the workshop, particularly Mrs. Ch. Weingartz for finalising the book manuscript.

Grenoble, Jülich
January 1986

C. Janot, W. Petry
D. Richter, T. Springer

Contents

Introductory Lecture. A Comparison of Neutrons and Synchrotron Radiation for Defect Studies. By G. Kostorz (With 6 Figures) 1

Part I **Short Range Order**

Monte Carlo Simulation of Alloy Phase Diagrams and Short-Range Order. By K. Binder (With 5 Figures) 12

The Determination of Atomic Interaction Energies in Solid Solutions from Short Range Order Coefficients – An Inverse Monte Carlo Method. By V. Gerold and J. Kern (With 5 Figures) 17

Short-Range Order and Atomic Interaction in $NiCr_x$
By W. Schweika and H.-G. Haubold (With 5 Figures) 22

High T-Dependence of Local Order in Magnetic FeM Alloys from Neutron Diffuse Scattering
By V. Pierron-Bohnes and M.C. Cadeville (With 4 Figures) 28

Short-Range Order in a $Ni_{63.7}\,Zr_{36.3}$ Amorphous Alloy. Effects of Structural Relaxation. By Y. Calvayrac, A. Quivy, S. Lefebvre, J. Bigot, and R. Bellissent (With 4 Figures) 32

A SANS Approach to Possible Hierarchical Structures of Defects in Amorphous Alloys. By C. Janot and B. George (With 5 Figures) 37

SANS Measurements on Metallic Glasses and the Contributions of Nuclear and Magnetic Structure Defects and of the External Surfaces
By R. Yavari, P. Desre, and P. Chieux (With 6 Figures) 42

Short-Range Order in Metallic Glasses
By P. Lamparter (With 4 Figures) 49

Part II **Precipitation and Growth; Time-Dependent Experiments**

Phase Separation in Alloys Using Low Q Neutron Scattering
By R.J. Stewart (With 7 Figures) 56

SANS-Investigation of Early Stage Precipitation Kinetics in Cu-2.9 at.% Ti. By H. Eckerlebe, R. Kampmann, and R. Wagner (With 10 Figures) 66

Phase Transformations in Fe-Cu-Alloys – SANS-Experiments and Theory. By R. Kampmann and R. Wagner (With 7 Figures) 73

Kinetics of Order-Disorder Phenomena in Some Intermetallic Platinum-Based Alloys. By M.C. Cadeville, C. Leroux, V. Pierron-Bohnes, and C.E. Dahmani (With 6 Figures) 78

Studies of Heterogeneities of Micrometer-Range with Double Crystal Diffractometer. By D. Schwahn and M.H. Yoo (With 6 Figures) 83

Decomposition Kinetics in the Superalloy Nimonic PE16 and the System NiAlTi Investigated by TEM, FIM and SANS By W. Hein, H.P. Degischer, D.J. Jensen, H. Strecker, W. Wagner, R.P. Wahi, and H. Wollenberger (With 7 Figures) 89

Comparative TEM, FIM and SANS Study of the Cu-2 at.% Co Decomposition. By H.P. Degischer, J. Piller, W. Wagner, and H. Wollenberger (With 5 Figures) 94

Quantitative Measurement of Anomalous Small-Angle X-Ray Scattering (ASAXS) in Some Al-Zn-Ag Alloys By O. Lyon and J.P. Simon (With 2 Figures) 99

Part III **Hydrogen in Metals**

Interstitial Point Defects (D,N,O) in Transition Metals By J. Peisl, H. Dosch, and A. von Schwerin (With 5 Figures) 106

Coherent Quasi-Elastic Scattering of Neutrons from Lattice Gas Systems. By D.K. Ross, D.A. Faux, M.W. McKergow, D.L.T. Wilson, and S.K. Sinha (With 1 Figure) 116

Mechanism of Hydrogen Diffusion in Intermetallic Hydrides By R. Hempelmann and D. Richter (With 5 Figures) 121

Hydrogen Diffusion in Amorphous Liquid-Quenched $Pd_xSi_{1-x}H_y$ Measured by QNS By G. Driesen, R. Hempelmann, and D. Richter (With 3 Figures) 126

Part IV **Diffusion in Alloys and Hydrogen in Metals**

Quasielastic Neutron Scattering Study of Fast Diffusion of Co in β-Zr. By W. Petry, G. Vogl, T. Flottmann, and A. Heidemann (With 2 Figures) 134

Systematic Investigation of Self-Correlation Functions in Cubic Lattices; Mono- and Divacancy Mechanisms
By K. Ruebenbauer and B. Sepioł (With 2 Figures) 139

Quasielastic Coherent Scattering for Atomic Diffusion via Vacancies
By K. Schroeder 143

Lattice Dynamics and Interstitial Motion
By H.R. Schober and A.M. Stoneham (With 1 Figure) 149

Interatomic Forces Near a Hydrogen Impurity in Metals
By G. Solt and H. Beck (With 4 Figures) 155

H Tunneling and Local Diffusion in NbO_xH_y
By H. Wipf, A. Magerl, and K. Neumaier (With 2 Figures) 161

Jahn-Teller Relaxation Effects in the Inelastic Neutron Scattering Spectra of Hydrogen Metal Systems. By A. Klamt (With 3 Figures) . 165

Localized Vibrations of Hydrogen in Metals
By D. Richter, R.Hempelmann, J.J. Rush, J.M. Rowe, J. Eckert, J.A. Goldstone, and D. Tonks (With 3 Figures) 170

One-, Two- and Three-Dimensional Disorder of D-Atoms in Non-Stoichiometric TaD_x. By J. Hauck and K. Parlinski (With 3 Figures) 176

Neutron Radiography Studies of Spinodal Decomposition in NbH_x
By G. Steyrer and J. Peisl (With 5 Figures) 181

Part V **Point Defects, Radiation Damage, Voids, and Bubbles**

Irradiation Damage in Metals: A Story of But Also's
By Y. Quéré (With 8 Figures) 188

Determination of Helium Densities in Bubbles Using a Contrast Variation Method. By D. Schwahn, W.Kesternich, S. Spooner, H. Schroeder, H. Ullmaier, and J. Schelten (With 6 Figures) 197

Studies of Stress- and Fatigue-Induced Defects Using Small-Angle Neutron Scattering. By J.R. Weertman (With 7 Figures) 204

Neutrons in Practical Metallurgy
By R.N. Sinclair and C.G. Windsor (With 4 Figures) 213

Point Defects in Metals in Thermal Equilibrium
By W. Trost, K. Differt, K. Maier, and A. Seeger (With 3 Figures) ... 219

Saturation of Irradiation-Induced Precipitation
By G. Martin (With 1 Figure) 225

Analysis of Defect Microstructures in Irradiated Ferritic Alloys
By P.A. Beaven, F. Frisius, R. Kampmann, and R. Wagner
(With 4 Figures) 228

Influence of Electron Irradiation on Decomposition Kinetics and
Morphology in CuNiFe Alloys
By W. Wagner, R. Poerschke, and H. Wollenberger (With 5 Figures) . 235

Index of Contributors 241

Introductory Lecture

A Comparison of Neutrons and Synchrotron Radiation for Defect Studies

G. Kostorz

Institut für Angewandte Physik, Eidgenössische Technische Hochschule, ETH-Hönggerberg, CH-8093 Zürich, Switzerland

1. Introduction

Defects in metals comprise a variety of features, and there have been several attempts to establish some systematic order among inherently "disorderly" items (see, e.g. [1,2]). As most metals commonly occur in their crystalline state, defects may be defined as any deviation from translational symmetry. This includes bulk and lower-dimensional inhomogeneities of structural or chemical nature as well as any excitations involving the basic building unit,or even only part of it (e.g. electronic excitations).

In the history of x-ray (and later neutron) diffraction, defects were present from the beginning. The experiment of von Laue and Knipping may have failed at the time if they had used a perfect crystal. In subsequent decades, crystal imperfections were taken into account in structural work in a very global way, e.g. as "grain size" or "mosaicity", "temperature factor" or "background". For an analysis of periodic structures, i.e. the measurement and evaluation of Bragg intensities, these concepts may be sufficient, as the final goal is a good "R-factor". With the steadily improving quality of x-ray and neutron sources and experimental equipment, however, important progress in the analysis of specific defects by scattering methods has resulted over the last 20-30 years, aided and complemented, wherever possible, by the simultaneous developments in transmission electron microscopy and other techniques.

Starting from a perfect crystalline structure, defects will lead to scattering at non-Bragg positions. Amorphous systems could be viewed as being in the extremely disordered, i.e. most defective state. However, the notion of "defects" in amorphous metals goes beyond this view and can be understood again as referring to local deviations from the average (bulk) amorphous state. In x-ray and neutron scattering, most of these cases can be treated on the basis of the kinematical scattering theory or its equivalent, the first Born approximation, and its validity is mostly implied in the following. A wealth of theoretical information can be found in the book of KRIVOGLAZ [3]; a detailed discussion of scattering methods and a review of x-ray and neutron results obtained until the early seventies has been given by SCHMATZ [4]. Since then, the availability of strong neutron beams - including especially cold neutrons - and of dedicated instruments for defect studies at the Institut Laue-Langevin, most of them based on pioneering work at Jülich, have confirmed the power of neutrons in defect research [5,6] and established neutron scattering as a genuine alternative and complement to x-ray scattering (see [7] for a review of recent results). Meanwhile, an increasing number of strong x-ray beams are becoming available [8,9] at synchrotrons, spallation neutron sources offer new possibilities [10], and new dedicated facilities are under construction or in the planning or discussion stage, e.g. [11,12]. In the light of these new scientific opportunities, it may be worthwile to compare the particular features of x-rays and neutrons, especially for defect studies in metals.

2. Basic Differences of Neutron and X-Ray Scattering

Both thermal neutrons and x-rays offer the appropriate range of wavelengths for diffraction studies in condensed matter. The fundamental differences in their interaction with fields and matter, however, assure that the two types of radiation can rarely be viewed as competitive, but rather frequently as complementary (see e.g. [13] for more details). Table 1 gives some properties of x-rays and neutrons. With $\underline{k}_0$ = wave vector of the incident radiation (modulus $k_0 = 2\pi/\lambda$ with λ = wavelength) and $\underline{k}$ = wave vector of the scattered radiation, the scattering vector $\underline{Q}$ is defined by $\underline{Q} = \underline{k}_0 - \underline{k}$. For elastic scattering, its modulus is $4\pi\sin\theta/\lambda$ with θ = half the scattering angle.

Table 1 Some properties of x-rays and neutron

Property	Value[a] for x-rays	Value[a] for neutrons
mass m, kg	0	1.675×10^{-27}
magnetic moment	0	$- 1.913\ \mu_n$
momentum p	$\hbar k$	$\hbar k = mv$
energy E	$\hbar ck$	$\hbar^2 k^2/2m$
velocity v (m/s)[b]	c	$437\ E^{1/2}$
wavelength λ(Å)[b]	12.4/E	3956/v
angular frequency $\omega = E/\hbar$ (rad/s)	$1.52 \times 10^{18} E$	$1.52 \times 10^{12} E$
scattering length	$Zf_0(Q \rightarrow 0)$	b (nuclear)
absorption	$\propto Z^4\lambda^3$(strong)	$\propto\lambda$ (mostly weak)

[a]with $\hbar$ = Planck's constant/2π, c = velocity of light, μ_n = nuclear magneton, Z = atomic number, f_0 = classical electron radius ($\sim 2.8 \times 10^{-15}$ m),
[b]for x-rays, E in [keV], for neutrons in [meV]

X-rays are scattered by electrons, neutrons "see" the nuclear potentials and local variations of magnetization, i.e. mainly unpaired spins. The scattering cross-sections depend strongly on Q for x-ray scattering (form factor) and for magnetic neutron scattering (magnetic form factor), but are constant in most cases for nuclear scattering of thermal neutrons. At intermediate values of Q and Z, the scattering cross-sections for neutrons and x-rays are comparable, but from element to element, they vary monotonically ($\propto Z^2$) for x-rays, whereas the neutron scattering cross-sections change in a non-systematic way from nucleus to nucleus. Absorption of x-rays depends quite strongly on Z and λ (see Table 1) - optimum sample thicknesses are roughly around 10 to 100 µm at 1 Å (see Fig.1). For thermal neutrons, 1 cm thick samples are mostly convenient. Thus, an x-ray experiment typically probes about 10^{20} atoms (but only $\sim 10^{11}$ may be sufficient for structure determination or absorption studies at a modern synchrotron), whereas neutrons often average over 10^{21} to 10^{24} atoms. (For comparison, in an analytical transmission electron microscope, as few as 10^4 atoms may be illuminated.) Depending on the type of inhomogeneity to be studied, one or the other radiation will be more suitable. Especially for scattering experiments in transmission (e.g. small-angle scattering), there are limitations to the use of x-rays of wavelengths above 2.5 Å, whereas neutrons of a wavelength of 20 Å can still penetrate rather thick samples (and sample containers etc.). If multiple Bragg scattering is to be avoided (as in many small-angle scattering studies), neutrons are the natural choice.

Collective excitations in condensed matter (phonons, magnons) are in the range of a few meV. Molecular excitations and electronic transitions extend to higher energies. In inelastic scattering, thermal neutrons ($\sim$ 1 to 400 meV) will experience appreciably higher relative energy changes than x-rays ($\sim$ 1-100 keV; in addition, optical selection rules apply). The separation of elastic and inelastic

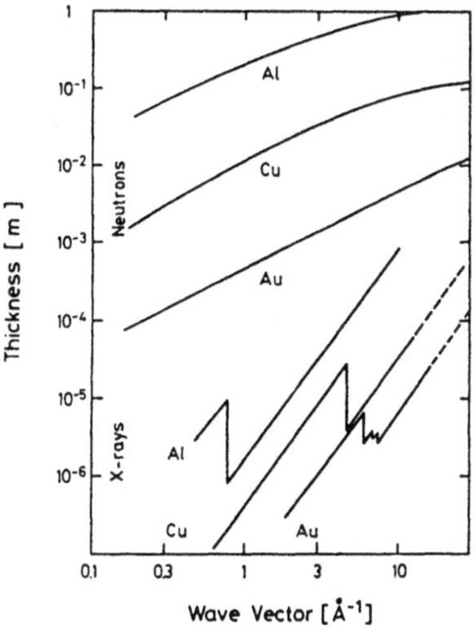

Fig.1 Optimum sample thickness D for x-rays and neutrons, calculated from $D = 1/\Sigma_t$ with Σ_t = macroscopic removal cross-section, as a function of wave vector k. (For neutrons, inelastic and Bragg scattering are not included in Σ_t as they will strongly depend on the experimental conditions.)

scattering is thus much easier with neutrons than with x-rays, and for inelastic scattering/quasielastic scattering below $\sim$ 1 eV (neutron spin-echo methods extend the resolution down to a few neV [14]) neutron scattering will be the proper tool. Figure 2 shows the ranges of $\Delta E(Q)$ currently accessible for neutrons at the ILL [15]. The simultaneous conditions of conservation of energy and momentum restrict these regions, e.g. large-ΔE transitions at small Q (optical phonons, crystal field transitions, "stiff" excitations) may not be reached by neutrons. Other limitations are, of course, given by the available beam intensity and resolution.

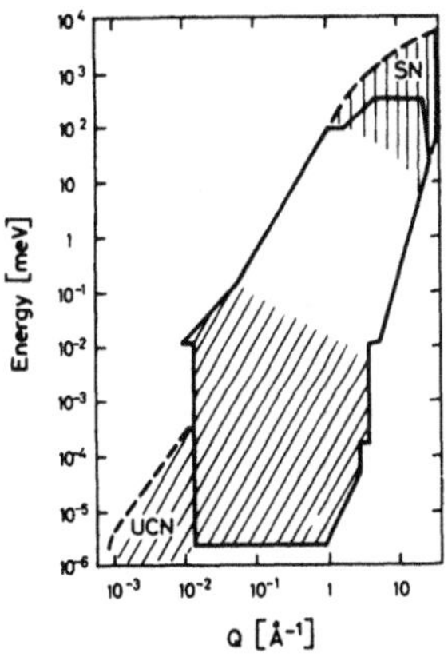

Fig.2 Energy and Q ranges accessible by neutrons at the Institut Laue-Langevin, Grenoble (heavy outline, after [15]). SN and UCN indicate the extension of the range expected from spallation neutrons and ultra-cold neutrons. Without particular energy resolution, Q may be smaller (e.g. $Q \gtrsim 10^{-4}$ Å^{-1} in small-angle scattering)

3. Properties of Neutron and Synchrotron Beams

As defect scattering cross-sections are much smaller than those for Bragg scattering, the intensity of the incident beam is an important parameter. According to MAIER-LEIBNITZ [13], the "momentum space density" $p(\underline{k}_o)$ is the adequate quantity for comparison of different sources of radiation, as the count rate $\dot{Z}$ at a detector in any scattering experiment (see [4] for details) is proportional to $p(\underline{k}_o)$ multiplied by the resolution elements $\Delta^3\underline{k}_o$ and $\Delta^3\underline{k}$ for incident and scattered radiation,

$$\dot{Z} \propto p(\underline{k}_o)S(\underline{Q},\omega)\Delta^3\underline{k}_o\Delta^3\underline{k} \qquad (1)$$

where $S(\underline{Q},\omega)$ is the scattering law to be studied.

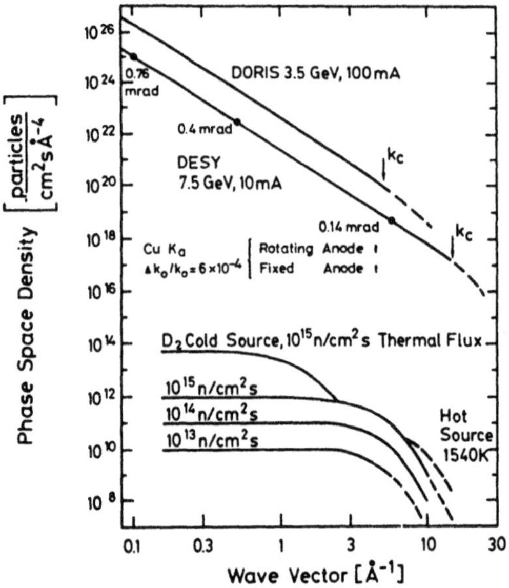

Fig.3 Momentum space density for different x-ray and neutron sources (after [4] and [17]). Source size 3 x 10 mm² for DESY and 1 x 10 mm² for DORIS. For $k \ll k_c$ the curve is proportional to $k^{-11/3}$ (and to the electron current). The characteristic momentum k_c is obtained from the characteristic energy $E_c = 2218\ E_s^3 R^{-1}$ where E_s is the energy of th synchrotron/storage ring, measured in GeV, and R the radius, measured in m

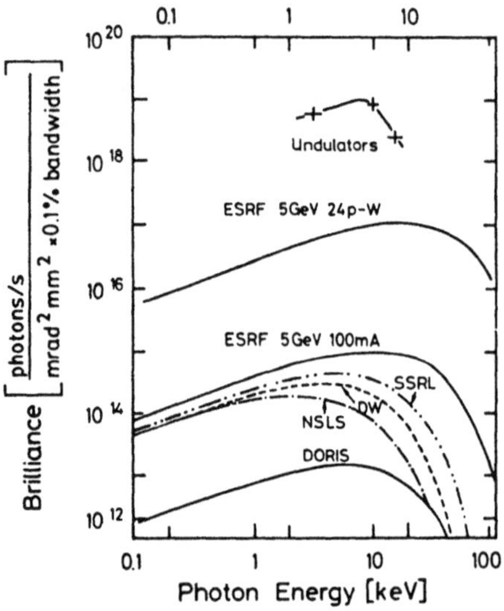

Fig.4 Spectral brilliance for the bending magnet, 24-pole wiggler and several undulators at the planned ESRF [11]. For comparison values for the bending magnets of DORIS (3.7 GeV, 100 mA) and the NSLS project (Brookhaven National Laboratory, 2.5 GeV, 300 mA), the 32-pole wiggler of DORIS (DW) and the 54-pole wiggler of the Stanford facility (SSRL) are also given

Figure 3 shows a comparison of continuous neutron sources and different x-ray sources. Spallation neutron sources may yield a peak intensity about a factor of ten higher than a high-flux reactor, but it is obvious that synchrotron radiation offers considerably higher values of $p(\underline{k}_0)$, over a very broad, continuous range of wavelengths from the optical to the short-wavelength x-ray range. Fig.4 shows the spectral brilliance (at equal relative energy resolution of 0.1% bandwidth) for some existing and planned synchrotron-radiation facilities [11], illustrating the gain to be expected from so-called insertion devices, i.e. wigglers and undulators [8,11].

Although Fig.3 suggests an immediate comparison on the basis of (1), there are quite important differences in the beam characteristics. Whereas thermal neutrons and also, to a large extent, spallation neutrons leave their sources (moderators) under all possible angles, the synchrotron beam is strongly collimated in the horizontal plane and directed tangentially to the electron orbit. The vertical divergence (FWHM) ψ_v (in mrad) is given by $\psi_v \simeq 0.68\ (E_s[\mathrm{GeV}])^{-1}(k_c/k)^{0.43}$, and typical values for DESY are given in Fig.2. The maximum brightness is exactly in the horizontal plane where photons are also 100% linearly polarized (electrical vector parallel to electron orbit plane). Although the magnetic scattering of photons is very small, the high brightness and well-defined polarization may be useful in the study of magnetic defects (magnetic satellite reflections with $\Delta Q/Q = 10^{-4}$, polarization analysis), especially in surface layers or thin films [11]. If extreme resolution is not required, a larger $\Delta^3\underline{k}_0$ for neutrons may compensate somewhat for the lower brilliance, but the flux on a sample will rarely exceed 10^6 neutrons/cm²s whereas an equivalent of 10^{14} photons/cm²s or more may easily reach the (smaller) sample at a synchrotron, posing heavy demands on the detector and counting chain. The difference becomes even more dramatic if the time structure of the beam is taken into account. Whereas neutrons from a reactor yield a continuous (generally stable and reliable) beam which must be chopped for time-of-flight experiments, spallation neutrons and synchrotron radiation emerge in pulses following the time structure of the accelerator. Table 2 gives some typical values.

Table 2 Time structure of synchrotron radiation and neutron beams

	synchrotron	spallation source	steady-state reactor
source pulse width	0.1 ns	0.4 μs	$\gtrsim 10^5$ s
repetition rate, Hz	$\gtrsim 10^6$	50	$\gtrsim 10^{-5}$
radiation pulse width	0.1 ns	2 - 100 μs	$\gtrsim 10^5$ s
average flux, part./cm²s	$> 10^{15}$	$\sim 10^{12}$	$\sim 10^{15}$
peak flux, part./cm²s	$> 10^{19}$	$\sim 10^{16}$	$\sim 10^{15}$

The time structure can be used in several ways. Apart from inelastic, time-of-flight spectroscopy with spallation neutrons, "time-slicing" on the time scale of the repetition rate of pulsed beams reveals the dynamical properties of the scattering systems which may be periodically excited to undergo reversible or irreversible changes (e.g. stimulation of molecular transitions, phase transformations, conformational changes etc.). For example, LARSON et al. [18] have synchronized laser pulses with synchrotron pulses and revealed details of the melting and crystallization process of a Si surface by time-resolved Bragg scattering.

It is not possible to mention all the possibilities and differences of synchrotron radiation and neutrons here. Table 3 gives a summary of techniques available at neutron and synchrotron facilities (restricted to the x-ray range). It may be stressed once more that the improved brightness of sources can be used to improve "resolution" in various ways, e.g. to study smaller objects or smaller parts of larger objects, including two-dimensional structures at surfaces or interfaces (see, e.g., [20] for an x-ray diffraction study of a large angle twist boundary in Au), to obtain shorter measuring times in order to follow rapid processes, to monitor smaller cross-sections or to study momentum or energy transfers with higher precision. Current and planned neutron and synchrotron radiation sources and facilities provide challenging opportunities for scientifically rewarding experiments, especially in the field of defects in metals.

4. Defects in Metals

The structural aspects of defects can be discussed quite generally under the term "local atomic arrangement", and some of the methods of studying these arrangements will be briefly assessed. Excitations and motion of defects can be followed primarily by neutron scattering techniques, but interesting spectroscopic methods, especially for surface defects, are also available at synchrotrons.

Table 3 Relative suitability of neutrons and synchrotron radiation in different techniques
V = very suitable, S = suitable, M = marginal

Technique	Neutrons	X-rays at synchrotron
Elastic scattering for		
- bulk structure	V	V
- surface structure	S	V
- magnetic structure	V	M
- inhomogeneities (small-angle scattering)	V	V
- defects (diffuse scattering)	V - S	V
Coherent inelastic scattering for		
- electronic excitations	M*	S
- phonons	V	M - S
- magnons	V	
Incoherent inelastic scattering	V	V
Quasielastic scattering	V	
Absorption and photoelectron spectroscopy		V
Chemical analysis (activation, fluorescence)	V	V
Topography	S	V
Radiography	S	S
Interferometry	S	S

* may be improved at spallation sources [19]

4.1. Local Atomic Arrangement

In the Ashcroft-Langreth notation (see, e.g. WASEDA [21]) the coherent scattering intensity of a completely disordered isotropic system containing n components with scattering lengths f_1, ... f_n can be written as

$$I(Q) = \sum_{i,j} (c_i c_j)^{1/2} f_i f_j^* \, S_{ij}(Q) \tag{2}$$

where the $S_{ij}(Q)$ are Fourier transforms of the radial distribution functions $\rho_{ij}(r)$ and c_i, c_j are the average concentrations. For n components, there are thus $n(n+1)/2$ independent partial "structure functions", and at least this number of independent measurements with different radiation(s) will be necessary to obtain a complete picture of the atomic arrangement. This remains valid in polycrystalline alloys, too, although reductions are possible if one is interested only in the occupational probabilities for fixed lattice sites r_m, e.g. expressed by the generalized Warren-Cowley short-range-order(SRO) parameters introduced by DE FONTAINE [22],

$$\alpha_{ii}(r_k) = \frac{P_{ii}(r_k)-c_i}{1-c_i} \quad \text{and} \quad \alpha_{ij}(r_k) = \frac{c_j-P_{ij}(r_k)}{c_j} \quad (i \neq j) \tag{3}$$

where P_{ij} is the conditional probability of finding an atom i at (r_m+r_k) if an atom j is at site r_m. There are only $n(n-1)/2$ different SRO parameters in an n-component alloy. Obviously, lattice displacements have been neglected, and only in this case will one measurement be sufficient for binary alloys, three for ternary alloys. A separation of SRO and displacements in crystals is possible if one makes use of the different symmetry and Q dependence of SRO and displacement scattering in reciprocal space. In the Borie-Sparks approximation [23,13], the displacements are considered to second order in Q, and variations of $\Delta f_{ij}/f_i$ with Q are ignored (the method is thus perfectly suited for neutron scattering). The minimum volume to be explored for fcc crystals corresponds to 1/8 of a unit cell in reciprocal space. If the variations of f_i with Q are taken into account [24,25], the volume and thus the number of points where measurements have to be taken is increased by a factor of 2.5. A recent x-ray study of aged Al-Cu (measurements are performed in the reflection mode) at the Cornell synchrotron source shows that reliable data for a binary alloy single crystal can be obtained in a few hours [26]. For comparison, measurements with neutrons for a complete Borie-Sparks analysis of SRO in Ni-25% Fe took several days at the ILL [27], but further gains are possible with improved instrumentation (see [28] for a recent experiment on ^{58}Ni-11% Cr at Jülich). It is not possible to express a general preference for x-rays (synchrotron radiation) or neutrons for these studies. In both cases, calibration and correction procedures, e.g. for background, incoherent scattering, absorption, thermal diffuse scattering (TDS) and Debye-Waller factors are tedious. With sufficient contrast, (f_i-f_j), neutrons may have a slight advantage despite the lower beam intensity, as TDS can essentially be eliminated by energy analysis and measurements can be performed in transmission on a more representative volume, and may include the small-angle scattering region where clustering and other inhomogeneities become visible.

For amorphous binary alloys and for ternary crystalline alloys, I(Q), (2), has to be measured for at least three different samples (isotopic substitution, neutrons) or at three different x-ray wavelengths such that the anomalous corrections to the scattering lengths f_i may be used to separate the partial structure functions or SRO parameters. As synchrotron radiation offers continuous wavelength tunability with high wavelength resolution, such experiments are now feasible [29]. Near an absorption edge, the x-ray scattering length varies according to

$$f = f_0 + f' + if'' \tag{4}$$

where f', f" are the real and the imaginary parts of the anomalous dispersion correction. Approaching an absorption edge from lower energies, f' decreases drastically. As there is no safe way to calculate f' so close to the edge, good measurements are necessary. They can often be performed on the same sample in transmission. From the absorption, one obtains f", and f' is calculated using the Kramers-Kronig dispersion relation. Fig.5 [30] shows a recent example for the K edge of Zn in two different Al-Zn-Ag alloys. The system of linear equations for

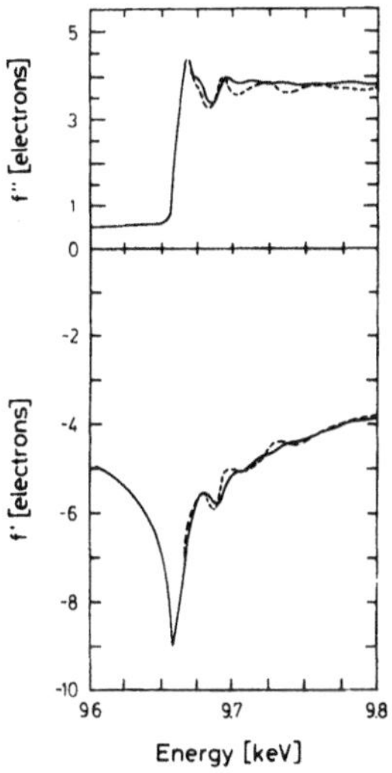

Fig.5 Anomalous dispersion corrections for Zn near the K edge, f" from extended absorption fine structure (EXAFS) measurements, f' calculated from f".
--- Sample C: Al-14 at.% Zn-4 at.% Ag
—— Sample E: Al-4 at.% Zn 14 at.% Ag
(After SIMON et al. [30])

the S_{ij} or α_{ij} can only be solved if the matrix of coefficients is not approaching singularity. This requirement restricts the choice of f_i, f_j. Statistical errors in intensity, uncertainties in f_i, f_j and variations of c_i, c_j all have large effects on the precision of the solutions. The situation has been discussed in detail [31] for Fe-23% Ni-21% Cr comparing analysis procedures for neutron scattering (experimental results were obtained and analysed from three different isotopic compositions) and x-ray scattering, assuming up to six different wavelengths. The authors conclude that each alloy requires careful, individual consideration for both types of radiation. Model calculations and simulated error analyses should help to decide about the feasibility of an experiment. Although differences in composition contribute to the overall error when different isotopic mixtures are prepared for neutron scattering, and further sources of error may be introduced by differences in the metallurgical states in the samples, at least the Q range is not restricted by the nature of the sample itself. For x-rays, the wave vector k_a corresponding to the absorption edge used for the anomalous scattering contribution also defines the maximum accessible Q, $Q_{max} \simeq 2k_a$. Fig.6 shows Q_{max} for the K and L edges as a function of Z.

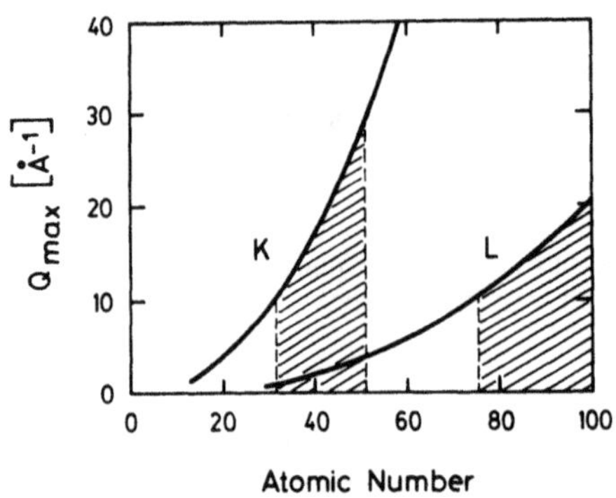

Fig.6 Maximum accessible scattering vector Q_{max} for incident x-ray wavelengths just below the K and L absorption edges. If one requires a certain Q_{max} (e.g. 10 Å^{-1}) and the shortest available wavelength is e.g. 0.4 Å, there may be intermediate elements (e.g. Z ≃ 50 -75) where anomalous scattering cannot be used

For decomposing systems (small-angle scattering) these restrictions are not important, and one might expect more frequent use of anomalous scattering in the study of precipitates etc. SIMON et al. [32] have compared the merits of small-angle neutron scattering (SANS) with isotopic substitution and of small-angle x-ray scattering (SAXS) using anomalous scattering contributions for Al-Zn-Ag alloys. SANS results using natural Ag, ^{107}Ag and ^{109}Ag in Al-5.2 at.% Zn-1.5 at.% Ag [33] had produced a negative S_{AgAg}. It was shown [32] that the remaining, though small, uncertainties in chemical composition and metallurgical state may account for this unphysical result. SAXS experiments, on the other hand, are not easier for this system. The final precision of the intensities after all corrections and relative calibration of different experiments must be around 1%. First SAXS measurements on Al-Zn-Ag alloys using f' near the Zn edge have been performed at Stanford [34,35].

As already visible in Fig.5, x-ray absorption measurements just above an absorption edge show a fine structure (EXAFS) which also contains information on the local atomic arrangement, mostly on nearest neighbours. This method is very powerful with synchrotron radiation for bulk as well as for surface atoms (see e.g. [36]). A combination with Auger electron spectroscopy may yield additional information on the binding state of specific atoms near the surface.

4.2. Defect Dynamics and Motion of Defects

Defect-induced changes in the phonon spectrum (see, e.g. NICKLOW [37]), motion and trapping of defects, especially of hydrogen dissolved in metals (e.g. SKOELD et al.

[38]) can be studied by inelastic and quasielastic neutron scattering, and activities in this field are extensively covered at this conference [39]. Synchrotron radiation may occasionally complement these studies, e.g. with topographic methods for lattice strains ($\varepsilon \simeq 10^{-9}$) and tilts ($\sim 10^{-9}$rad) or macroscopic phase separation, or with surface-sensitive methods to study segregation effects. Apart from surface EXAFS and surface diffraction, the method of standing x-ray waves may be useful, as it becomes possible to not only excite fluorescence of atoms segregating to a surface, but also to identify the sites they occupy [40].

5. Conclusion

Many more features of neutrons and synchrotron radiation deserve detailed discussion in the context of defect studies. A complete comparative assessment will never be possible, as new ideas, experiments and techniques occur at any moment. With the available and forthcoming synchrotron radiation and neutron facilities, appropriate tools for many defect problems are and will become accessible. Considerable efforts will still be necessary in each case to prepare good samples and to chose the best instrument.

References

1. F. Seitz: "Imperfections in Nearly Perfect Crystals: A Synthesis", in Imperfections in Nearly Perfect Crystals, W. Shockley, J.H. Hollomon, R. Maurer, F. Seitz, eds. (Wiley, New York 1952) pp. 3-76
2. A. Seeger: "Theorie der Gitterfehlstellen", in Handbuch der Physik, VII-1, S. Flügge, ed. (Springer, Berlin 1955) pp. 383-665
3. M.A. Krivoglaz: Theory of X-Ray and Thermal Neutron Scattering by Real Crystals, translated from Russian, S.C. Moss, ed. (Plenum, New York 1969)
4. W. Schmatz: "X-Ray and Neutron Scattering Studies on Disordered Crystals", in Treatise on Materials Science and Technology, H. Herman, series ed., (Academic, New York 1979) pp. 105-229
5. W. Schmatz: "Disordered Structures", in Neutron Diffraction, H. Dachs, ed. (Springer, Berlin 1978) pp. 151-196
6. G. Kostorz, ed.: Neutron Scattering, Vol.15 of Treatise on Materials Science and Technology, H. Herman, series ed., (Academic, New York 1979)
7. G. Kostorz: "X-Ray and Neutron Scattering", in Physical Metallurgy, 3rd ed., R.W. Cahn and P. Haasen, eds. (North-Holland, Amsterdam 1983) pp. 793-853
8. C. Kunz, ed.: Topics in Current Physics - Synchrotron Radiation, Techniques and Applications (Springer, Berlin 1979)
9. A. Bienenstock and H. Winick: Physics Today, June 1983, pp. 48-58
10. G.H. Lander and D.L. Price: Physics Today, January 1985, pp. 38-45
11. B. Buras and S. Tazzari: European Synchrotron Radiation Facility (ESRF), (ESRP, Geneva 1984)
12. Major Facilities for Materials Research and Related Disciplines (National Academy Press, Washington D.C. 1984)
13. L.H. Schwartz and J.B. Cohen: Diffraction from Materials (Academic, New York 1977)
14. F. Mezei: Physica 120B, 51 (1983)
15. R. Pynn and B.E.F. Fender: Physics Today, January 1985, pp. 47-53
16. H. Maier-Leibnitz: Nukleonik 8, 61 (1966)
17. Synchrotron Radiation - a Perspective View for Europe, Report of the working party on synchrotron radiation (ESF, Strasbourg 1977)
18. B.C. Larson, C.W. White, T.S. Noggle and D. Mills: Phys. Rev. Letters 48, 337 (1982)
19. S.W. Lovesey: Comments on Solid State Phys. 10, 45 (1981)
20. J. Budai, P.D. Bristowe and S.L. Sass: Acta Metall. 31, 699 (1983)
21. Y. Waseda: Novel Application of Anomalous (Resonance) X-Ray Scattering for Structural Characterization of Disordered Materials (Springer Lecture Notes in Physics 204, Berlin 1984)

22. D. de Fontaine: J. Appl. Cryst. 4, 15 (1971)
23. B. Borie and C.J. Sparks: Acta Cryst. A27, 198 (1971)
24. J.E. Tiballs: J. Appl. Cryst. 8, 111 (1975)
25. P. Georgopoulos and J.B. Cohen: J. Physique 12, C7-191 (1977)
26. E. Matsubara and P. Georgopoulos: J. Appl. Cryst., to be published
27. S. Lefebvre, F. Bley, M. Fayard and M. Roth: Acta Metall. 29, 749 (1981)
28. W. Schweika and H.G. Haubold: this conference
29. P.H. Fuoss, P. Eisenberger, W.K. Warburton, A. Bienenstock: Phys. Rev. Letters 46, 1537 (1981)
30. J.P. Simon, J.J. Hoyt, O. Lyon, R. Pro, B.E.C. Davis and D. de Fontaine: J. Appl. Cryst. 18, 181 (1985)
31. P. Cénédèse, F. Bley, S. Lefebvre: Acta Cryst. A40, 228 (1984)
32. J.P. Simon, O. Lyon, D. de Fontaine: J. Appl. Cryst. 18, 230 (1985)
33. A. Salva-Ghilarducci, J.P. Simon, P. Guyot, I. Ansara: Acta Metall. 31, 1705 (1983)
34. O. Lyon, J.J. Hoyt, R. Pro, B.E.C. Davis, B. Clark, D. de Fontaine, J.P. Simon: to be published
35. O. Lyon, J.P. Simon: this conference
36. T.M. Hayes and J.B. Boyce: Solid State Phys. 37, 173 (1982)
37. R.M. Nicklow: "Phonons and Defects", in Ref. [6], pp. 191-226
38. K. Sköld, H. Mueller, T.O. Brun: "Hydrogen in Metals", in Ref. [6], pp. 423-460
39. H. Peisl: this conference
- and subsequent papers
40. J.A. Golovchenko, J.R. Patel, D.R. Kaplan, P.L. Cowan and M.J. Bedzyk: Phys. Rev. Letters 49, 560 (1982)

Part I

Short Range Order

Monte Carlo Simulation of Alloy Phase Diagrams and Short-Range Order

K. Binder

Institut für Physik, Johannes-Gutenberg-Universität,
Postfach 3980, D-6500 Mainz, and
Institut für Festkörperforschung, Kernforschungsanlage Jülich,
Postfach 1913, D-5170 Jülich, F. R. G.

Abstract: As a prototype model for order-disorder phenomena in binary alloys, a face-centered cubic lattice is considered, the sites of which can be taken by either A-atoms or B-atoms, assuming pair-wise interactions between nearest (J_{nn}) and next nearest neighbours (J_{nnn}). The phase diagram is constructed from Monte Carlo calculations. Some technical aspects essential for the success of such calculations are briefly mentioned (use of grand-canonical rather than canonical ensemble, how to obtain the free energy needed to locate first-order phase transitions, etc.). It is shown that the topology of the phase diagram changes when the ratio $R = J_{nnn}/J_{nn}$ is varied, and this behaviour is discussed in the context of approximate calculations (mean field, cluster variation method) and renormalization group predictions. The temperature variation of some Cowley short-range order parameters is compared to pertinent experiments on the Cu-Au system.

1. Introduction

Order-disorder phenomena on fcc lattices have been a challenge for both metallurgy and statistical physics for many years [1-17] . A typical example is Cu-Au (Fig.1) [18]; its long- and short-range order (SRO) parameters, internal energy etc. have been extensively studied [19-24] ; one would like from these measurements to infer the effective interaction parameters of the system. This knowledge then could be used to predict other physical properties, and to check first-principle calculations of these parameters, which is a difficult problem [25].

In this work we study this problem using Monte Carlo simulations; parts of the results have been described earlier [5-7,10] . We study very simple models, including interactions between nearest and next-nearest neighbours only; we use the results both for a check of approximate calculations (such as the cluster variation method [1,4,8] performed for exactly the same model, and for a comparison with experiment [19-24].

2. Model and Simulation Technique

We use a finite lattice (usually 16^3 unit cells with periodic boundary conditions, and introduce a pseudospin $S_i = +1$ if site i is taken by an A-atom, $S_i = -1$ if it is taken by a B-atom. We assume an Ising Hamiltonian

$$\mathcal{H} = -J_{nn} \sum_{<nn>} S_i S_j - J_{nnn} \sum_{<nnn>} S_i S_j - H \sum_i S_i \ , \quad R \equiv J_{nnn}/J_{nn} \ ; \qquad (1)$$

here <nn>, <nnn> means nearest (next nearest) neighbour pairs; the "exchange" J is related to the standard pair-wise interactions v^{AA}, v^{AB}, v^{BB} as $J = -(v^{AA}+v^{BB}-2v^{AB})/4$; the "field" H is related to the chemical potential difference between A and B (see refs. 6,7 for more details).

Note that unlike the experiment where the concentration $c_B=(1-m)/2$, $m=\langle S_i \rangle_T$ is fixed (canonic ensemble), we work in the grand canonic ensemble; the final results are translated back into the canonic ensemble. The use of the grand canonic ensemble is

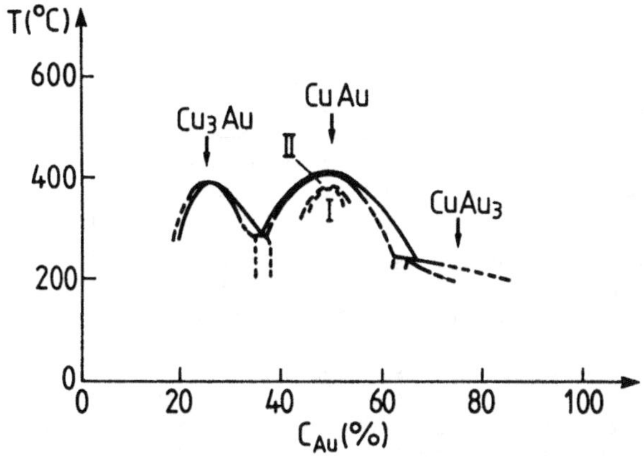

Fig.1: Partial phase diagram of copper-gold alloys in the temperature-concentration plane, indicating the existence regions of the three ordered phases. From Ref. 18.

necessary since (i) equilibrium is reached quicker, particularly in the ordered phase, and (ii) finite size effects are easier to handle. E.g., crossing a two-phase region shows up as a "jump" in an isotherm rather than a shallow van der Waals-like loop, etc.(Fig.2A). Due to hysteresis, locating the first-order phase boundaries often is difficult; we solve this problem by comparing the free energies of the various phases {F is obtained by integrating the internal energy $U=-\partial(F/k_BT)/\partial(1/k_BT)$ or "magnetization" $m=-(\partial F/\partial H)_T$ along a suitable path in the T-H plane, see [7,11]} (Fig.2B).

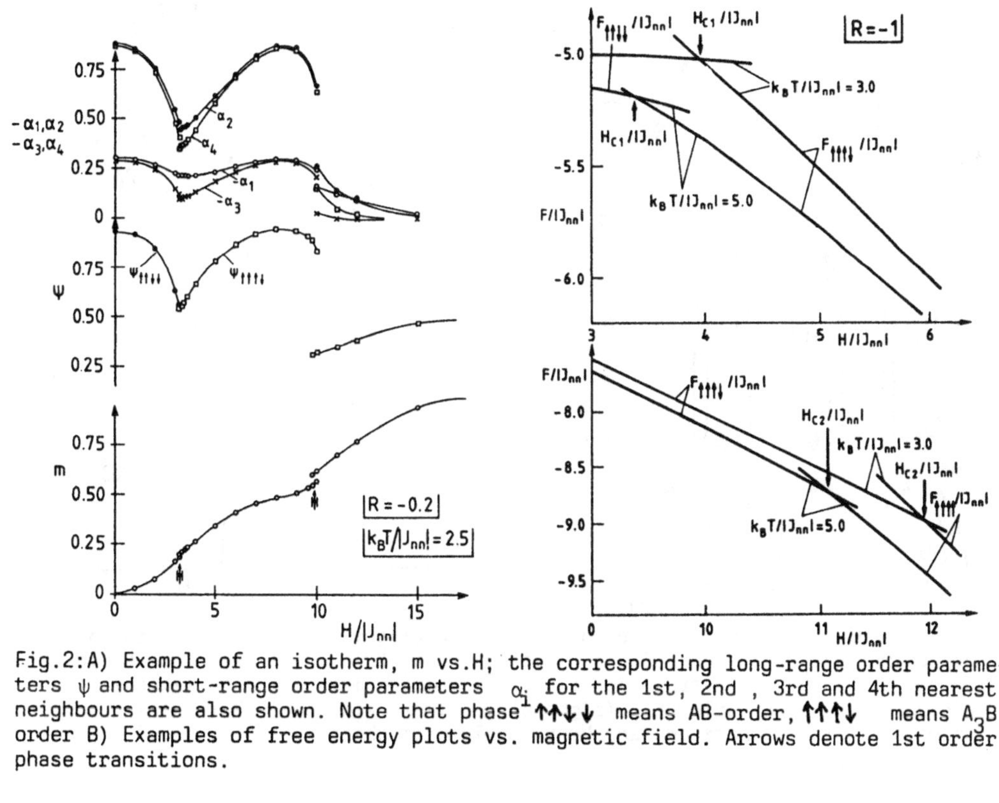

Fig.2:A) Example of an isotherm, m vs.H; the corresponding long-range order parameters ψ and short-range order parameters α_i for the 1st, 2nd , 3rd and 4th nearest neighbours are also shown. Note that phase ↑↑↓↓ means AB-order, ↑↑↑↓ means A_3B order B) Examples of free energy plots vs. magnetic field. Arrows denote 1st order phase transitions.

3. Results and Discussion

The phase diagram for the nearest-neighbour model (R=0) is shown in Fig.3 and compared to various approximations. Bragg-Williams approximation [2], quasichemical approximation [3] and the cluster variation method [4] (CVM, in the tetrahedron approximation give dramatically different predictions. Hence the question must be asked: which of these methods - if any at all - is reliable, what is the true phase diagram of the model? Fig.3B compares the Monte Carlo results to the CVM phase diagram (tetrahedron approximation): it is seen that the ordering tendency is overestimated by the CVM: at stoichiometric compositions the error is only a few percent, while in the region from $c_B \approx 0.35$ to $c_B \approx 0.40$ the situation is much worse. Ref. 4 predicts a triple point at a rather high temperature T_t, while according to Monte Carlo it is very low [5,6,12], possibly even $T_t = 0$ [5]. This discrepancy has stimulated further efforts to improve the CVM approach [8]: in fact, using the more complicated tetrahedron-octahedron approximation [8] the results are in nearly complete agreement with the Monte Carlo data (apart from T_t, which now is lowered from about $k_BT/|J_{nn}| \approx 1.65$ to $k_BT/|J_{nn}| \approx 1.2$). This is a fine example of how Monte Carlo results can be useful for asserting the reliability of various analytical approximations.

Fig. 4 shows the behaviour for attractive next-nearest neighbour interaction, R <0. This interaction removes the high ground-state degeneracy (see [5,6,13], and references therein), and also has a pronounced effect on the phase diagram. With -R increasing the two triple points move upwards, until for $-R=R_m$ ($-R_m \approx 0.25$-0.3 [7,12]) they merge (at $c_B = 1/2$) in a multicritical point. For $-R>-R_m$ the transition from AB order to the disorder phase is second order. This behaviour is in accord with a

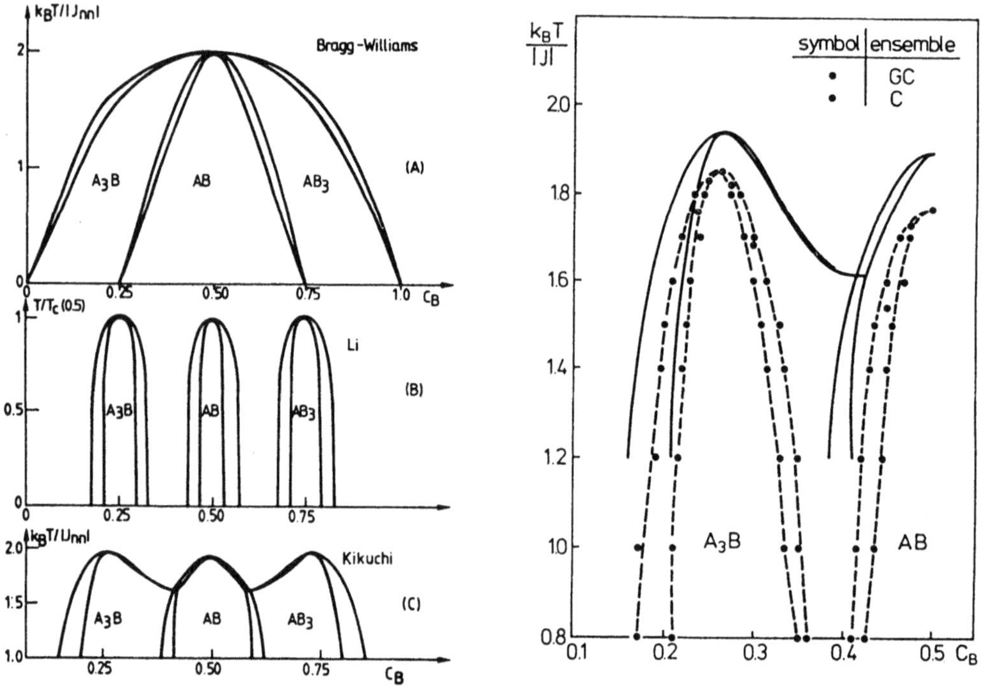

Fig.3: Phase diagram of the nearest-neighbour fcc binary alloy AB according to various approximations (left part: Bragg-Williams method [2], quasi-chemical approximation due to Li [3], tetrahedron cluster variation method due to Kikuchi [4]), and according to Monte Carlo [6] (right part). There also the CVM results [4] are shown again for comparison, and Monte Carlo data obtained from runs in the canonic ensemble are included.

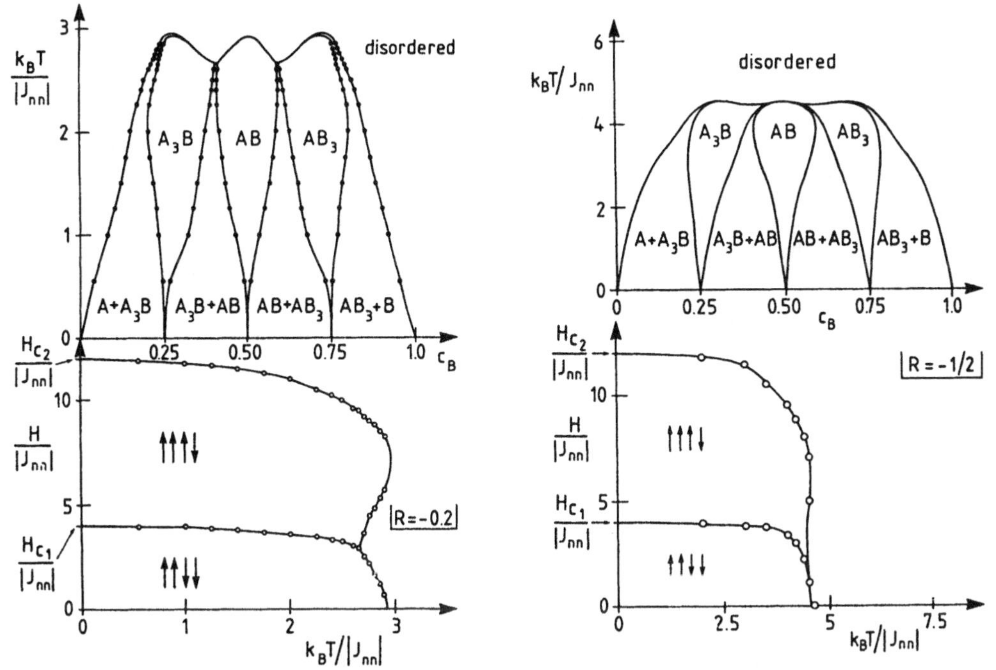

Fig. 4: Phase diagram of the fcc Ising antiferromagnet in a field and corresponding alloy phase diagrams for R = -0.2 (left) and R = -0.5 (right). From Ref.10.

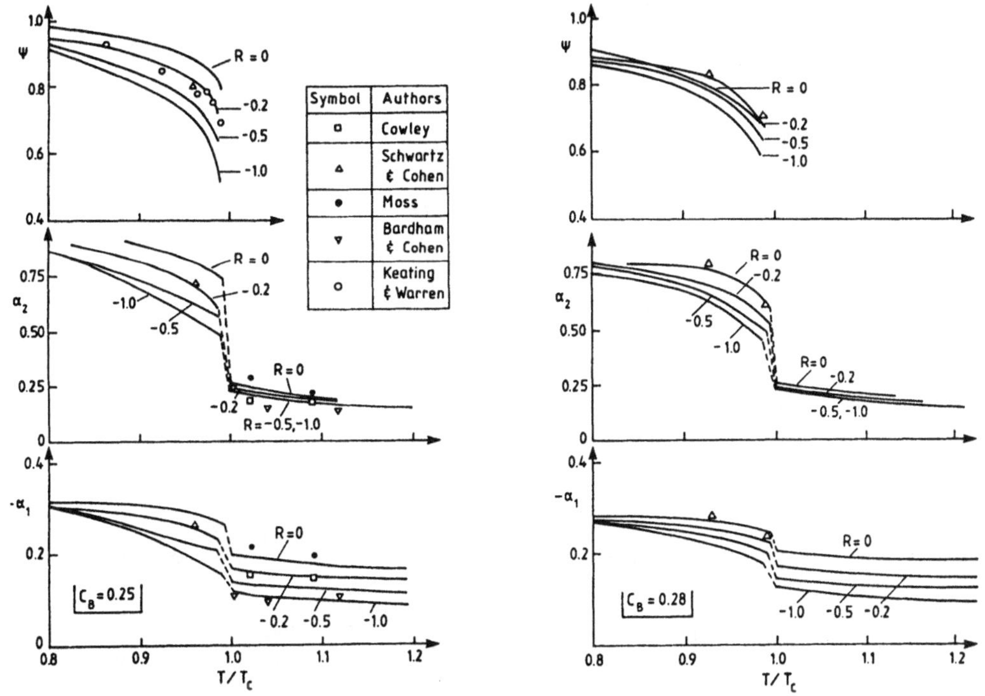

Fig. 5: Temperature dependence of the long-range orderparameter ψ of the Cu_3Au structure and of the Cowley short-range order parameters α_1, α_2 for c_B=0.25 (right part) and c_B=0.28 (left part). Curves are present Monte Carlo calculation, points show experimental data of Refs. 19-24 as indicated.

recent renormalization group analysis [26] . For $R \approx R_m$ we thus encounter a new kind of multicritical point, the behaviour of which has not yet been studied in detail [26].

Finally Fig. 5 shows the temperature dependence of the order parameter of the A_3B-structure as well as of the short-range order parameters α_1, α_2 at two concentrations, together with experimental data [19-23]. Of course, to account for the asymmetry of the real phase diagram (Fig. 1) we either would need concentration-dependent interactions or higher than pairwise interaction terms [13,14], and thus we should not expect the model to account for the real system very well. While the long-range order data favour $R \approx -0.2$, for both $c_B = 0.25$ and $c_B = 0.28$, consistent with the structure of the phase diagram on the Cu-rich side, the short-range order data do not allow any clear conclusion: the scatter of data taken from different authors is larger than the range of variation of α_1, α_2 with R from R=0 to R = -1! This clearly shows that to make further progress, what is most urgently needed are more reliable short-range order data.

Finally we mention that the system, Eq.(1), may model an interstitial alloy rather than a sustitutional one as well: we re-interpret the lattice sites as the interstitial sites, $S_i=-1$ corresponds to a site taken by an interstitial, $S_i=+1$ represents an empty site (=lattice gas model). In this case, the diffusion of the interstitials also is of great interest, and can conveniently be studied by simulations as well [27].

References

1. D. de Fontaine, in Solid State Physics (edited by H. Ehrenreich F.Seitz and D. Turnbull), Vol. 34, p.73, Academic Press, New York 1979
2. W. Shockley, J.Chem. Phys.6, 130 (1938)
3. Y.Y. Li, J.Chem. Phys. 17, 447 (1949)
4. R.Kikuchi, J.Chem. Phys. 60, 1071 (1974)
5. K. Binder, Phys. Rev.Lett. 45, 811 (1980)
6. K. Binder J.L. Lebowitz, M.K.Phani, and M.H.Kalos, Acta Metall.29,1655 (1981)
7. K. Binder, Z.Phys. B45, 61 (1981)
8. J.M. Sanchez, D.De Fontaine and W. Teitler, Phys.Rev.B26, 1465 (1982)
9. U.Gahn, J.Phys.Chem.Solids 43, 977 (1982)
10. K. Binder, W. Kinzel, and W. Selke, J.Magn.Mag.Mat. 31-34,1445 (1983)
11. T.L.Polgreen, Phys.Rev.B29, 1468 (1984)
12. J.L.Lebowitz, M.K.Phani and D.F.Styer,J.Stat.Phys.38,413 (1985)
13. D.F.Styer, M.K.Phani and J.L.Lebowitz, preprint
14. R.Kikuchi and D.de Fontaine,in Applications of Phase Diagrams in Metallurgy and Ceramics, Vol.2,p.967.NBS Special Publication 496 (1978)
15. R.C.Kittler and L.M.Falicov, Phys.Rev.B18, 2506 (1978); 19, 291 (1979)
16. W.Schweika, Dissertation RWTH Aachen 1985 (unpublished)
17. J.Kern, Dissertation Univ. Stuttgart 1983 (unpublished);J.Kern and V.Gerold, to bepublished
18. M.Hansen, Constitution of Binary Alloys, McGraw Hill, New York 1958
19. D.T.Keating and B.E.Warren, J.Appl.Phys.22, 286 (1951)
20. S.C.Moss, J.Appl. Phys.35, 3547 (1964)
21. J.M.Cowley, J.Appl.Phys. 21, 24 (1950)
22. L.H.Schwartz and J.B.Cohen, J.Appl.Phys.36, 598 (1965)
23. P. Bardhan and J.B.Cohen, Acta Cryst.A32, 597 (1976)
24. C.Sykes and F.W.Jones,Proc.Roy.Soc. London, Ser.A157, 213 (1938); R.L.Orr,Acta Metall. 8, 498 (1960); R.L.Orr, J.Luciat-Labry, and R.Hultgren, Acta Metall. 8 431 (1960)
25. A.Bieber and F.Gautier, J.Phys.Soc.Japan 53, 2061 (1984); Z.Phys.B57, 335 (1984)
26. E.Domany, Y.Shnidman and D.Mukamel, J.Phys.C15, L 495 (1982)
27. R. Kutner, K. Binder and K.W. Kehr, Phys.Rev.B26, 2967 (1982);B28, 1846 (1983)

The Determination of Atomic Interaction Energies in Solid Solutions from Short Range Order Coefficients – An Inverse Monte Carlo Method

*V. Gerold and J. Kern**

Max-Planck-Institut für Metallforschung, Institut für Werkstoffwissenschaften, D-7000 Stuttgart 1, F. R. G.

1. Introduction

There exist two ways in the literature to analyse quantitatively short-range order in a solid solution. One way is the measurement of the modulated monotonic Laue scattering of X-rays or neutrons from a real solid solution crystal [1,2]. The results are used to evaluate the short-range order coefficients α_j where j is the number of the coordination shell (Fig. 1, case a). From these coefficients it is possible to create a model crystal in the computer (Fig. 1, solid arrow) which has the same or nearly the same set of α_j (j = 1 to 6, for example) [3,4]. The other way (Fig. 1, case b, dashed arrows) is a purely theoretical one. The starting point is the energetic interaction between the atoms in a solid solution, the simplest assumption being pair interaction energies V_j. These energies (usually the first two) or better the ratios V_j/kT are used to simulate pair exchanges in a model solid solution crystal containing A and B atoms [5]. These exchanges finally lead to a state of thermal equilibrium for the temperature T with some short-range order which can be determined in this model crystal and can be compared with the reality (Fig. 1). So far only a very approximative method exists to deduce the interaction energies V_j/kT immediately from the diffraction experiment [6,7]. This approximation is only valid for very high temperatures, i. e., for small values of V_j/kT.

The aim of the present paper is to suggest another way which should be more generally valid. The starting point is the model crystal obtained by the Gehlen-Cohen method from a set of experimentally determined α_j values from a crystal in thermodynamic equilibrium (Fig. 1, solid arrow). The procedure uses a Monte Carlo method and will be called inverse Monte Carlo method (IMC) and is presented in Fig. 1 as open arrow (case c). As will be shown, this method is able to deduce

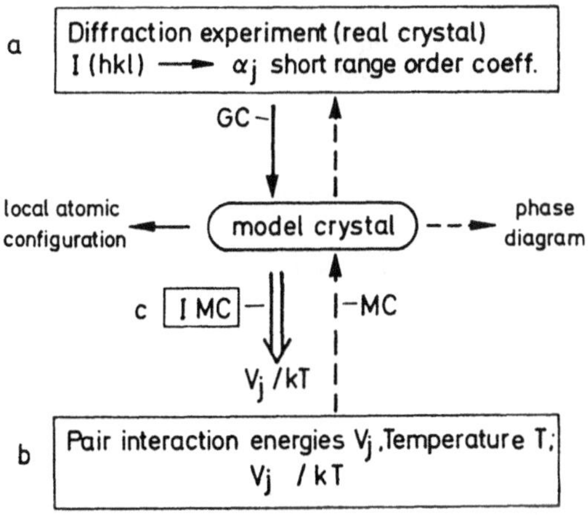

Fig. 1 Schematic sketch of the different methods used to study short-range order (see text). GC = Gehlen and Cohen method [3]; IMC = inverse Monte Carlo method

*Now at McKinsey & Company, Inc., D-4000 Düsseldorf 30, FRG

reliable V_j/kT values up to j = 8. The method would be also applicable for other types of interactions besides pair interaction.

2. The Monte Carlo Method (MC)

The inverse Monte Carlo method is based on equations which describe the MC method. In that method an AB nearest neighbour pair of the model crystal is selected randomly (running number k), and the change of the configuration energy, ΔU_k, is calculated for a site exchange of this pair:

$$\Delta U_k = \sum_{j=1}^{j_{max}} \Delta N_{j,k} V_j \tag{1}$$

In this equation $\Delta N_{j,k}$ is the change in number of AA pairs in the j th coordination shell of the crystal. The probability p_k for site exchange is then given by [5]

$$p_k = \frac{\exp(-\Delta U_k/kT)}{1+\exp(-\Delta U_k/kT)} \tag{2}$$

After a large number k_{max} of such exchange trials the total change of AA pairs in the i th coordination shell is given by

$$\Delta N_i = \sum_{k=1}^{k_{max}} \Delta N_{i,k} p_k \qquad \text{for } i = 1 \text{ to } i_{max} \tag{3}$$

As a criterion for thermal equilibrium, a further set of site exchanges must fulfill the condition

$$\Delta N_i \approx 0 \qquad \text{for } i = 1 \text{ to } i_{max} \tag{4}$$

3. The Inverse Monte Carlo Method (IMC)

The IMC makes use of equations (3) and (4); the starting point is a simulated model crystal (13824 lattice sites in our case) in thermal equilibrium evaluated from known short-range order data α_i (Fig. 1). In order to determine a set of unknown interaction energies V_j the following steps are undertaken [8]:

i. As in the Monte Carlo method an AB nearest neighbour pair is selected randomly (running number k).
ii. The changes $\Delta N_{i,k}$ are calculated for the exchange of this pair and registered in the computer.
iii. In order to keep the correct short-range order the exchange is made only virtually.
iv. Steps one to three are repeated 10 000 times (k_{max}).
v. The registered numbers $\Delta N_{i,k}$ are inserted into the following set of equations:

$$\sum_{k=1}^{k_{max}} \Delta N_{i,k} p_k = 0 \qquad \text{for } i = 1 \text{ to } i_{max} \tag{5}$$

vi. The unknown interaction energies V_j which enter Eq. (5) via Eqs. (2) and (1) are evaluated by solving the set of equations, Eq. (5). This has been done by a library program of the computer.

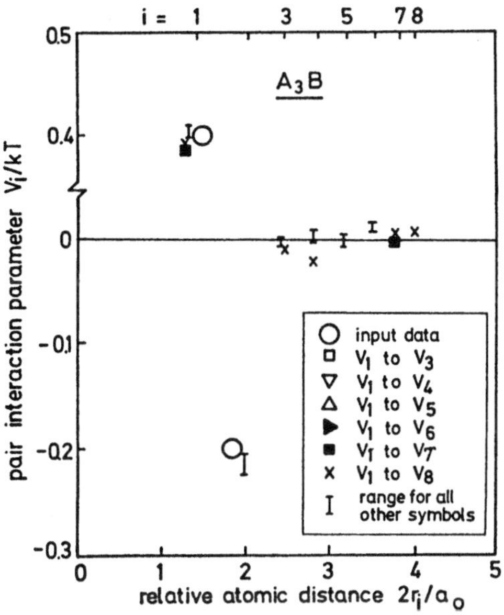

Fig. 2 Test of the IMC method

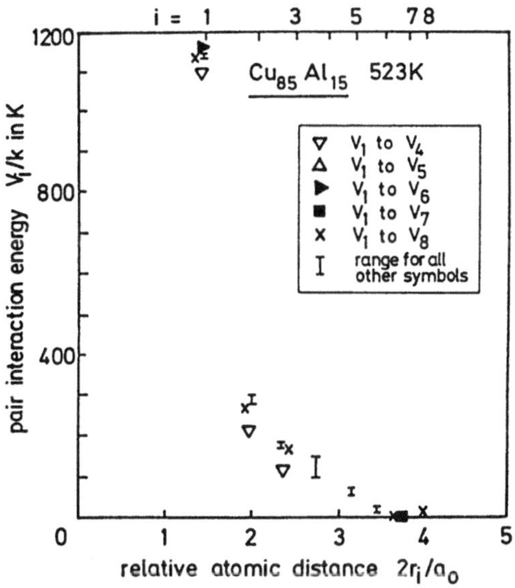

Fig. 3 Pair interaction energies for Cu-15at%Al at 523K

This procedure was repeated 5 times in each of three model crystals simulated in a similar way. The V_j values reported are averages of all 15 runs. In addition, in step (v.) i_{max} was chosen to run from 3 to 8. Thus 6 independent determinations of the V_j's (with j_{max} also running from 3 to 8) were possible.

4. Test of the Inverse Monte Carlo Method (IMC)

The IMC method has been tested using a fcc A_3B model crystal simulated by the MC method using parameters $V_1/kT = 0.4$ and $V_2/kT = -0.2$ (circles in Fig. 2). This model crystal in thermal equilibrium was either used directly or the short-range order coefficients $\alpha_j (j = 1 \text{ to } 6)$ were determined and a new model crystal was simulated using these coefficients as criteria. The result of the first of these tests (i_{max} = 3 to 8) is plotted in Fig. 2. For each i_{max} a different symbol has been used. Since very often these symbols cluster in a small energy range a single sign "I" covering this range is used instead of the many symbols.

Surprisingly, the method gave consistent results for all six independent calculations. The values are $V_1/kT = 0.4 \pm 0.015$, $V_2/kT = -0.21 \pm 0.01$ and all other $V_j/kT \approx 0$. The other tests did show that in order to simulate a crystal from the α_i's at least 6 coefficients have to be taken into account in order to get reliable results. In addition, since in these tests the V_j/kT values were predetermined, each simulated MC crystal had a different set of α_i values.

The best results were obtained if several (at least 3) MC crystals were created and the average α_i values were evaluated and used to simulate the crystals for the IMC method. In that case the results for the V_j/kT were comparable to those where the MC crystals were used immediately. From these results the general procedure was adopted to simulate always 3 crystals and apply 5 runs to each of them. The final V_j (as shown in Fig. 2) are the averages of all 15 tests.

5. Application of the IMC Method

The IMC method has been applied to real fcc solid solutions where the short-range order coefficients in thermal equilibrium have been reported in the literature. The first solid solution was $Cu_{85}Al_{15}$ measured by EPPERSON et al. [9] at 423 and 523 K. As an example, Fig. 3 shows the result of the IMC analysis for 523 K. In this figure the data for i_{max} = 2 and 3 have been omitted, because their results

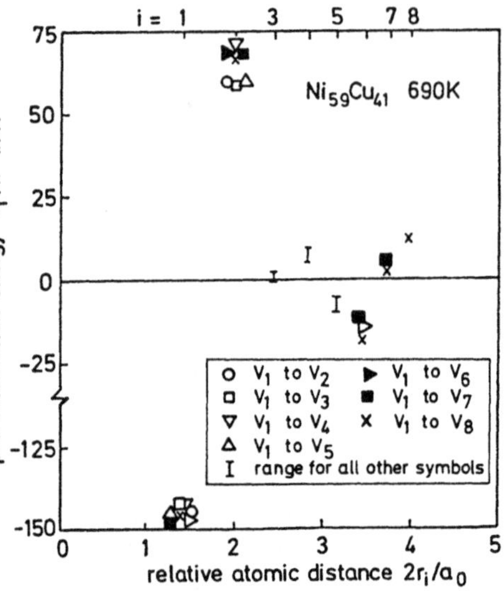

Fig. 4 Pair interaction energies for Ni-41at%Cu at 690K

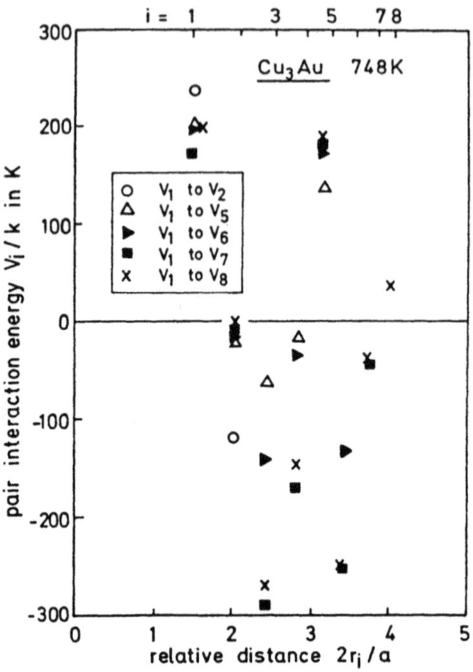

Fig. 5 Pair interaction energies for Cu-25at% Au at 748K

differ from the other V_j evaluations. Obviously, in this solid solution the first four to five energy terms V_j are important and therefore necessary to describe the V_j values correctly. Similar results were obtained for 423 K. Only V_1/k = 1550 K has a distinctly higher value compared to that at 523 K.

The next case studied was the solid solution Ni-41at%Cu at 690 K. The short-range order coefficients were taken from the paper by WAGNER et al. [10]. The V_j values obtained are shown in Fig. 4. In this case all evaluated values are shown for i_{max} = 2 to 8. Most data fit well to each other. Only the value for V_2 depends on the choice of i_{max}. Since this system shows short-range clustering ($\alpha_1 > 0$) the first term V_1 is negative. Again, the method gives reasonable results.

The third case studied was the solid solution Cu_3Au. The experimentally determined short-range order coefficients were taken from the paper by BARDHAN and COHEN [11]. In that case the method gave less reliable values. As an example, Fig. 5 shows the result for 748 K. The points scatter more widely than in the previous cases. Only the data for 1023 K seem to show better agreement. Again, a larger number of interaction energies of i_{max}=5 is necessary to get more consistent V_j values. The data for i_{max} = 2 to 4 differ considerably from the others. Altogether, five temperatures were investigated. The V_j's always oscillated with increasing j giving positive values for j = 1, 5 and 8. All others were negative or close to zero.

Since the system Cu-Au has a large atomic size effect compared to the other systems Cu-Al and Cu-Ni it is felt that the reliability of the α_j-values is less than in the other two systems. In order to get them from diffraction data one has first to eliminate the influence of the size effect from these data, which becomes more and more inaccurate with increasing size effect. On the other hand, the size effect influences the short-range order by internal strain, which cannot be described by pair interactions. Thus a model with pair interaction energies is less suitable for this case.

Very recently the inverse Monte Carlo method has been successfully applied by SCHWEIKA [12,13] in his study of short-range order in Ni-Cr solid solutions.

6. Conclusions

It has been shown that the inverse Monte Carlo method introduced in this paper is capable to evaluate interaction energies responsible for the equilibrium short-range order in solid solutions provided the short-range order coefficients α_i are available. The efficiency of the method has been demonstrated for pair interaction energies, where it gave reliable results up to 8 energy terms. The method can be applied as well to evaluate any other kind of interactions, such as four-body interaction energies. Such energy terms have been introduced by FONTAINE and KIKUCHI [14] by their cluster variation method.

References

1. B. Borie and C.J. Sparks: Acta Crystallogr. A27, 198 (1971)
2. G. Georgopoulos and J.B. Cohen: J. de Physique, Colloque C7 38, C7-191 (1977)
3. P.C. Gehlen and J.B. Cohen: Phys. Rev. 139A, 844 (1965)
4. J.E. Epperson: J. Appl. Cryst. 12, 51 (1979)
5. K. Binder: Monte Carlo Methods in Statistical Physics, in Topics in Current Physics (Springer, Berlin 1979)
6. P.C. Clapp and S.C. Moss: Phys. Rev. 142, 418 (1966)
7. M.A. Krivoglaz: Theory of X-Ray and Thermal Neutron Scattering by Real Crystals (Plenum Press, New York 1969)
8. J. Kern: Dissertation Universität Stuttgart (1983)
9. J.E. Epperson, P. Fürnrohr and C. Ortiz: Acta Crystallogr. A34, 667 (1978)
10. W. Wagner, R. Poerschke, A. Axmann and D. Schwahn: Phys. Rev. B21, 3087 (1980)
11. P. Bardhan and J.B. Cohen: Acta Crystallogr. A32, 597 (1976)
12. W. Schweika: Dissertation R.W.T.H. Aachen (1985)
13. W. Schweika and H.-G. Haubold: This conference
14. D. de Fontaine and R. Kikuchi: NBS Publication SP 496, 999 (1978)

Short-Range Order and Atomic Interaction in $NiCr_x$

W. Schweika and H.-G. Haubold

Institut für Festkörperforschung, Kernforschungsanlage Jülich, Postfach 1913, D-5170 Jülich, F.R.G.

1. Introduction

Even at high temperatures the atoms of an alloy will not be distributed purely at random on the lattice sites: The atomic interactions between the alloying atoms always cause at least some short-range order (= SRO) or, below critical temperatures, long-range order or decomposition into precipitates of fully ordered phases. An exact knowledge of the atomic interaction thus is of basic interest and, furthermore, a requirement for the calculation of realistic phase diagrams.

Usually, the atomic interaction energies are calculated from the Warren-Cowley SRO-parameters |1| which are measured in a short-range ordered single crystal by diffuse scattering of either X-rays or neutrons. In this procedure (KRIVOGLAZ |2| and CLAPP, MOSS |3|) a high-temperature approximation is applied. However, more precise atomic interaction energies can be calculated from the measured α's by the use of a new inverse Monte Carlo method |4|. With such precisely determined energy values, the coherent phase diagrams can be calculated by direct Monte Carlo methods |5|. This allows to study the phase stabilities in the not-measured low-temperature regions of the phase diagrams and to check the interaction potential's validity by comparison of the predictions with the known parts of the phase diagram.

In the present study, we have applied this procedure for the first time. A NiCr alloy was chosen for several reasons. First, because of the small lattice parameter change with increasing Cr concentration |6|, lattice distortions were expected to be small. Hence their contribution to the diffuse scattering intensity is small and the SRO diffuse-scattering intensity can be separated accurately. Secondly, the atomic interaction is expected to be long range, a necessary condition for the stability of the Ni_2Cr long-range ordered structure |7| as observed by X-ray |8| and neutron |9| diffraction. In addition, $Ni_{0.89}Cr_{0.11}$ is a technically interesting material for thermocouples. Residual resistivity changes were observed after aging and thermal treatments which was attributed to changes of the SRO |10,8|.

The diffuse scattering experiment was performed with neutrons since compared to the alternative of an X-ray study, the scattering cross-sections are significantly larger. Furthermore, the inelastic thermal diffuse background-scattering can be experimentally separated by an energy analysis of the scattered neutrons.

2. Experiment

The equilibrium SRO-state at 833 K was studied at room temperature in a quenched $^{58}Ni_{0.89}Cr_{0.11}$ single crystal. The diffuse scattering was measured on a new neutron spectrometer in Jülich, which is especially designed for diffuse scattering studies. Inelastic background scattering was separated by the time-of-flight method. The use of the ^{58}Ni isotope improves the signal-to-background ratio between the coherent SRO- and the incoherent elastic background scattering by a factor of 50. Measurements were performed in the (110) and (100) reciprocal lattice planes. For the (100) plane the measured coherent elastic scattering is shown in fig. 1a. The absolute calibration with a vanadium standard is accurate to 5 %.

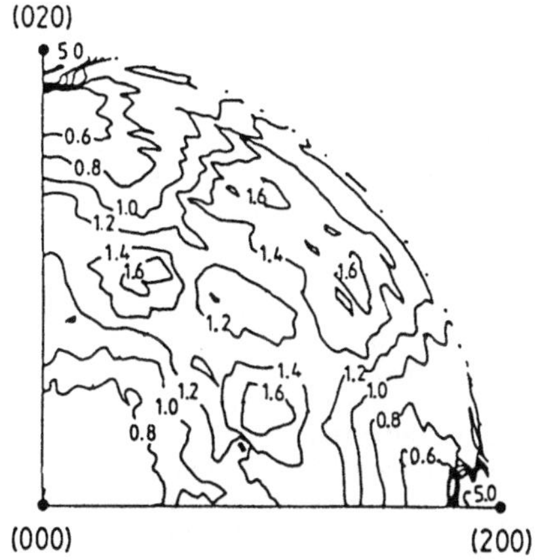

Fig. 1a: Measured iso-intensity contours of coherent elastic scattering on the (100) plane

in Laue units: $c_{Ni}c_{Cr}\ |f_{Cr}-f_{Ni}|^2$

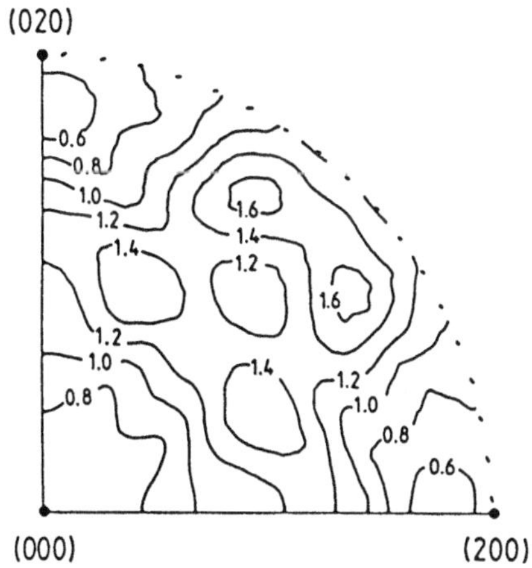

Fig. 1b: Reconstructed diffuse scattering intensity due to SRO and lattice displacements

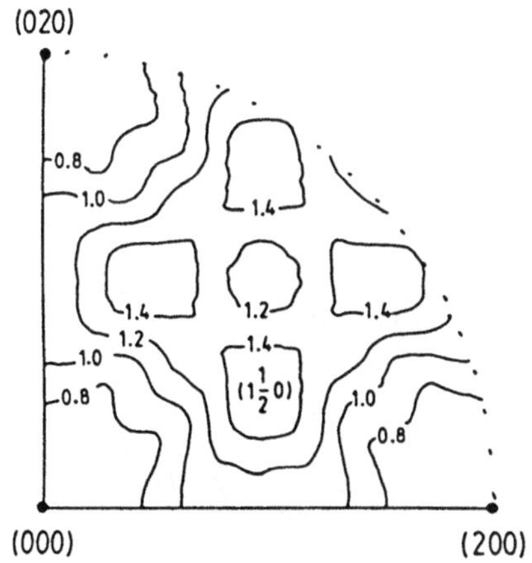

Fig. 1c: SRO contribution to the elastic diffuse scattering intensity.

3. Data Analysis and Results

3.1 Short-Range Order

The total coherent elastic diffuse scattering due to SRO and lattice displacements can be expressed as /1/:

$$\frac{d\sigma}{d\Omega}(\underline{Q}) = c_{Ni}\, c_{Cr}\, |f_{Cr}-f_{Ni}|^2 \,\{ (\sum_i \alpha_i \cos(\underline{Q}\,\underline{R}_i) + \sum_i \delta_i\, \underline{Q}\,\underline{R}_i \sin(\underline{Q}\,\underline{R}_i) \} \qquad (1)$$

where c is the atomic fraction, f is the scattering amplitude, α_i are the Warren-Cowley short-range order parameters and δ_i are the lattice distortion parameters. The expression is linear in atomic displacements and is obtained in the approximations of small distortions and elastic isotropy.

Both short-range order and lattice distortion parameters α and δ were determined by a weighted linear least squares fitting procedure to 1737 data points using the algorithm of Singular Value Decomposition |11|. Values for 35 short range order parameters and 15 lattice distortion parameters were calculated. As expected, the contribution of lattice distortion although very small is not negligible. The diffuse intensity as reconstructed from these parameters is shown in Fig. 1b, the separated contribution due to short-range order scattering is given in Fig. 1c. The SRO intensity maximum at the (1 1/2 0) position is only 50 % larger than the Laue scattering. The values of the corresponding short-range parameters are given in Fig. 2. The value of $\alpha_0 = 0.998 \pm 0.008$ is very near unity, the theoretical value.

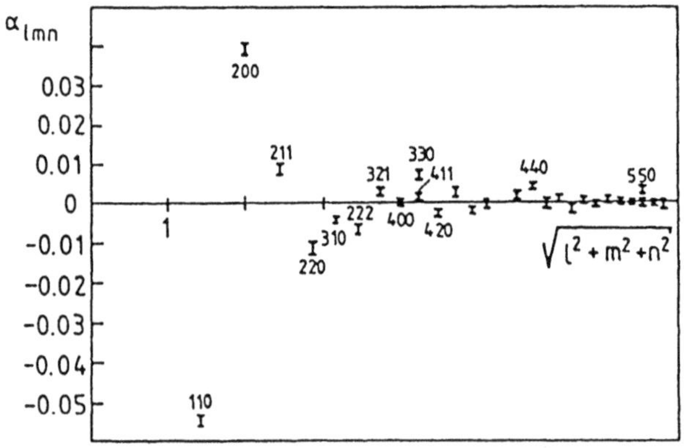

Fig. 2: SRO parameters α for $Ni_{0.89}Cr_{0.11}$ as quenched from thermal equilibrium at 833 K. α's are labelled by Miller indices ℓ, m, n.

3.2 Interaction energies

The atomic pair interaction energies V_i were determined from the short-range order parameters with two different methods: From the usually applied Clapp-Moss |3| high temperature approximation, and the new inverse Monte Carlo method |4|. Both results are given in Fig. 3.

Using the inverse Monte Carlo Method, the pair interaction energies were calculated employing a computer model of the short-range ordered crystal which was created with 22 parameters α_i and about 200 000 atoms. The V_i results of both methods are in very good agreement for $i > 1$ and show an oscillating long-range behavior especially in <110> and <100> directions. Slightly different values are calculated for V_1 because of the fact that V_1 is not really small in comparison to the thermal energy of 72 meV: $V_1/kT = 0.36$! This, however, is a necessary condition for the validity of the Clapp-Moss approximation.

The inverse Monte Carlo Method allows the calculation of many-body interactions. A small contribution was found to arise from the three-body interaction between nearest Cr neighbors (Fig. 4). This is in agreement with theoretical predictions for transition metal alloys |12,13| and reflects the local and angular dependent character of d-electron interaction.

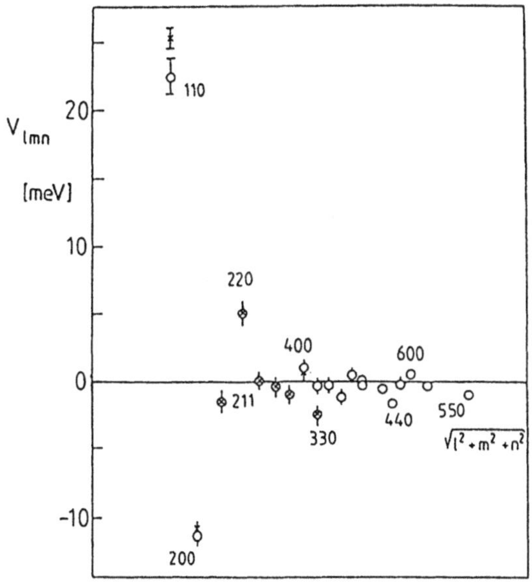

Fig. 3: Atomic pair interaction energies $V_{\ell mn}$
ϕ : High temperature approximation; ⸸ : Inverse Monte Carlo Method

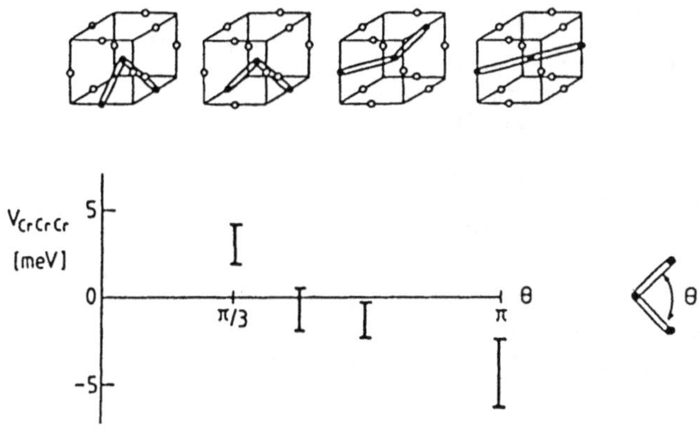

Fig. 4: Three-body interaction energies of the four Cr triplet configurations between nearest neighbours. Stretching angle $\theta = \pi/3$, $\pi/2$, $2\pi/3$ and π.

3.3 Phase Diagram

The coherent phase diagram of the fcc NiCr alloy was determined with the Monte Carlo Method by using a grand canonical ensemble |5|. The simulation was performed on a Cray-XMP computer using a new parallel algorithm within a total computing time of about 30 min. A minimum of 5 pair interaction energies is necessary to define the ground state of the known Ni_2Cr phase uniquely |7|. Thus in a first attempt, only the five most contributing pair interactions were used for the Monte Carlo simulation. The calculated concentration and temperature-dependent phase-diagram is shown in Fig. 5.

Three long-range ordered phases are found to be stable: Ni_2Cr_2, Ni_2Cr and Ni_3Cr. Presumably the Ni_2Cr_2 phase will not exist because in this region a coexistence between fcc and bcc structures is observed. The Ni_2Cr phase is known to exist. The small remaining difference between the calculated and the experimental va-

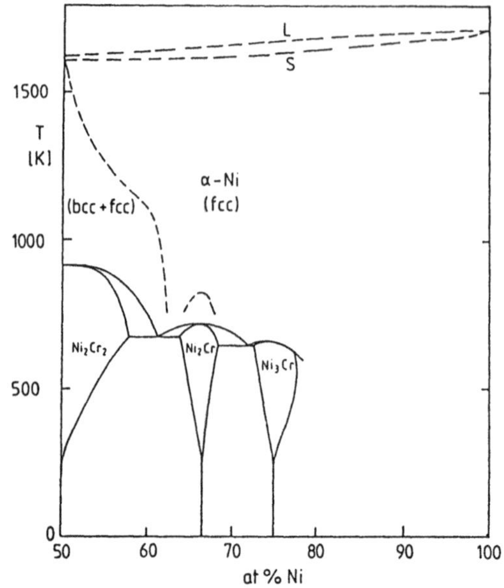

Fig. 5: Coherent phase diagram as calculated from V_{110}, V_{220}, V_{211}, V_{220}, V_{330} (solid lines). Broken lines are the experimental phase boundaries after Shunk |14|.

lues of T_c can be explained either by the neglected longer-ranging part of the interactions, or by a change of the interaction which may be caused either by the lattice contraction accompanying the ordering process |8|, or by a concentration dependence of the interaction energy. According to the simulation, a previously undetermined long-range ordered phase Ni_3Cr should exist (DO_{22}).

To judge the results of this first attempt one has to be aware of some complicating facts. The analysis of all possible ground state structures could not be performed completely. Therefore, further long-range ordered structures might exist, particularly for lower Cr-fractions. Because of metastability problems (first order phase transitions) such structures will not necessarily be found by the Monte Carlo simulation. The experimental errors in the interaction energies and the restriction in their number lead to an uncertainty of about 100 K for the critical temperatures. Hence, it cannot be excluded that the Ni_3Cr phase might be suppressed by a possible further long-range ordered structure.

4. Conclusions

The application of the Inverse Monte Carlo Method to neutron diffuse scattering from short-range ordered alloys has yielded accurate values for the long-range pair interactions and for the first time the three-body interactions. These results will be useful for a quantitative comparison with future ab initio calculations in the coherent potential approximation |12,13,15|. Moreover, for the first time a realistic phase diagram has been calculated with the accurate interaction energies determined from diffuse scattering data. This offers a promising technique to extend phase diagrams into the low-temperature region and might become a basis for future systematic studies of phase diagrams.

References

1 B.E. Warren: "X-Ray Diffraction", p. 229, Addison-Wesley, Reading, Massachusetts, USA (1969)
2 M.A. Krivoglaz: "Theory of X-Ray and Thermal Neutron Scattering by Real Crystals", Plenum, New York, (1969)
3 P.C. Clapp, S.C. Moss: Phys. Rev. Vol. 142, No. 2, 418 (1966)

4 V. Gerold, J. Kern: this conference
5 K. Binder, J.L. Lebowitz, M.K. Phani, M.H. Kalos: Acta Met., Vol. 29, 1655-1665 (1981)
6 L. Karmazin: Czech. J. Phys. B29 (10), 1184 (1979)
7 J. Kanamori, Y. Kakehashi: J. Phys. 38, C7, 274 (1977)
8 H.G. Baer: Z. Metallkde, Bd. 49, 12, 614-623 (1958)
9 Ye.Z. Vintaykin, A.A. Loshmanov: Fiz. met. metalloved 27, 7, 754 (1967)
10 H. Heidsiek, Thesis Aachen, 1979
11 C.L. Lawson, R.J. Hanson, "Solving Linear-Least Squares Problem", Englewood-Cliff, New Jersey (1974)
12 A. Bieber, F. Gautier: J. Phys. Soc. Jap., Vol. 53, No. 6, 2061-2074 (1984)
13 A. Bieber, F. Gautier: Z. Phys. B - (Cond. mat.) 57, 335-334 (1984a)
14 F.A. Shunk (1969) "Constitution of Binary Alloys", McGraw Hill, New York
15 B.L. Gyorffy, G.M. Stocks: Phys. Rev. Lett., Vol. 50, No. 5, 374-377 (1983).

High T-Dependence of Local Order in Magnetic FeM Alloys from Neutron Diffuse Scattering

V. Pierron-Bohnes and M.C. Cadeville

L.M.S.E.S. (UA CNRS n° 306), Université Louis Pasteur,
4, rue Blaise Pascal, F-67070 Strasbourg Cedex, France

We present in this paper the first "in situ" at high temperature investigation of the atomic short-range order (SRO) T-dependence in magnetic alloys by neutron diffuse scattering (NDS). Up to now analogous experimental studies were done on quenched samples by either NMR in dilute alloys [1-3] or NDS in concentrated ones [2, 4, 5]. NDS was generally done at room or at 4 K temperature in alloys annealed and quenched. In the large samples necessary for neutron experiments, the equilibrium state at the annealing temperature T_a can be frozen in by a quench only for low values of T_a. At such temperatures, the annealing time necessary to attain equilibrium makes it impossible to vary T_a on a large range. This limitation can be overcome by doing "in situ" at temperature measurements. NDS is then the most suitable technique because it allows one to separate the elastic contribution in contrast to X-rays. In this work, we focus on the high-temperature aspects of these NDS measurements : metallurgy and Debye-Waller factors ; the paramagnetic contribution evaluation will be detailed elsewhere. We show that in spite of all these difficulties and in spite of the increased noise due to the oven and the smoothing of SRO at high temperature, it is possible to perform such measurements and so to obtain SRO parameters at temperatures as high as 1500 K.

1. EXPERIMENTAL DETAILS

The NDS measurements have been performed on D7 instrument of I.L.L. with an unpolarized neutron beam of 4.83 Å wavelength. The furnace has Nb screens and an Al vacuum box which give rise to 2 Bragg peaks in the observed q-range (0.21 to 2.5Å^{-1}). Its temperature is checked against a W-Rh thermocouple. D7 is built with a time-of-flight analysis which is essential at high temperature to separate the elastic scattering from inelastic ones.

Measurements were done on two polycrystalline bcc alloys : $\underline{Fe}Co_{0.2}$ and $\underline{Fe}V_{0.2}$. The experimental mounting did not allow to separate the magnetic cross-section contribution, so we had to avoid it. The ferromagnetic contribution is in first approximation proportional to $(\mu_{Fe}-\mu_M)^2$. In $\underline{Fe}Co$ this quantity is small, the ferromagnetic contribution is only few percents of the nuclear one and can be neglected [4]. $\underline{Fe}Co_{0.2}$ has been studied between the order-disorder transition and the martensitic transformation (800 - 1150 K). In $\underline{Fe}V_{0.2}$ the ferromagnetic contribution cannot be neglected, so we have been restricted to the paramagnetic domain : 1125 to 1500 K (T_{CM} = 1048 K from [6]). The paramagnetic contribution has been evaluated from Fe results [7] and found negligible under our measurement conditions.

2. METALLURGICAL ASPECTS

A SRO study has to be performed on alloys in equilibrium state, so metallurgical aspects are of prime importance [1, 3, 4]. In metallic alloys, equilibrium order kinetics proceeds through a vacancy jump process [8, 9] and is relatively slow. Atomic relaxation can be simply described on a small T-range by a phenomenological Arrhenius law : $\tau(T) = \tau_o \exp E_{SD}/kT$. The necessary annealing times ($\sim 10\,\tau$) can be calculated at each temperature when the self-diffusion energy E_{SD} and the factor τ_o are known. For $\underline{Fe}V_{0.2}$, we used $\tau_o = 10^{-15}$ s and E_{SD} = 2.66 eV determined through kinetics measurements in a $\underline{Fe}V_{0.1}$ alloy [3] and for $\underline{Fe}Co_{0.2}$, $\tau_o = 10^{-14}$ s and E_{SD} = 2.8 eV from tracer diffusion results [10].

3. DEBYE-WALLER FACTOR CORRECTIONS :

Debye-Waller factors (DWF) have been measured in a $FeCo_{0.15}$ alloy at I.L.L. on D1B spectrometer which covers a 80° angle in 2θ, so that 6 Bragg peaks can be observed with the 1.28 Å wavelength neutron beam. DWF are plotted on Fig. 1. Moreover DWF have been evaluated in the quasi-harmonic approximation (QHA). As Fe, Co and V are neighbours in the periodic table, all physical quantities have been taken as a linear combination of those of pure metals. The QHA is only valid below $2\theta_D$ (θ_D : the Debye-temperature ∿ 470 K in Fe), a deviation of the experimental data is observed at high T (Fig. 1) ; it can be approximated by $B - B^{QHA} = 1.585.10^{-2}(T/\theta_D)^3$ (inset of Fig. 1) and was assumed to be transferable to $FeCo_{0.2}$ and $FeV_{0.2}$ alloys.

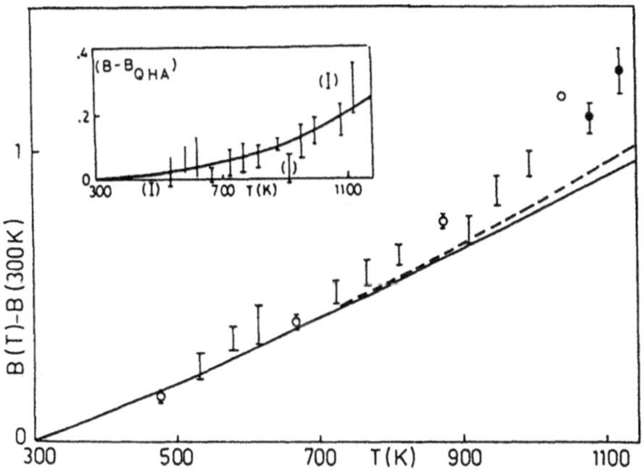

Fig. 1 : Debye-Waller factor measurement in $FeCo_{0.15}$. Comparison of the measured Debye-Waller factors with those calculated in the harmonic (—) and quasi-harmonic (---) approximations (measurements at increasing temperature (o) ; decreasing temperature (.) ; in a two-phase-state (•)) Inset : T-dependence of the departure of the Debye-Waller factors from the QHA ; —— : its T³-fit as mentioned in the text

4. ANALYSIS of DATA - DETERMINATION of SRO PARAMETER α_i

In bcc alloys, the first two shells of neighbours have very close radii (0.87 and $1\times a_o$), like the 4th and 5th shells (1.55 and 1.73 a_o). Resolution and sensitivity of NDS measurements in polycrystals are not sufficient to determine separately the corresponding SRO parameters : in general, when doing so [2, 4, 5] the results are out of their physical range : $\alpha_i \geq (-c/1-c)$. On can tide over this difficulty by regrouping these shells to obtain significant values of the average SRO parameters $\bar{\alpha}_{12}$ and $\bar{\alpha}_{45}$ (with $\alpha_{ij} = (N_i\alpha_i + N_j\alpha_j)/(N_i + N_j)$). This is justified by the fact that in these alloys α_1 and α_2 (as the corresponding energies) have the same sign and order of magnitude [1-4, 11, 12]. So the following expansion of $\alpha(\vec{q})$'s spherical average is used to fit the data with a least square method :

$$\alpha(q) = 1 + \bar{\alpha}_{12}(S_1 + S_2) + \alpha_3 S_3 + \bar{\alpha}_{45}(S_4 + S_5) + \sum_{i=6}^{i_\ell} \alpha_i S_i$$

where $S_i = Z_i(\sin qR_i)/qR_i$; R_i and Z_i are respectively the radius and number of atoms of the ith shell. In $FeV_{0.2}$, the α_i are stable for $i_\ell \geq 5$. Thus we obtain significant values of the 4 SRO parameters : $\bar{\alpha}_{12}$, α_3, $\bar{\alpha}_{45}$ and α_6 ($\alpha_6 \cong 0$). In $FeCo_{0.2}$, the modulation of the cross-section is weaker and its maximum badly defined, being located near the limit of the explored q-range (Fig. 2). $\bar{\alpha}_{12}$ is the only significant result in this case.

The resolution, the accuracy and the number of SRO parameters determined through this investigation of polycrystalline alloys are clearly lower and smaller than those usually given by experiments using single crystals. The comparison between both has not been possible yet in these systems. However, in the present study, we are interested by showing a physical phenomenon, say the influence of magnetism

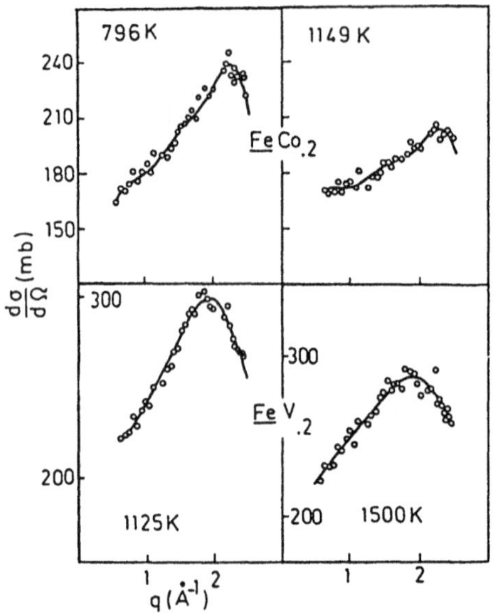

Fig. 2 :
NDS cross-section of $\underline{Fe}Co_{0.2}$ and $\underline{Fe}V_{0.2}$ at the extreme measurement temperatures
o : experimental points ; — fit with i_ℓ = 6 ($\underline{Fe}$Co) or 7 ($\underline{Fe}$V)

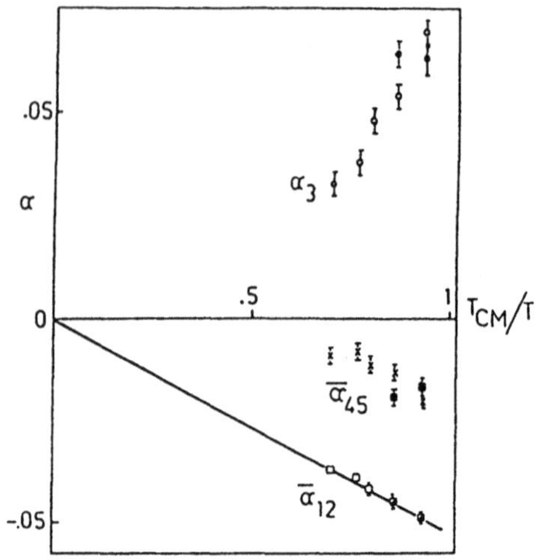

Fig. 3 :
T_{CM}/T dependence of SRO parameters in $\underline{Fe}V$
□: $\overline{\alpha}_{12}$; o : α_3 ; × : $\overline{\alpha}_{45}$ at increasing T ; ■ ● ⊠ at decreasing T

on the atomic distribution, and not by determining a large number of accurate SRO parameters, as necessary for example to calculate pair interaction energies. Thus we believe that the accuracy and the number of significant SRO parameters so determined are sufficient to illustrate our purpose as discussed in the following. Moreover, our values of α_i are in good agreement in sign and order of magnitude with our previous NMR and NDS results in the same systems [1-4]. Measurements on single crystalline alloys are also in progress.

5. DISCUSSION of RESULTS

In the two systems, we find a strong repulsive SRO between M atoms (M = Co or V) as first and second neighbours, in $\underline{Fe}$V a strong attractive interaction between third neighbours.

In $\underline{Fe}$V, the SRO parameters show the usual alternating and damping variation with the radius of the shell. Strong position correlations still exist at 1500 K. The 1/T variation of α_i is linear (Fig. 3), but only $\overline{\alpha}_{12}$ is proportional to 1/T as predicted by the usual non-magnetic mean field models at T high compared to the ordering temperature [13].

In $\underline{Fe}$Co, the T-dependence of $\overline{\alpha}_{12}$ displays a rapid decrease of SRO when T approaches the Curie temperature (Fig. 4). It is in agreement with the variation we have observed by NMR on dilute alloys [1] : the existence of an abrupt change of behaviour of $\overline{\alpha}_{12}$ at T_{CM} has been attributed to the effect of ferromagnetism on SRO and can be well described by a Bethe model, which takes into account both magnetic and chemical interactions through a double Ising Hamiltonian [1, 12]. The rapid decrease of SRO near T_{CM} is related to the disappearance of the average magnetic moment. The enhancement of SRO by its coupling with magnetism is due to a strong Fe-Co magnetic interaction, larger than those between FeFe or CoCo atoms, as can be expected in this system.

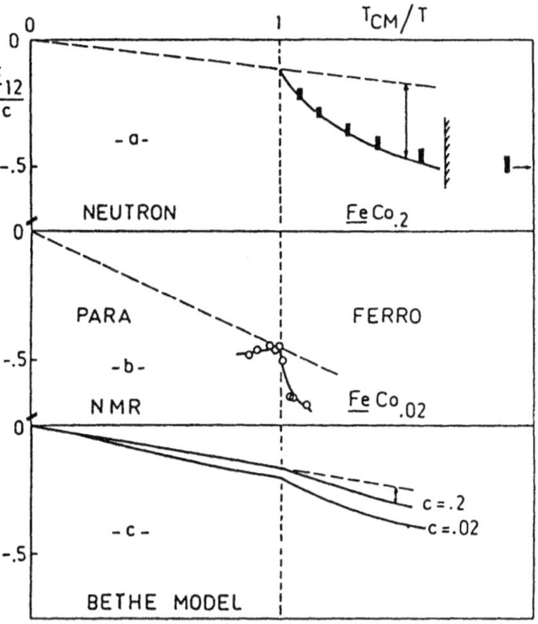

Fig. 4 : T_{CM}/T-dependence of SRO parameters in FeCo : $\bar{\alpha}_{12}/c$ from NDS (a), NMR (b) ; α_1/c from Bethe model (c)

The error bars on experimental points in (a) are given by the heights of the symbols.

ACKNOWLEDGEMENTS

The authors wish to thank Prof. O. Schaerpf and Dr. Fujara for their valuable help during the measuring time on D7, MM Rebesco and Andant for their effective technical assistance and Dr. Mangin for his active participation in measurements on D1B.

1. V. Pierron-Bohnes, M.C. Cadeville, F. Gautier: J. Phys. F : Metal Phys. 13, 1689 (1983) & "Phase Transformation in Solids", Symp. Maleme, Creete, 1983, p. 401, ed. by Tsakalakos (North-Holland).
2. I. Mirebeau, Cadeville M.C., Parette G., Campbell I.A.: J. Phys. F : Metal Phys. 12, 25 (1982).
3. V. Pierron-Bohnes, I. Mirebeau, E. Balanzat, M.C. Cadeville: J. Phys. F : Metal Phys. 14, 197 (1984).
4. V. Pierron-Bohnes, M.C. Cadeville, G. Parette: J. Phys. F : Metal Phys. 15, 1441 (1985).
5. I. Mirebeau, M. Hennion, G. Parette: Phys. Rev. Lett. 53, 687 (1984).
6. I. Mirebeau:, Third Cycle Thesis, Orsay (1982).
7. P.J. Brown, J. Deportes, D. Givord, K.R.A. Ziebeck: Proc. of Workshop on 3d Metallic Magnetism, Grenoble (August 1983).
   G. Shirane, O. Steinwell, Y.J. Uemura, J. Wicksted: J. Appl. Phys. 55 (6) 1887 (1984).
8. S. Radelaar: J. Chem. Solids 27, 1375 (1966).
9. K. Gschwend, H. Sato, R. Kikuchi: J. Chem. Phys. 69, 5006 (1978).
10. S.G. Fishman, D. Gupta, D.S. Lieberman; Phys. Rev. B 2, 1451 (1970).
    G. Henry, G. Barreau, G. Cizeron; C.R. Hebd. Seances Acad. Sci. Ser. C, 280, (16) 1007 (10) (1975).
11. F. Gautier, A. Bieber: private communication.
12. V. Pierron-Bohnes, M.C. Cadeville, A. Bieber, F. Gautier: ICM'85, San Francisco, to be published in J. Mag. Mag. Mat.
13. P.C. Clapp, S.C. Moss: Phys. Rev. 171, 535 (1968).

Short-Range Order in a $Ni_{63.7}Zr_{36.3}$ Amorphous Alloy. Effects of Structural Relaxation

Y. Calvayrac, A. Quivy, S. Lefebvre, and J. Bigot

CECM-CNRS, 15 rue G. Urbain, F-94400 Vitry, France

R. Bellissent

LLB, CEN Saclay, F-91191 Gif sur Yvette Cedex, France

Introduction

Amorphous metallic alloys undergo changes in many physical properties upon annealing at temperatures well below the crystallization temperature. The different phenomena observed are often attributed to changes in the nature and degree of short-range order, but reliable direct evidence by diffraction techniques are very few. Differential Scanning Calorimetry (DSC) is particularly suited to study the enthalpic relaxation of the glass towards the equilibrium liquid configuration and hence to monitor a diffraction study. Two relaxation effects have been observed by DSC [1] :

- An exothermic phenomenon, extending over a broad range of temperature, occurs during the first scan of the as-quenched sample. This effect is irreversible.
- After prolonged annealing in a critical temperature range below (but close to) the glass transition temperature T_g, an endothermic peak is observed at T_g upon reheating : in the glass transition region the structural relaxation rate increases and the specimen rapidly readjusts towards the equilibrium state. The effect of initial annealing is eliminated but can be reproduced by repeated annealing.

We have planned a complete study of the changes in topological and chemical short-range order produced by these two stages of structural relaxation in the amorphous alloy $Ni_{63.7}Zr_{36.3}$. We present here preliminary results for an isothermal heat-treatment which just eliminates the broad (irreversible) exothermic phenomenon. The three partial structure factors (PSF) were determined from neutron diffraction experiments on three isotopically substituted samples : $^{nat}Ni_{63.7}Zr_{36.3}$, $^{60}Ni_{63.7}Zr_{36.3}$ and $^{62}Ni_{39.2}{}^{60}Ni_{24.5}Zr_{36.3}$ ($\langle b \rangle = 0$, "zero alloy"). The normalised determinant of the equations to be solved to obtain the PSF is exceptionally high,[1] allowing a direct determination of the three PSF from the data without using constraints. The "zero alloy" yields a direct determination of the Bhatia-Thornton partial structure factor $S_{cc}(q)$ which describes the chemical short-range order. Reports of the results obtained on the as-quenched samples have been published elsewhere [2,3]: We indeed observe the existence of strong chemical short-range order and important size effect.

Experimental

The amorphous ribbons were obtained by planar flow casting under a helium atmosphere. These are typically 6 mm wide and 35 µm thick. The specimen for neutron diffraction consisted of between 50 and 100 portions of ribbon (5 cm long) stacked together. These were annealed in a separate furnace, under ultra-high vacuum, for 8 hours at 430°C. Accurate repositioning of the sample in the spectrometer after the anneal was therefore required.

The experiments have been carried out on the spectrometer 7C2 on the hot source of the Orphée reactor using a wavelength of 0.704 Å and a 640-cell position sensitive detector described elsewhere [4] . Data reduction consisted of correction for

1. The value of the normalised determinant is -0.615 for the Faber-Ziman PSF and 0.9 for the Bhatia-Thornton PSF.

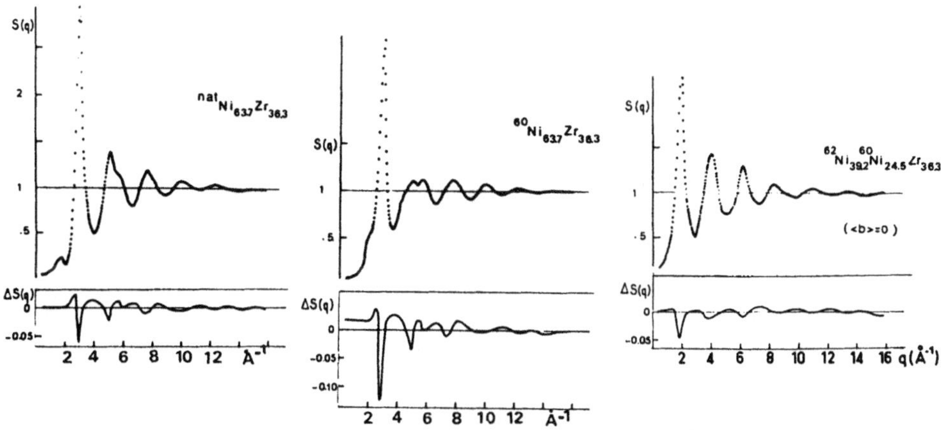

Figure 1. Total structure factors S(q) and their change due to annealing at 430°C for 8 hours. ($\Delta S(q) = S(q)_{as\text{-}quenched} - S(q)_{annealed}$)

transmission [5] , incoherent, multiple scattering |6| and inelastic scattering |7| following classical procedures.

Results

The total structure factor $S_{(q)}$ may be defined by the relation $S_{(q)} = I_{coh}(q)/\langle b^2\rangle$ which is particularly appropriate for zero-alloys. According to the Bhatia-Thornton formalism [8] $S_{(q)}$ for a binary AB alloy can be expressed as

$$S_{(q)} = \frac{\langle b\rangle^2}{\langle b^2\rangle} S_{NN}(q) + \frac{2\Delta b\langle b\rangle}{\langle b^2\rangle} S_{NC}(q) + \frac{\Delta b^2}{\langle b^2\rangle} S_{CC}(q) \qquad (1)$$

where $S_{NN}(q)$, $S_{NC}(q)$ and $S_{CC}(q)$ are the number-concentration partial structure factors. For the three isotopic compositions the factors in equation (1) are given in table 1.

Table 1

	$\langle b\rangle^2/\langle b^2\rangle$	$2\Delta b\langle b\rangle/\langle b^2\rangle$	$\Delta b^2/\langle b^2\rangle$
$^{nat}Ni_{63.7}Zr_{36.3}$	0.973	- 0.667	0.117
$^{60}Ni_{63.7}Zr_{36.3}$	0.815	+ 1.614	0.801
$^{62}Ni_{39.2}{}^{60}Ni_{24.5}Zr_{36.3}$	0	0	4.329

Fig. 1 shows $S_{(q)}$ for the three samples. The structure factor for the ^{nat}Ni-Zr sample, although dominated by the topological short-range order $S_{NN}(q)$, shows a small pre-peak. For the zero alloy this pre-peak becomes the main peak. Fig. 2 is an enlargement of the small q range for the three samples. It can be clearly seen that, for the ^{nat}Ni-Zr sample, the pre-peak is split. The number-concentration PSF are given in Fig. 3. The behaviour of $S_{NC}(q)$ and $S_{CC}(q)$ in the small q range,

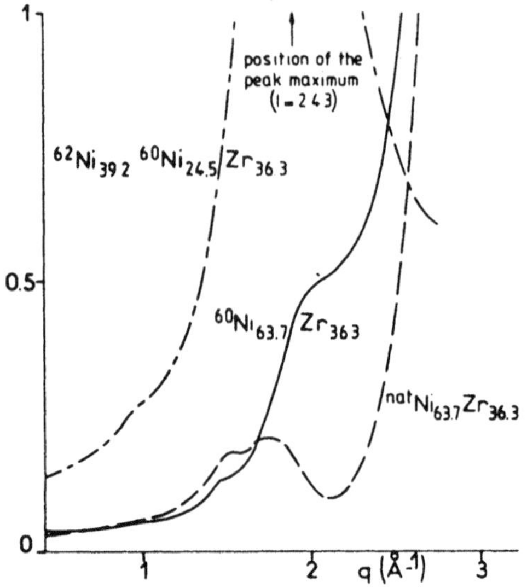

Figure 2. Enlargement of the total structure factors $S_{(q)}$ of figure 1. (small q range).

together with an examination of Table 1, show that the split pre-peak of the structure factor of ^{nat}Ni-Zr clearly reveals the existence of both chemical short-range order and size effect. For the ^{60}Ni-Zr sample the weighting factors of S_{NC} (q) and S_{NC} (q) increase, but the change in the sign of this factor for S_{NC} (q) strongly modifies the shape of the pre-peaks by a partial compensation of S_{CC} (q) and S_{NC} (q) in this range of q (Fig. 2).

Fig. 1 shows the variation, after annealing, of the structure factor for the three samples. The differences are very small, and the observed effects are at the limit of detectability, but a striking feature is that the shape and the position of the oscillations observed in the difference curves for both the ^{nat}Ni-Zr and the ^{60}Ni-Zr samples are the same. The differences on the number-concentration PSF are shown in Fig. 3. The effects on S_{NN} (q) are clearly stronger than those on S_{CC} (q) and S_{NC} (q) and they are similar to those obtained on the structure factors of the ^{nat}Ni-Zr and ^{60}Ni-Zr samples. This is only to be expected for ^{nat}Ni-Zr since the weighting factors are such that S_{NN} (q) largely dominates. The variations of S_{CC} and S_{NC} have no effect on the total structure factor for ^{60}Ni-Zr (despite stronger weighting factors) since these variations are too small. Therefore the observed changes are essentially due to variations in topological SRO.

The global effect observed corresponds to a sharpening of the peaks with no change in the angular position of the maxima. A similar effect has also been observed for Cu-Zr glasses by X-ray diffraction [9]. Such small effects do not lead to changes in the coordination numbers obtained from the partial radial distribution functions. There are no detectable differences for the first neighbours peak in the partial pair correlation functions g_{NiNi} (r) and g_{ZrZr} (r) calculated in the Faber-Ziman formalism. In g_{NiZr} (r) a very small increase of the first maximum is observed. The changes in the number-concentration correlation functions are given in figure 4. The small differences of topological short-range order which appear up to $r \simeq 7$ Å cannot be quantitatively described. The use of a "zero alloy" allows a tiny increase of chemical short-range order to be detected.

During the annealing treatment carried out, an irreversible relaxation occurs with a heat evolved corresponding approximately to 20 % of the heat of crystallization : this effect is considerable. On the other hand, the diffraction experiments just described allow a precise determination of the state of topological and chemical short-range order. Good conditions existed therefore to measure the variation of short-range order on annealing. Qualitatively, significant effects

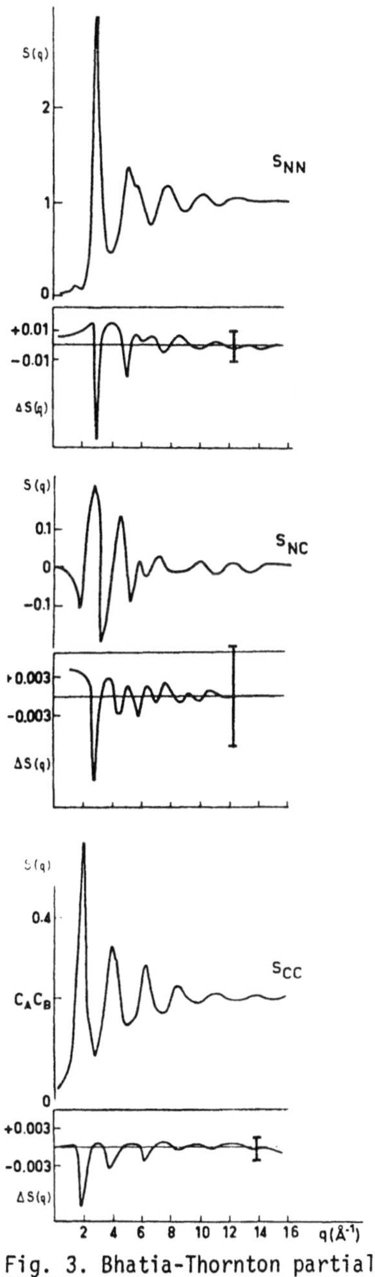

Fig. 3. Bhatia-Thornton partial structure factors and their change due to annealing at 430°C for 8 hours.

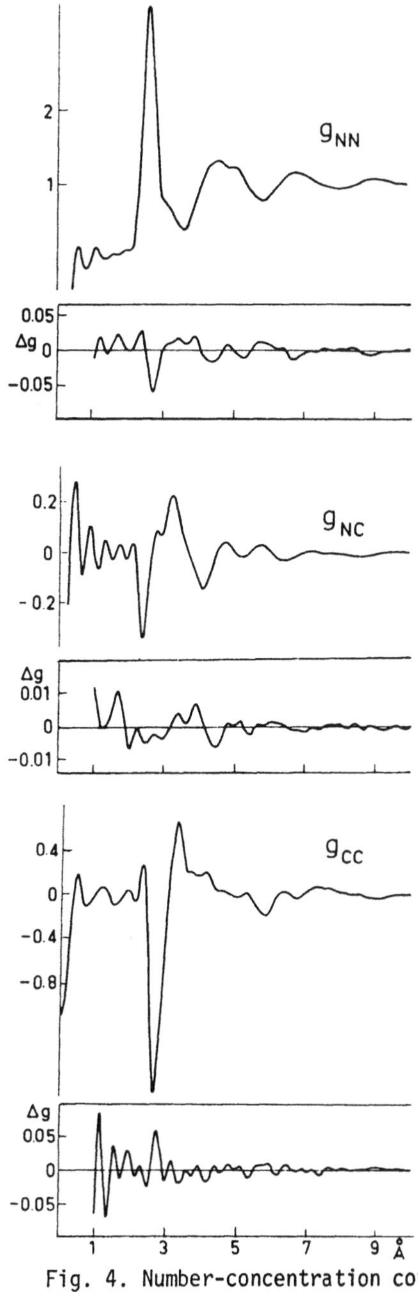

Fig. 4. Number-concentration correlation functions and their change due to annealing at 430°C for 8 hours.

$$(\Delta S(q) = S(q)_{\text{as-quenched}} - S(q)_{\text{annealed}})$$

were observed, leading to a more ordered atomic distribution in a wide sense. However, these effects are very small. With an experimental accuracy of 0.4 % on the total structure factors the error bars associated with the FSP are as large as the changes observed (Fig. 3). Hence, before attempting an even qualitative description of structural changes, the reproducibility and the reliability of the measurements have to be fully guaranteed.

References

1 M. Harmelin, Y. Calvayrac, A. Quivy, J. Bigot, P. Burnier and M. Fayard : J. Non-Cryst. Solids 61/62, 931 (1984)
2 S. Lefebvre, A. Quivy, J. Bigot, Y. Calvayrac and R. Bellissent : J. Phys. F : Met. Phys. 15, L 99 (1985)
3 R. Bellissent, J. Bigot, Y. Calvayrac, S. Lefebvre and A. Quivy : Third Int. Conf. on the Struct. of Non Cryst. Mat, Grenoble (1985)
4 J.P. Ambroise and R. Bellissent : P.S.D. of Thermal Neutrons (Academic, London 1983)
5 H.H. Paalman and C.J. Pings : J. Appl. Phys. 33, 2635 (1962)
6 I.A. Blech and B.L. Averbach : Phys. Rev. A 137, 1113 (1965)
7 J.L. Yarnell, M.J. Katz, R.G. Wenzel and H.S. Koening : Phys. Rev. A 7, 2130 (1973)
8 A.B. Bhatia and D.E. Thornton : Phys. Rev. B 2, 3004 (1970)
9 P. Garoche, Y. Calvayrac, W. Cheng and J.J. Veyssié : J. Phys. F : Met. Phys. 12, 2783 (1982)

A SANS Approach to Possible Hierarchical Structures of Defects in Amorphous Alloys

C. Janot[1] *and B. George*[2]

[1]Institut Laue-Langevin, 156 X, F-38042 Grenoble Cedex, France
[2]Physique du Solide, B.P. 239, F-54506 Vandoeuvre Cedex, France

1. Introduction

Amorphous systems are structurally characterized by a lack of long-range periodicity and the presence of a rather well defined short-range order. In amorphous metallic systems one satisfactory approximation proposed for the structure corresponds to close packing tetrahedra on the ground that a regular tetrahedron is the densest configuration for the packing of four equal spheres. As the dihedral angle of a tetrahedron is not a submultiple of 2π, a perfect tiling of the 3D Euclidean space with regular tetrahedra is not possible. SADOC [1] has first proposed a definition of an ideal amorphous structure by allowing for space curvature so that the local configuration can propagate without defect (a 2D example of this model is the perfect tiling of the surface of a sphere with regular pentagons). The so-called real amorphous structure would then be obtained by projection into our ordinary Euclidean 3D space. Obviously enough projection means distortions and introduces topological defects, among which disclination lines play an important role. Projection from curved to Euclidean space means also decurving. MOSSERI and SADOC [2] have shown that it is possible to remove the curvature using an iterative procedure which produces disclination networks, step by step. The key idea is to introduce, at each step, a disclination network (instead of successive single disclination lines) whose symmetry group is contained in the so-called polytope symmetry group G ({3,3,5} polytope for instance for a perfect icosahedral environment). The final model is a hierarchical structure with regions of positive curvature where the polytope {3,3,5} is maintained and regions of negative curvature associated with the line defects, arranged in such a way that the mean curvature is zero. The successive disclination networks are interlaced so that each site of the network reached at a given step is used to generate the next step network by application of the G group symmetry operations as sketched in Fig. 1.

The purpose of this paper is to report SANS data, measured on amorphous alloys, which may be understood in terms of such a hierarchical structure description of topological inhomogeneities.

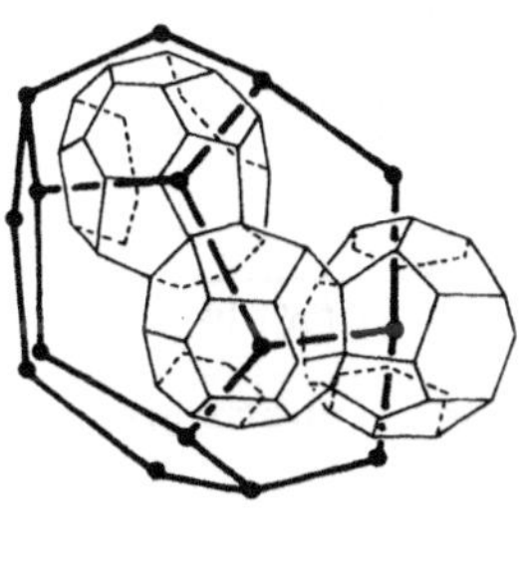

Fig. 1: Local view of the model of interlaced hierarchical networks of disclinations corresponding to a polytope {3,3,5}

2. Sample description and characterization

As the purpose was to observe possible hierarchical microstructural features in non-crystalline systems, optimized conditions would correspond to a random close packing of atoms of a pure metal. Although strictly not achievable, such an ideal situation has been reasonably approximated by amorphous alloys of the $(Fe_{1-x}Mn_x)_{75}P_{15}C_{10}$ systems prepared by the melt spinning technique, with x = 0, 0.26, 0.70 and 0.84. The quenched alloys were checked [3] [4] for the amorphous structure and the Mn-Fe isomorphicity using X-ray diffraction, differential scanning calorimetry and Mössbauer spectroscopy. Within the consistency of these data with an amorphous state and the isomorphous random substitution of Fe by Mn over the same subnetwork [3], the particular alloys with x = 0.26 and 0.84 (refered to as M26 and M84 hereafter) can be suitably studied using neutron diffraction measurements. Due to contrast effects associated to the negative scattering length of manganese (b_{Mn} = - 0.373in 10^{-12} cm) M26 is an alloy without any contrast between the average metal and non-metal atoms. The scattered neutrons are then only sensitive to topological aspects and give the S_{NN} structure factor which Fourier transforms into the $G_{NN}(R)$ density fluctuations at atomic level. Similarly M84 is an alloy with a zero average scattering length which allows direct neutron diffraction measurement of the diffuse scattering S_{CC} (Fourier transformed into G_{CC}) related to possible chemical order. A detailed analysis of the whole set of neutron diffraction data in terms of atomic structure has been previously produced elsewhere [4]. The main conclusions restricted to the present case of interest are illustrated by the curves of Fig. 2 which show that chemical order is practically absent, except perhaps in the first coordination shell, and that rather well-defined topological distances exist at R_1 = 2.53 Å, $R_2 = 1.68R_1$, $R_3 = 1.95R_1$, $R_4 = 2.53R_1$, $R'_4 = 2.94R_1$, $R_5 = 3.40R_1$ ... very comparable to polytetrahedral structure [5]. Finally it can be considered that the M26 alloy is an acceptable real case for the typical topological inhomogeneities of an ideal dense random packing of atoms to be observed.

3. Small-angle neutron scattering study of the M26 alloy

The diffraction of neutron with small scattering vector Q provides information about the state of the specimen over distances larger than several atomic diameters. The limiting value of the structure factor as Q → O simply measures the macroscopic variance of the density of the medium which, for monatomic systems in thermodynamic

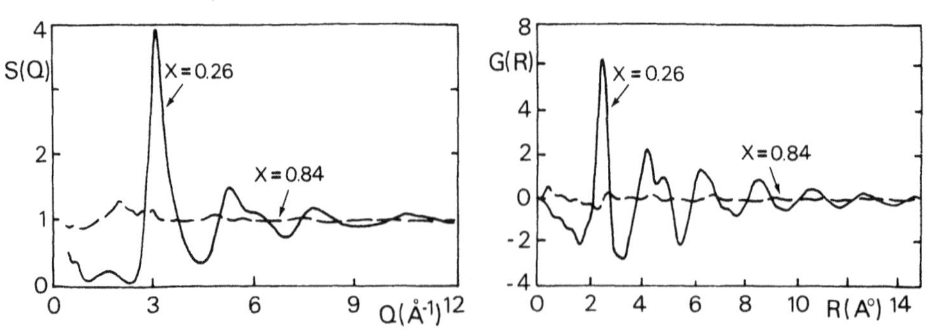

Fig. 2: Structure factors and reduced pair correlation functions of the M26 and M84 alloys

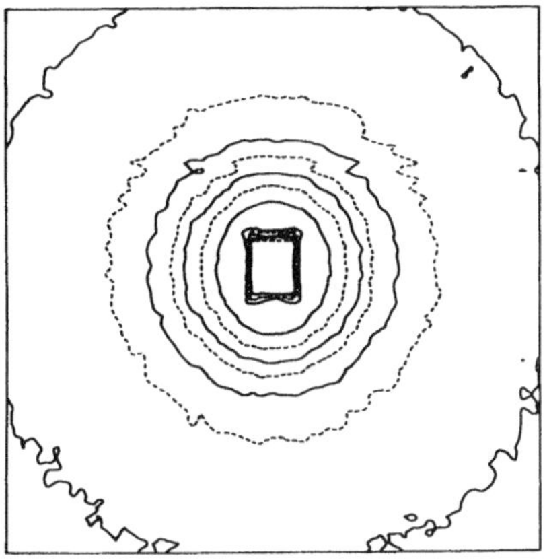

Fig. 3: A tow-D small - angle scattering pattern measured at 4.2K on the M26 sample. Lines are drawn through the equal intensity scattering points.

equilibrium, is simply proportional to isothermal compressibility χ_T (nuclear scattering) or isothermal susceptibility (magnetic scattering). However, amorphous materials are not in thermodynamic equilibrium, nor in an homogeneous single-phase state, so that there is no reason for this proportionality law to be satisfied.

On the ground that the M26 alloy may be typical for neutron studies of a possible hierarchical microstructure, small - angle neutron scattering patterns have been measured on this alloy, at temperatures ranging from 2K to RT, using the D11 small-angle camera at the ILL (Grenoble). With a sample-to-detector distance of 10.6 meters and a neutron wavelength of 7 Å, a Q range from 0.4 to 3.2 (10^{-2} Å^{-1}) has been covered. Surface imperfections effects have been removed using a contrast variation method [6], namely the immersion of the sample in a deuterated liquid whose scattering length has been adjusted to about 5 x 10^{10} cm/cm^{-3}. The 2D-scattering patterns (see Fig. 3) do not show any evidence for anisotropic effects despite the samples being made of about 50 pieces of ribbons carefully aligned to the vertical position with their shiny side perpendicular to the incident beam. Then, it is assumed hereafter that the presence of anisotropic scattering particles can be ruled out and that isotropic data regrouping is likely to be sensible. The regrouped and corrected I(Q) patterns have been fitted to a set of spline functions [7] which artificially allows for extrapolation of the data toward both small and large Q sides. This set of spline functions can then be Fourier transformed to reach a description, in real space, of the inhomogeneities which are responsible for the measured SANS signals. A typical fitted I(Q) curve is shown on Fig. 4 along with a real space distance correlation function P(r) obtained in a simple two-phase model with identical particles and without contribution from interparticle interferences. As the displayed P(r) is not the one expected from isotropic particles, an alternative analysis has been carried out in terms of scattering volume distribution V(d) over diameter d of nonidentical spheres, interparticle interferences being still neglected. A selection of V(d) curves are displayed in Fig. 5 corresponding to SANS data measured at different temperatures. Differences in these curves are due to magnetic transitions (at about 25K and 170K) and neutron multiple refraction effects on domain walls as explained elsewhere [4]. All these curves have in common very well defined features, whose positions are little affected by temperature changes. These features correspond to d_o = 85 Å, d_1 = 208 Å, d_2 = 338 Å = 1.63 d_1, d_3 = 440 Å = 2.12 d_1, d_4 = 570 Å = 2.73 d_1, d_5 = 720 Å = 3.46 d_1. Such peaks in the performed size distribution

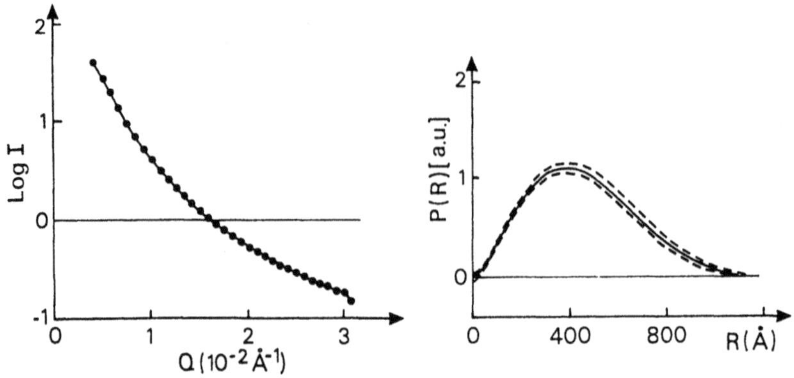

Fig. 4: Regrouped I(Q) data (in Log scale) corresponding to the isotropic pattern of Fig. 3 and distribution function P(R) for the correlation distances

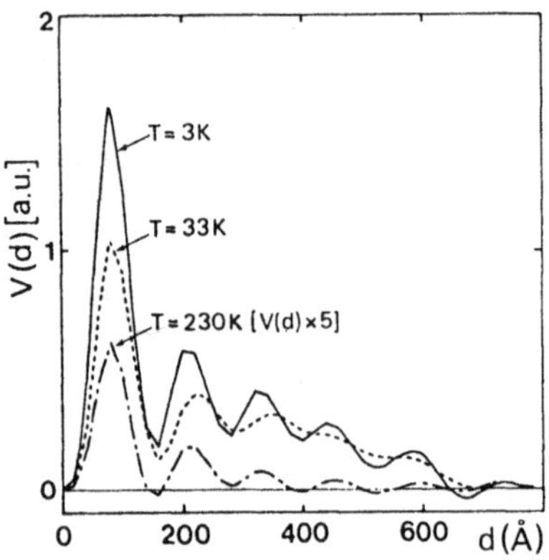

Fig. 5: Distributions of the volume of the scattering particles into spheres of different diameters d, for the M26 alloys, as measured at different temperatures (the amplitudes shown are multiplied by 5 when data have been collected in the paramagnetic state

are obviously intended to be linked to the non-interacting families of different spherical particles which contribute to the SANS signal. However, assuming that we would have made the wrong approach and that scattering would come from, say, identical spheres with interparticle interference, the calculated V(d) distributions would anyway show maximum values at d_o for the sphere diameter and d_1, d_2, etc. corresponding to the successive maximum values of the spheres pair correlation function. Then it may be interesting to point out that, on the one hand, the first larger diameter d_1 is only a little above 2.225 d_o as if d_1 were the diameter of a sphere containing four d_o spheres not quite closely packed and, on the other hand, d_2, d_3, d_4, d_5 are related to d_1 by factors very close to those measured for the short range atomic correlations (see Fig. 2 and R_i values in section 2). Such a result is understandable in a model of atoms icosahedrally packed together into 85 Å diameter "spheres", without significant coherency losses within the measured Q range, these 85 Å diameter spheres being in turn packed into some sort of polytetrahedral structure.

4. Conclusion

Contrast methods in neutron diffraction and small-angle scattering have proved to be very suitable to simulate chemically homogeneous amorphous alloys and to eliminate surface imperfection effects. SANS intensity due to topological bulk inhomogeneities have been conveniently analysed in real space in terms of correlated scattering spheres. The experimental results are consistent with a russian-doll-like microstructural model, in which the atomic polytetrahedral packing would be infinitely reproduced at larger scales within given coherency limits.

5. Acknowledgements

The authors are very grateful to R. MAY for assistance in data treatment.

References

1. J.F. Sadoc: J. Non-Cryst. Solids 44, 1 (1981)
2. R. Mosseri and J.F. Sadoc: J. Phys. Lett. 45, L827 (1984)
3. Chr. Janot, B. George, C. Tête, A. Chamberod and J. Laugier: J. Phys 46, 1233 (1985)
4. Chr. Janot, B. George and C. Tête: Proceedings NCM-3. To be published (1985)
5. Ph. Mangin: "Les Amorphes métalliques", (Les Editions de Physique, Paris 1984), p.175
6. Chr. Janot and B. George: J. Phys. Lett. 46, L85 (1985)
7. O. Glatter: J. Appl. Cryst 10, 415 (1977)

SANS Measurements on Metallic Glasses and the Contributions of Nuclear and Magnetic Structure Defects and of the External Surfaces

R. Yavari and P. Desre

L.T.P.C.M., U.A. CNRS n° 29, Institut National Polytechnique de Grenoble, BP 75, Domaine Universitaire, F-38402 Saint Martin d'Hères Cedex, France

P. Chieux

Institut Laue-Langevin, BP 126 X, F-38042 Grenoble Cedex, France

1. Introduction

We are interested in the study of structures or heterogeneities in the medium-size range (10 to 500 A) that may exist in our metallic glasses. Small-angle neutron scattering (SANS) is an appropriate tool for studying such defects, because unlike for SAXS, sample thicknesses of the order of 1 mm can be used to increase the irradiated volume and the signal and because low scattering vectors $Q = 4\pi\sin\theta/\lambda$ and scattering angles 2θ are easily accessible with long neutron wave lengths $\lambda \simeq 10$ A. Among the defects we might observe, we will discuss here : quenched-in density fluctuations, regions of compositional heterogeneities, magnetic structural defects such as Bloch walls in ferromagnetic glasses and external surfaces and surface defects.

2. Density Defects

When liquid alloys undergo rapid quenching, the melt must undergo thermal contraction during the quench and a net excess "free-volume" must be evacuated to the external surfaces and might be partially quenched in as density defects.

The kinetics of atomic mobility and of relaxation towards the equilibrium density must somehow depend on the melt fluidity and viscosity. The viscosity of a pure metal rises rapidly with undercooling. That of a glass-forming alloy increases even more sharply with reducing temperature T [1] and rises to 10^{10} poise at Tg where a glass-transition is observed on reheating. Thermal contraction is quickly accommodated by relaxation as long as the fluidity is high, but as the melt becomes viscous near T_g, and for the very fast quench-rates (10^5 to 10^6 K/S) required for alloy glass formation, the higher temperature topology and specific volume of the melt are frozen into the glassy state. For cooling between T_g and the substrate temperature $T_o \simeq 300$ K, this results in excess atomic volume fraction :

$$\Delta V/V \simeq 3\alpha_l\ (T_g - T_o) \simeq 0{,}9\ \% \qquad (1)$$

where $\alpha_l \simeq 10^{-5}$/k is a linear thermal expansion coefficient and as is usually the case, we have taken T_g to be between 0.4 and 0.6 times the melting temperature T_m of for example 1200 K.

$\Delta V/V$ corresponds to the volume fraction that must be evacuated from the bulk of the liquid towards the surfaces in a time $t = 6.10^{-4}$ seconds or else it will become part of the quenched-in free volume. For a sample of thickness $\varepsilon = 50\ \mu$m, this would require a free-volume diffusion coefficient

$$D_v \simeq (\varepsilon/2)^2/t \qquad (2)$$

or about 10^{-2} cm^2/s. The fastest possible non-correlated transport mechanism for a defect in a liquid is random displacements at the speed of sound $V_s \simeq 10^5$ cm/s over mean-free-paths of length d_o. Using the interatomic spacing for d_o ($\simeq 10^{-8}$ cm) such a diffusion mechanism for a hole (or free-volume element) will give a coefficient :

$$D_v \simeq V_S a_o \simeq 10^{-3} \mathrm{cm}^2/\mathrm{s} \qquad (3)$$

We would therefore need this upper-limit maximum diffusion coefficient or more to evacuate to the surface all of the free volume produced by the thermal contraction of the bulk during the quench as given by equation(1). While such D_v values are not unexpected above the melting temperature ($T > T_m$), the diffusion of a hole is also expected to be temperature dependent (thermally activated) and somehow dependent on the melt viscosity, which attains high values at high undercoolings in the range $T \lessdot T_g$ and values of $D_v << 10^{-3} cm^2/s$ are expected below T_g. In the disordered state of a glass, if hole diffusion coefficients were near $D_v \approx 10^{-3} cm^2/s$, all glasses would immediately relax to their maximum density ideal glass-state while in fact, annealing over several hours is needed to obtain densification $\Delta V/V \approx 0.4$ % (for a recent article, see for example [2]) which implies by equation (2), D_v values of the order of $10^{-8} cm^2/s$ less than 100 K below T_g. Both these relaxation kinetics and the observed densification support the idea that some of the volume-contraction during the quench is trapped as free-volume in the glassy state.

In a crystal, such a volume excess would be like a massive vacancy supersaturation of $\Delta V/V$ = 4000 ppm. Short of annealing-out at the surfaces, grain-boundaries or dislocation-cores, excess vacancies in crystals quickly form clusters in the bulk, clusters that grow and produce a variety of defects such as stacking-fault tetrahedra, voids or dislocation loops, and eventually anneal-out on reheating at sufficiently high T of the order of 0.4 to 0.6 T_m. Their formation, growth and disappearance results in simultaneous dramatic changes in mechanical and other properties [3]. In a glass, excess free-volume is not expected to necessarily take the form of atomic-sized vacancies, but qualitatively, their annihilation process other than at surfaces and heterogeneities, can hardly be other than through clustering and collapse of the larger clusters or the formation of voids. Such defects would be expected to be detectable by SANS measurements.. Any observed SANS however, could also be attributed to possible compositional heterogeneities of fluctuations. We therefore conducted the following experiment to distinguish between these interpretations.

3. The $Pd_{80}Si_{12}Ge_8$ Experiments

We have previously shown that $Pd_{80}(Si_xGe_{1-x})_{20}$ alloys can be treated as quasi-binary Palladium-Metalloid alloys [4,5]. This allows us to prepare a quasi-binary $Pd_{80}Si_{12}Ge_8$ alloy with zero contrast for composition fluctuations as follows. With coherent scattering lengths b_{Pd} = 0.59, b_{Ge} = 0.84 and b_{Si} = 0.42, (all in 10^{-12} cm units), for the composition $Pd_{80}Si_{12}Ge_8$, we obtain $\Delta b = b_{Pd} - b_{metalloid} = 0$. Thus, any observed SANS due to metal-metalloid compositional heterogeneities or fluctuations should disappear in the case of $Pd_{80}Si_{12}Ge_8$ provided that compositional fluctuations do not result in large internal stresses. This is likely to be the case, as the alloy atom density is rather insensitive to composition between 0 and 33 % metalloid content [5].

Most SANS data presented here and in the rest of this paper have been obtained at ILL (D17 and D11 spectrometers) with some having been obtained on the PAXY spectrometer at LLB-Saclay. Corrections such as for background, sample-absorption and incoherent scattering were done as before [5,6] with normalisation done using the incoherent scattering of water. The glassy ribbons are placed with their thickness perpendicular to the neutron beam and the ribbon long axes parallel to one of the edges of the 2-dimensional square area-detectors as shown in figure 1. Figure 2 shows I(Q) versus Q curves obtained in this manner for amorphous $Pd_{80}Si_{12}Ge_8$ prepared in fine narrow ribbon form (about 1 mm wide) and 20 mm wide ribbons which are also somewhat thicker. Also shown is I(Q) versus Q for the narrow

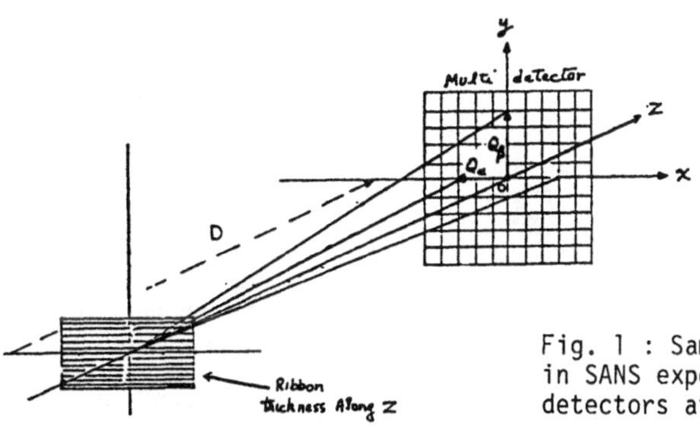

Fig. 1 : Sample-detector geometry used in in SANS experiments on 2-dimensionnal detectors at ILL- Grenoble and LLB-Saclay

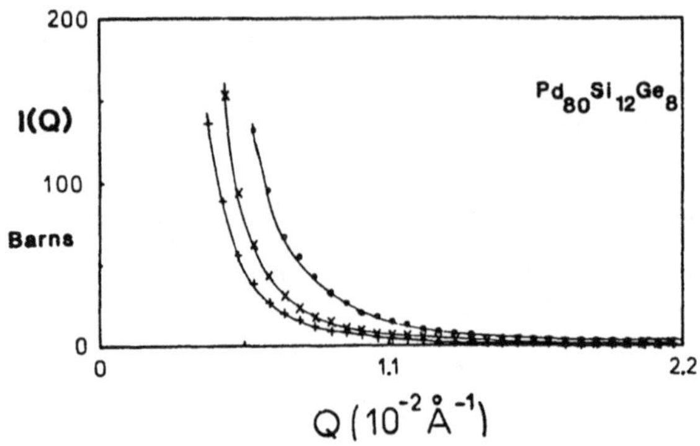

Fig.2 : SANS intensities for (top) narrow, (bottom) wide and (middle) narrow annealed glassy ribbons.

ribbon after a relaxation annealing at 523 K well below T_g (some anisotropy [7] in the SANS has been observed and attributed to a shape anisotropy of the scatterers induced by substrate shearing of the solidifying liquid [8]).

Clearly, the curve for the narrow glassy ribbon which is quenched more rapidly and contains more quenched-in defects is well above that of the more slowly cooled wide ribbon, but upon relaxation annealing drops towards the latter [9] as the tail of I(Q) moves towards lower Q.

However, we have not been able to deduce the exact nature of the quenched-in defects. Guiniers plots do not yield linear regions in the Q-range corresponding to $|R(Q).Q|<2$ with the I(Q) data fitting better to a Q^{-3} law which has been attributed to edge dislocations in crystals [10]. Although such dislocations cannot exist in amorphous metals, we can imagine regions ressembling dislocation loops which would have an edge-type character, and which would form by the agglomeariation and collapse of quenched in excess free volume $\Delta V/V \simeq 0.4\%$ as in eq(1).

4. The Contribution of the External Surfaces to SANS

When the population of appropriately sized quenched-in structural defects is very low, as in the more slowly cooled wide ribbon of Fig. 2, we must naturally begin to see the contribution of the last remaining defects, which are the sample surface and its structure. Indeed log I - Log Q plots of low-intensity SANS curves for slowly quenched wide ribbons exhibit a slope of -4 which is typical of abrupt interfaces such as the external surface, as we have previously reported [4]. Note that in the configuration of Fig. 1 and if the surface are flat, the discontinuity that is the surface will be an extended defect of dimensions $d = 2\pi/Q$ much larger than the range $d \ll 1500$ A detectable in a range of $Q \gg 4.10^{-3}$A as in our experiments. Thus surface defects and not the surface itself will contribute to SANS in this configuration (sharp ribbon edges could contribute but for the wide ribbons the edges are not in the neutron beam). There are usually many defects on the substrate side of roller-quenched metals, as shown in the SEM picture of Fig. 3. These defects, mostly due to gas bubbles trapped between the substrate and the solidifying ribbon, are somewhat elongated along the casting direction, due to substrate shearing of the solidifying liquid [8]. These defects are cavities with depth of a few μm and their inner concave surfaces and their edges represent, depending on the curvature (or depth to width ratio), surface components perpendicular to the detector plane of Fig. 1 which scatter in the Q-plane of the detector. We propose therefore that when the bulk defects detectable in the SANS range are absent, the remaining intensity due to surfaces [4] is more particularly due to the components of cavity surfaces perpendicular to the detector plane.

More recently, Rodmacq et al [11] have measured SANS due to surfaces in wide (thus slowly quenched) $Pd_{80}Si_{20}$ ribbons submerged in mixtures of CH_3OH and CD_3OD with increasing D/H ratios. They found for as-quenched, mechanically polished and etched ribbons that each time SANS goes through a minimum for a given D/H ratio in the solution used for wetting its surfaces and that the minimum SANS values correspond to less than 25 % of the SANS measured in air. This brought these authors to conclude that more than 75 % of all SANS in metallic glasses is due to surface scattering, which is of course not the case, as such SANS is very sensitive to quench-rate (see previous section). Their results, however, do confirm that

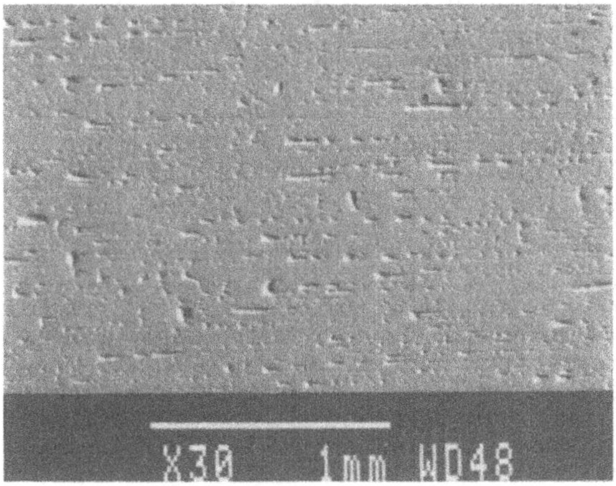

Fig. 3 : SEM micrograph of substrate side of glassy ribbon

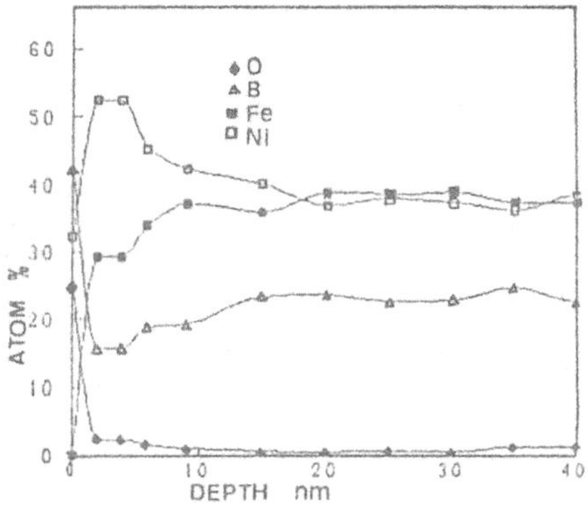

Fig. 4 : Surface elemental depth profiles in glassy $(Fe\text{-}Ni)_{83}B_{17}$ by A.E.S.

surface scattering is present and becomes important in the absence of significant quenched-in defect populations. By way of example, the SANS value these authors measure in air for as-quenched $Pd_{80}Si_{20}$ at $Q = 8.10^{-3}$A is about 5 to 10 barns. At this same Q value for $Pd_{80}Si_{12}Ge_8$ we measured 13 barns for the more slowly quenched wide ribbons and ≈50 barns for the more rapidly quenched narrow ribbons of Fig.2.

Another important result of Rodmacq et al [11] which, however, they did not explain, is that the D/H ratio in the $CH_3OH\text{-}CD_3OD$ mixture at which the minimum in SANS is observed is higher for the polished than the as-quenched sample, and is highest for the etched sample. Furthermore, only the average coherent scattering length per volume of the solution yielding the SANS minimum for the etched sample corresponds to that of $Pd_{80}Si_{20}$. These interesting results can be explained as follows : metal surfaces are in reality almost always covered by oxide layers. Figure 4 shows, by way of example, a composition profile we have recently obtained on the extreme surface of an amorphous $(Fe\text{-}Ni)_{83}B_{17}$ ribbon [12] by Auger Electron spectroscopy. It can be seen that even below the oxide layer the surface composition does not correspond to that of the bulk. More particularly since oxides have lower atom-densities than metals, they reduce any SANS due to surfaces of most metals and alloys. XPS analysis on our $Pd_{80}Si_{20}$ glassy ribbons [13] has equally shown that the surface is almost exclusively covered by SiO_2 and Palladium oxide. This explains why in the experiment of [11] with $Pd_{80}Si_{20}$ submerged in a CH_3OH/DH_3OH solution, as the surface (oxide) layer is removed, partially by polishing and completely by chemical etching, the solution having the same coherent scattering length per unit volume as the sample is obtained at increasing D/H ratios.

Concerning SANS due to surfaces in as-quenched glassy alloys, there are two ways that an upperlimit to this contribution can be defined. Firstly, for a given morphology, the surface contribution should scale with the coherent scattering length, such that whether covered by an oxide layer or not, it cannot be increased by say more than a factor of 2 in going from $Pd_{80}Si_{20}$ to another alloy composition. Secondly, in a given composition, only a range of morphologies are consistent with glass-formation by liquid quenching, which requires very high quench rates and good thermal contact at the alloy-substrate interface. For example, if instead of the morphology of Fig.3, all of the ribbon's face was covered with very fine, deep cavities, although as per our discussion this would strongly increase SANS due to surfaces, the as-quenched alloy would no longer be amorphous due to insufficient metal-metal contact on the surface and a reduced quench-rate. Therefore SANS due to surfaces of as-quenched metallic glasses will always be of the same order as that measured by Rodmacq et al [11] and the lower I(Q) versus Q curve of our Fig. 2.

Separately from the effect of the discontinuity that is the surface, one must also consider the effect of the surface layers, such as in Fig. 4, which are compositional heterogeneities with 50 to 100 A thickness and likely to evolve with annealing in different ways under different conditions.

5. The Contribution of Bloch Walls to SANS in Ferromagnetic Glasses

Normally in soft ferromagnetic metallic glasses, all magnetic scattering of neutrons at small angles should be suppressed if a moderately intense magnetic field (H ≃ 5 to 10 KOe) is applied in the plane of the glassy foil. There is however some controversy on this matter, with some experimentalists reporting disappearance [14] and others the persistance [15] of SANS in similar alloys when such magnetic fields are imposed. We have previously discussed some of the reasons for these differences [16]. We can add here that although most of the magnetic structure disappears with such saturating magnetic fields, some may remain. Firstly and in analogy with crystalline materials and the discussion concerning quenched-in defects in this paper, if there exist free volume agglomerates, quasi-dislocations or other density defects, the magnetic moments of atoms around them will be much more difficult to allign in an arbitrary direction, and saturation may be incomplete. Secondly, slower quench-rates result in stress zones with perpendicular magnetisation and micro-domains [17] covered on the surface with arrays of closure domains [18]. These structures are more difficult to remove by imposing H in the ribbon plane, and some may remain and contribute to SANS. A better way of separating magnetic and nuclear SANS is to measure SANS well below and well above the Curie temperature T_c. We are therefore conducting such experiments on Fe-Ni-based glasses with $T_x - T_c > 100$ K and the results will be published elsewhere.

A simple way of studying qualitatively the magnetic and nuclear contributions to SANS is by measuring at ambient temperature and with no magnetic field, the variation in SANS with composition in glasses whose T_c are very sensitive to small changes in composition. We have done such experiments on $Fe_xNi_{80-x}P_{14}B_6$ glassy alloys with x = 0.40, 0.31 and 0.24 having Curie temperature near 500, 400 and 300 K respectively. Figure 5 shows I(Q) versus Q for these three alloys in the low Q range. It can be seen that for $Q < 10^{-2}$A, I(Q) rises sharply but is strongly dependent on the composition (and T_c) falling rapidly with decreasing X and T_c. A further run with Fe content X = 0.16 atomic fraction and T_c ≃ 200 K yielded a I(Q) curve nearly identical to that of X = 0.24 and T_c = 300 K and is therefore not shown here. As we have discussed elsewhere [16], $b_{Ni} = b_{Fe}$ and Fe-P and Fe-B alloys have short-range-order and crystalline structure similar to Ni-P and Fe-B. It is

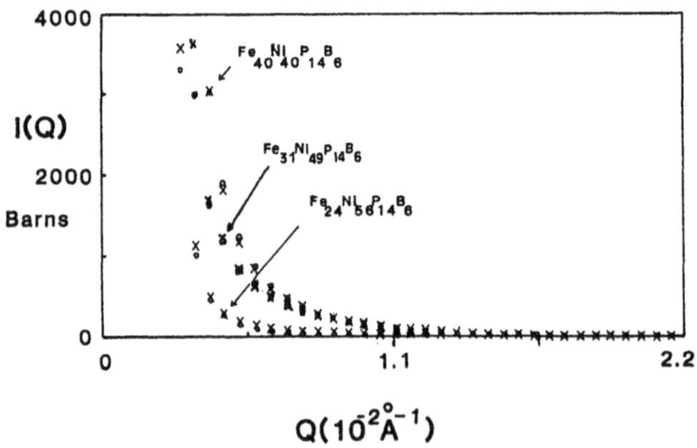

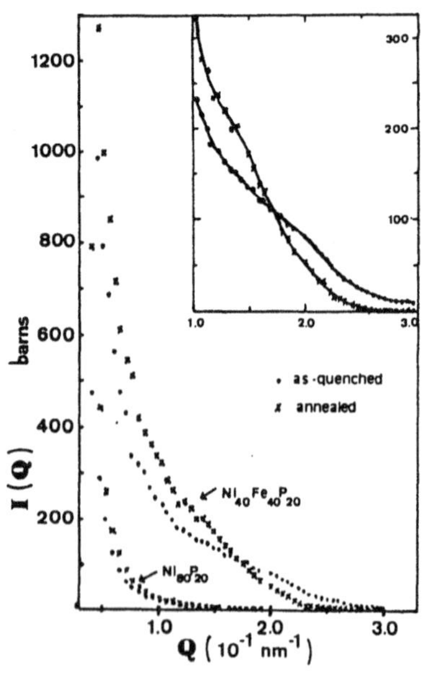

Fig. 5 : (top) SANS in $Fe_xNi_{80-x}P_{14}B_6$ glassy ribbons

Fig. 6 : (right) SANS in $Fe_{40}Ni_{40}P_{20}$ as-quenched and annealed glassy wide ribbons.

therefore not expected that a 9 % increase in Ni content (as is the case for $Fe_{24}Ni_{56}P_{14}B_6$ compared to $Fe_{31}Ni_{49}B_{14}P_6$) will cause a drastic change in the nuclear SANS. The strong dependence of I(Q) on T_c and X may thus be attributed to the magnetic structure and magnetic SANS. There may however exist a magnetic SANS due to magnetic structure around nuclear structural defects (e.g. voids or quasi-dislocations). Such magnetic coupling with nuclear structural defects can only increase the effective scattering cross-section σ of unpolarised neutrons by [19] :

$$\sigma/\sigma_{nucl} = (b^2 + P^2 \sin^2 \alpha)/b^2 \qquad (4)$$

where is the angle between the scattering vector Q and the direction of atomic magnetic spin. Now p depends on the magnetic form factor and the magnetic moments and for natural mixtures of isotopes of Ni and Fe we have $P < 1.46$ [16]. Thus $\sigma/\sigma_{nucl} < 3$ for magnetic SANS contrast due to nuclear structural defects in the alloys of Fig.5 while experimentally, at lower Q values $Fe_{40}Ni_{40}P_{14}B_6$ with $T_c \approx 500$ K scatters more than 7 times that of $Fe_{24}Ni_{56}P_{14}B_6$ with $T_c \approx 300$ K. These considerations all converge towards the reasoning that the observed scattering is due to Bloch walls : such Bloch walls disappear with the domains at $T > T_c$. Furthermore, the walls represent regions of abrupt atomic magnetic spin rotation over thicknesses d of the order of 1000 A that should be detectable near $Q = 2\pi/d = 6 \times 10^{-3} A^{-1}$ which indeed corresponds to the Q-window of Fig.5.

6. Applications of SANS in Metallic Glasses : the Case of Embrittlement

We have seen that for certain alloy compositions, SANS can be used to monitor relaxation and the annealing-out of quenched-in density defects in metallic glasses. However, SANS intensity is much higher when magnetic scattering of neutrons is present, as in ferromagnetic metallic glasses. This renders the measurements more efficient, thus requiring less beam time. Furthermore, Fe-based ferromagnetic glasses are of technological importance due to their soft-magnetic properties. However, optimisation of these same properties requires relaxation annealing at temperatures which result in embrittlement. Recently we have studied the embrittlement of $Fe_{40}Ni_{40}P_{20}$ by SANS at the LLB Saclay [20]. We will present preliminary results at this workshop, while complete results will be published elsewhere.

Figure 6 shows SANS in $Fe_{40}Ni_{40}P_{20}$ wide ribbons before and after annealing for 2 hours at 473 K which results in embrittlement. It is seen that I(Q) evolves sensitively with annealing and a "hump", already present in the I(Q) versus Q curve of the as-quenched sample, is displaced towards lower angles in the embrittled state. The development of phosphorous-rich zones has been previously observed by Auger-electron spectroscopy on fracture surfaces of embrittled phosphorous containing Fe-based metallic glasses [21]. The displacement with annealing of the hump in the SANS in $Fe_{40}Ni_{40}P_{20}$ towards lower Q-values is consistent with an increase in a correlation distance between scatterers. Further observations using SANS and STEM on this glassy alloy are in progress.

In conclusion, although the application of SANS to metallic glasses is still under development, it is increasingly clear that SANS will contribute as a powerful metallurgical tool for studying the medium-range structure of metallic glasses.

References.

[1] A.R. Yavari, Amorphous Metals and Non-Equilibrium Processing, pp.31-43, Les Editions de Physique 1984.
[2] T. Komatsu, K. Matusita, R. Yokota, J. Non. Cryst. Solids, 69 (1985) 347.
[3] A.R. Yavari, Phys. Stat. Sol.-a, 78 (1983) 583.
[4] A.R. Yavari, J.C. Joud, M.C. Bellissent, Phys. Lett. 105 A (1984) 88.
[5] A.R. Yavari, M. Maret, C.R. Acad. Sci. 296 (1983) 1637.
[6] A.R. Yavari, D. Bijaoui, M. Maret, P. Chieux, Amorphous Metals and Non-Equilibrium Processing, pp; 281-6, Les Editions de Physiques 1984.
[7] A.R. Yavari, P. Desré, J. Phys. F.: Met. Phys. 14 (1984) 291.
[8] A.R. Yavari, P. Hicter, P. Desré, Proc. Rapid. Quench. Metals, vol. I, pp. 713-6, Japan Metal Society, Sendaï 1981.
[9] A.R. Yavari, J. de Physique Lett. 46 (1985) L-189.
[10] H.H. Atkinson, P.B. Hirsch, Philos. Mag. 3 (1958) 213.
[11] B. Rodmacq, Ph. Mangin, A. Chamberod, Phys. Rev. B. 30 (1984) 6188;

[12] D. Bijaoui, A.R. Yavari, J.C. Joud, Proc. 3rd Int. Conf. Struct. Non-Cryst. Mat. Grenoble, July 1985 in press.
[13] G. Jorgé, Diplôme d'Etudes Approfondies, ENSEEG, 1985 ; G. Jorgé, R. Faure, R. Durand, A.R. Yavari, Congrès MALISORA, Toulouse, juin 1985.
[14] G.E. Fish, H.R. Child, J. Appl. Phys. 52 (1981) 1880.
[15] E. Nold, S. Steeb, P. Lamparter, G. Rainer, J. de Physique 41 (1980) C8-186.
[16] A.R. Yavari, Proc. 5th Int. Conf. Rapidly Quenched Metals, Vol. I, pp. 495-500 Wurzburg, Sept. 1984
[17] A.R. Yavari, P. Desré, J. Mat. Sci. Lett. 2 (1983) 516.
[18] H. Kronmuller et al, J. Mag- Mag : Mat. 13 (1978) 53.
[19] G.E. Bacon, Neutron Scattering, Clarendon Press, Oxford 1975.
[20] D. Bijaoui, A.R. Yavari, M.C. Bellissent, Rap. Exp. LLB-Saclay, Sept. 1985.
[21] J.L. Walter, F. Bacon, F.E. Luborsky, Mat. Sci. Eng. 24 (1976) 239.

Short-Range Order in Metallic Glasses

P. Lamparter

Max-Planck-Institut für Metallforschung, Institut für Werkstoffwissenschaften, D-7000 Stuttgart, F.R.G.

1. Introduction

In recent years considerable progress has been made in understanding the atomic-scale structure of binary metallic glasses. On the experimental side, the three partial pair correlation functions have been determined for a variety of binary metallic glasses. On the theoretical side models have been constructed, mostly by computer calculations, which fit the experimental data quite well. However, in spite of this bulk of information there is still a lot of controversy about the construction schemes of the structure of metallic glasses.

Some general structural features of all the glasses investigated up to now are well known, as for example the chemical short-range ordering (CSRO), due to the chemical interaction between unlike atoms, and also the size effect which is caused by the different atomic diameters of both constituents. However, some questions cannot be answered finally up to now: To what extent does the short-range order in glasses resemble that of related crystalline phases? Or, is the structure rather governed by quite general features which may even be the same in both groups of metallic glasses, namely the metal-metal (M-M) alloys and the metal-metalloid (M-m) alloys? The size effect is obviously one of those features.

In the following, some of our present knowledge on the structure of M-M and M-m glasses will be briefly illustrated. It should be noted that some correlation functions used in this work have been taken from plots in publications.

2. Metal-metalloid (M-m) glasses

Since the experimental evaluation of the complete set of the three partial pair correlation functions of amorphous Ni(81)B(19) [1] a couple of models for its structure have been developed which, however, are based on quite different ideas. In Fig.1 three selected examples are presented.

Figure 1a shows the result of a simple hard sphere computer model developed in [2] and applied to this glass in [1] where the short-range order was simulated by packing of hard spheres with two different appropriate diameters, hereby observing the rule that the smaller atoms (m) never touch each other. This model already describes the general features of the G(R) quite well. Even the double peak structure of the B-B correlation near 3.5 Å, which has been the most interesting and crucial experimental result, shows up in the model curve.

Figure 1b shows the result of a molecular dynamics model calculation [3] where the short-range order results from relaxation of the system under the influence of appropriate pair potentials. Besides out-of-phase oscillation of the B-B correlation at larger distances, the agreement with the experimental data is very good.

Figure 1c shows a completely different approach. In this trigonal prismatic packing model [4] the structural units are not single atoms but trigonal prisms formed by the Ni-atoms and containing a B-atom in their center and packed together with following relaxation of the system. This model is based on the idea that in related crystalline phases of M-m glasses, the occurrence of trigonal prisms is a characteristic structural feature, and that the short-range order in the glasses is governed by the same feature. Again we find very good but not complete accordance with the experimental data.

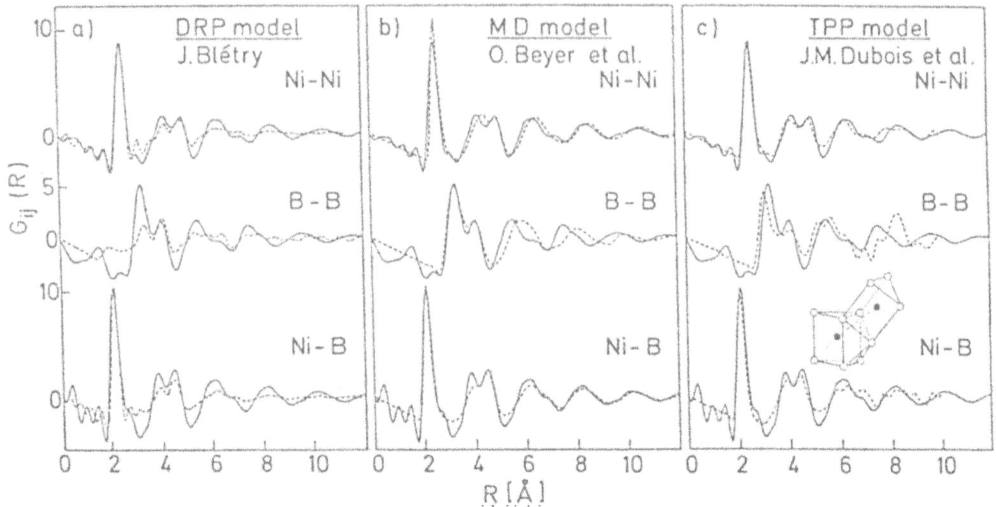

Fig. 1) Ni(80)B(20): Partial pair correlation functions G(R)
—— experimental, ---- theoretical models. For details see text

Table 1. Structural parameters of metallic glasses. M-m glasses: 1=M,2=m. M1-M2 glasses: 1=M1, 2=M2. R-ij=distance between i-atoms and j-atoms. Z-ij=number of j-atoms around a central i-atom. η=short-range order parameter. ri=atomic radius of i-atoms. Δ(R-12)=(r1+r2)-(R-12)[pct]. Data for crystalline alloys in parentheses.

a:amorphous c:crystalline	Ref.	R-11 R-21 R-22		Z-11 Z-21 Z-22		η r1/r2 Δ(R-12)
		a:	c:	a:	c: :	
a:Fe(80)B(20)	[7]	2.57		12.4	(10.7)	0.148
c:Fe(3)B		2.14		8.6	(9)	1.29
		3.57			(4)	5.0 pct
a:Ni(81)B(19)	[1]	2.52	(2.59)	10.8	(11.3)	0.168
c:Ni(3)B		2.11	(2.04)	9.3	(6)	1.27
orthorhombic		3.29;4.02	(3.06;3.55)	3.6;3.7	(2;2)	5.0 pct
a:Ni(64)B(36)	[8]	2.55	(2.59)	9.2	(11)	0.235
c:Ni(2)B		2.12	(2.14)	8.7	(8)	1.27
		1.72		1.1		4.5 pct
a:Ni(80)P(20)	[6]	2.56	(2.68)	9.4	(10)	0.198
c:Ni(3)P		2.28	(2.28)	9.3	(9)	0.97
tetragonal		3.73;4.30	(3.44)	5.3;3.3	(4)	9.5 pct
a:Ti(84)Si(16)	[9]	2.90		11.5		0.135
		2.64		9.4		1.11
		-		-		5.4 pct
a:Y(67)Ni(33)	[10]	3.56				0.15
		2.87				1.44
		2.51				5.0 pct
a:Dy(69)Ni(31)	[11]	3.50		14.7		0.03
		2.85		11.0		1.45
		2.54		3.2		6.3 pct
a;Ti(60)Ni(40)	[12]	3.01	(2.91-2.99)	8.1	(9)	0.169
c:Ti(2)Ni		2.60	(2.49-2.89)	7.9	(9)	1.19
		2.63	(2.87)	2.3	(3)	4.1 pct
a:Zr(50)Ni(50)	[13]	3.32	(3.26-3.46)	7.8	(8)	0.132
c:ZrNi		2.73	(2.66-2.87)	6.7	(7)	1.29
		2.63	(2.49;3.26)	3.3	(2;2)	3.9 pct
a:Zr(64)Ni(36)	[14]	3.30	(2.98-3.43)	11.0	(11)	0.03
c:Zr(2)Ni		2.85	(2.76)	8.6	(8)	1.29
		2.45	(2.62)	3.3	(2)	0 pct?

At this point, where so different models almost give the same result, the situation seems to be somewhat hopeless. This problem arises from the fact that experimentally derived pair correlation functions present a one-dimensional picture of a three-dimensional structure, that is, it contains no information about, e.g., triplet correlations. Further enlightenment of this situation is expected from a three-dimensional analysis of dense random-packing models and of molecular dynamics models. Which structural units, maybe trigonal prisms, are favoured in the resulting structures? Do the different models, despite their different starting points, end up in a similar three-dimensional structure? Indeed, in [5] it was shown that relaxation of a dense packing model for a Fe-B glass leads to an increase of the population of distorted trigonal prisms.

Another progress is expected from experimental data for further M-m glasses. In Fig. 2 the partial pair correlation functions of amorphous Ni(80)P(20) [6] are shown as solid lines and compared to those of the Ni-B glass, dotted lines. Whereas the main features are similar in both alloys, especially the double peak of the m-m correlation, there are also quite different details, e.g., in the m-m curve near 6 Å. Application of existing theoretical models to these recent experimental data will be very revealing.

Structural data of M-m glasses are compiled in Table 1. The most interesting feature of the figures for these glasses is the fact that independent of the two constituents, of the ratio of their atomic diameters, of the concentration, and of the different structures of related crystalline phases, the metalloid atoms (m) always have - within the experimental accuracy - nine metallic (M) neighbors (Z-mM = 9). Further common features are a narrow m-M correlation in G(R) which is sharper than the M-M correlation (see, e.g. [1,6,7]) and the avoidance of nearest m-m neighbors. The latter feature is reflected in the maximum possible value (100 %) of the short-range order parameter η for the case of M(80)m(20) glasses, which, according to [15,16] follows from the coordination numbers. Obviously, the m-atoms play a dominant role in the construction of the amorphous structure, dictating their own neighborhood by chemical interaction with the M-atoms.

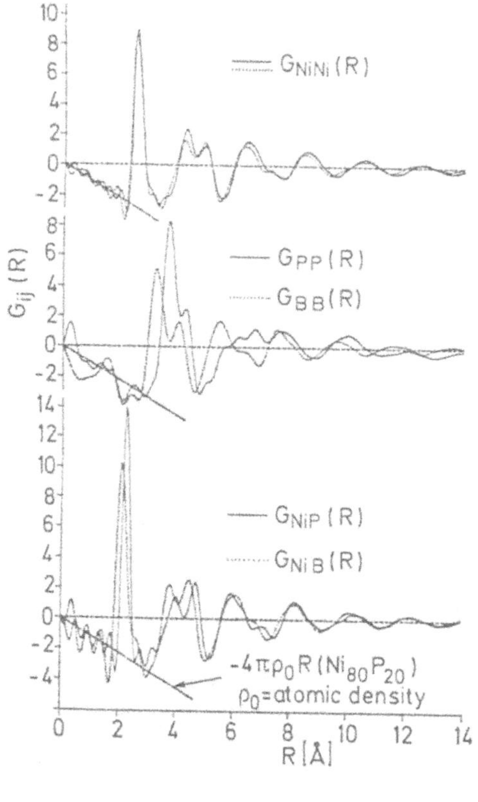

Fig. 2) Ni(80)P(20), Ni(80)B(20): Partial pair correlation functions. The curves for Ni-B(•••••) are the same as the experimental curves (——) in Fig. 1

However, the details of the structures also show differences (see Fig. 2, e.g.), and the question whether these details are correlated with the short-range order in crystalline phases needs further investigations (see [17] and [18], e.g.).

3. Metal-metal (M1-M2) glasses

Table 1 also shows structural parameters of metal-metal (M1-M2) glasses, where M2 = nickel is always the smaller constituent, whose partial pair correlation functions have been determined recently. Concerning model caculations, for this type of glasses it should be noted that to the knowlege of the author of the present work no theoretical models have been tested using these data up to now.

Comparison of the data of the glasses with those of related crystalline phases in Table 1 shows that there exists some degree of accordance, which usually is stressed by the respective authors. However, there are also significant differences, especially in the Ni-Ni correlations, which make it difficult to recognize general relationships between the short-range order in M1-M2 glasses and crystalline compounds.

On the other hand we find, however different the structural details may be, also within this group remarkable common features. It will be interesting to compare these features with those of the M-m glasses: First we observe also with these glasses a pronounced size effect in the ratio r1/r2 of the atomic radii of M1 and M2 which is displayed in the larger values of R-11 compared to R-22. The chemical affinity between unlike neighbors shows up in the smaller distance R-12 between unlike neighbors compared to the sum of their atomic radii r1+r2, where the difference Δ(R-12) in the order of 5 pct is as large as that found in M-m glasses (see Table 1).

Before the publication of partial correlation functions, the glasses belonging to the M1-M2 group have been regarded as in principle substitutional alloys, in contrast to the M-m glasses, maybe with some degree of CSRO. Recent results, however, show that this does not hold, as can be seen in Fig. 3 for the case of amorphous

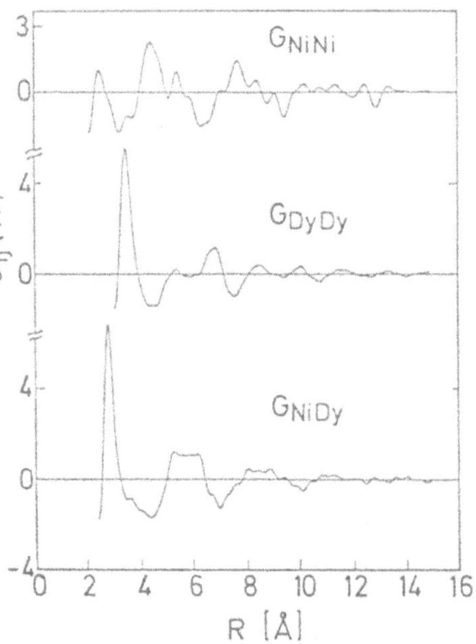

Fig. 3) Dy(69)Ni(31):
Partial pair correlation functions

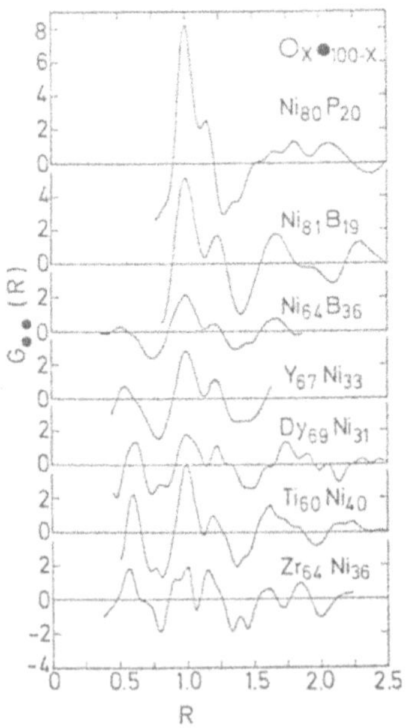

Fig. 4) M-m and M1-M2 glasses: Partial pair correlation between the smaller atoms. The distances are normalized to the position of the larger subpeak of the splitted second maximum

Dy(69)Ni(31) [11] (compare also the discussion in [19]). The small first peak of the Ni-Ni correlation in Fig. 3 compared to those of the Ni-Dy- and the Dy-Dy correlations shows that the structure of Dy(69)Ni(31) is asymmetric with respect to both constituents in that sense, that close contact between the smaller Ni-atoms is much more unlikely than contact between the larger Dy-atoms. Following this small peak, we observe a double peak structure near 5 Å, where correlation between the Ni-atoms, far away from close contact, is favoured. This behaviour, as a trend, is in surprising accordance with the m-m correlations in M-m glasses, and it is interesting to compare the M2-M2 correlations of M1-M2 glasses with the m-m correlations of M-m glasses in Fig. 4: We generally observe a rather small first peak, if at all, followed by a larger second maximum which exhibits a double peak structure.

From this we conclude the unlikelihood of close contact between the smaller atoms, not occurring at all in M(80)m(20) glasses and being restricted to one nearest neighbor in the Ni(64)B(36) glass, is a common feature of metallic glasses, whatever the group they belong to. Also, the structure of the split maximum at larger distances with a higher subpeak at R = 1.0 in Fig. 4 seems to be a general feature. One may note that the occurrence of those common features is not very encouraging to the idea that the short-range order in metallic glasses is close to that in their crystalline counterparts.

On the other hand, one should keep in mind the differences in the structures of M-m- and M1-M2 glasses regarding, e.g., the short-range order parameter η (see Table 1). Whereas in M(80)m(20) glasses the absence of m-m nearest neighbors consequently results in the maximum possible value of η the corresponding value in M1-M2 glasses where M2-M2 nearest neighbors do occur is always lower than the respective maximum value. A quantitative comparison, however, is not possible at present, because of the difficulties in defining the maximum possible value of η in concentrated M1-M2 glasses, as discussed in detail in [16].

References

1 P. Lamparter, W. Sperl, S. Steeb, and J. Blêtry: Z. Naturforsch. 37a, 1223 (1982)
2 J. Blêtry: Z. Naturforsch. 33a, 327 (1977)
3 O. Beyer and C. Hoheisel: Z. Naturforsch. 38a, 859 (1983)
4 J.M Dubois, P.H. Gaskell, and G. Le Caer: Int. Conf. Rapidly Quenched Metals, Würzburg 1984, Proc., pp. 567-572
5 J.L. Finney, G. Bushnell-Wye, and J.P. Bouquiere: ibid, pp. 563-566
6 P. Lamparter and S. Steeb: ibid, pp. 459-462
7 E. Nold, P. Lamparter, H. Olbrich, G. Rainer-Harbach, and S. Steeb: Z. Naturforsch. 36a, 1032 (1981)
8 N. Cowlam, Wu Guoan, P.P. Gardner, and H.A. Davies: J. Non-Cryst. Sol. 61, 62, 337 (1984)
9 P. Lamparter, A. Habenschuss, and A.H. Narten: Submitted to J. Non-Cryst. Sol. June 1985
10 M. Maret, P. Chieux, P. Hicter, M. Atzmon, and W.L. Johnson: Int. Conf. Rapidly Quenched Metals, Würzburg 1984, Proc., pp. 521-524
11 A. Wildermuth: Thesis work, University of Stuttgart, 1985, and A. Wildermuth, P. Lamparter, and S. Steeb: Z. Naturforsch. 40a, 191 (1985)
12 T. Fukunaga, N. Watanabe, and K. Suzuki: J. Non-Cryst. Sol. 61, 62, 343 (1984)
13 T. Fukunaga, N. Hayashi, N. Watanabe, and K. Suzuki: Int. Conf. Rapidly Quenched Metals, Würzburg 1984, Proc., pp. 475-478
14 T. Mizoguchi, S. Yoda, N. Akutsu, S. Yamada, J. Nishioka, T. Suemasa, and N. Watanabe: ibid, pp. 483-486
15 G.S. Cargill III and F. Spaepen: J. Non-Cryst. Sol. 43, 91 (1981)
16 F. Spaepen and G.S. Cargill III: Int. Conf. Rapidly Quenched Metals, Würzburg 1984, Proc., pp. 581-584
17 P. Lamparter, W. Sperl, E. Nold, G. Rainer-Harbach, and S. Steeb: Int. Conf. Rapidly Quenched Metals, Sendai 1981, Proc., pp. 343-346
18 S. Aur, T. Egami, and I. Vincze: ibid, pp. 351-354
19 P. Lamparter, S. Steeb, and E. Grallath: Z. Naturforsch. 38a, 1210 (1983)

Part II

Precipitation and Growth; Time-Dependent Experiments

Phase Separation in Alloys Using Low Q Neutron Scattering

R.J. Stewart

J.J. Thomson Physical Laboratory, University of Reading, Whiteknights, Reading, RG6 2AF, England

Introduction

Small-angle scattering has been used extensively to study the phase separation in alloys since the pioneering x-ray experiments of Guinier [1]. The first tentative neutron experiments were begun in the late 1960's [2,3,4,5,6], however, the depth and range of the investigation of the phase segragation processes have benefited immensely from small-angle neutron scattering (SANS) studies over the last decade or so. This is mainly due to the availability of intense long wavelength neutron beams at the Institut Laue-Langevin and the development of large area position-sensitive neutron detectors. I will begin with a brief outline of phase separation in alloys and then discuss the background to the SANS technique before illustrating what information can be obtained from such experiments. The early work on phase separation has been surveyed in the excellent detailed reviews of Gerold and Kostorz [7] and Kostorz [8]. More recent work is covered in the reviews of Kostorz [9,10,11].

Phase separation

Consider two metals A and B which are mixed to form a condensed system, the equilibrium state of the resulting alloy corresponds to that in which the Gibbs free energy is a minimum. In the simplest case, two forms for the free energy dependence on composition at a given temperature are found. Firstly, a curve with a single minimum (i.e a "U-shaped" curve); this indicates that the solid solution has the lowest free energy and thus a single phase will be stable at all compositions of the alloy. Put in another way, the components of the alloy $A_{1-x}B_x$ are completely miscible over the entire composition range x = 0 to 1. The second form for the free energy curve consists of a maximum positioned between two minima at compositions c_1 and c_2 (i.e. a "W-shaped" curve). For compositions lower than the first minimum ($<c_1$) or higher than the second minimum ($>c_2$) the lowest free energy is obtained with a single phase (a solid solution of B in A for $x<c_1$ or a solid solution of A in B for $x>c_2$). However, for compositions $c_1<x<c_2$ the lowest free energy is obtained with a mixture of two phases whose constituents have compositions c_1 and c_2 in the most stable condition. On a phase diagram a gap in miscibility is thus found between c_1 and c_2 for a "W-shaped" free energy curve. If an alloy $A_{1-x}B_x$ with $c_1<x<c_2$ is cooled (or quenched) from the single phase region into the two phase region (where one of the phases is usually of the same form as the parent solid solution), phase separation will occur (e.g. $\alpha \rightarrow \alpha+\beta$). The rate at which phase segragation proceeds depends on the concentration x, the presence of trace impurities, the stress level and the vacancy concentration. The vacancy concentration controls the rate of diffusion of the atoms in a substitutional alloy, and is of course temperature dependent, and thus the rate of decomposition of a supersaturated solid solution into two phases is temperature dependent and at sufficiently low temperatures phase separation can be inhibited. An excess vacancy concentration can be achieved by quenching rapidly from the single phase region into the two phase region. The magnitude of the excess vacancy concentration depends on the homogenisation temperature in the single phase region, the quench rate and the final ageing temperature in the two phase region. Thus, diffusion rates very much faster than those characteristic of the ageing temperature can be achieved.

The mechanisms involved in the early stages of decomposition within the miscibility gap regime are conceptually easy to describe, but in practice there is only limited agreement between such models and the experimental results. In simple terms, the mechanism can be divided into two types: nucleation followed by growth and spinodal decomposition. For a specific alloy composition, x ($c_1<x<c_2$), the dividing line between the two modes of decomposition is termed the spinodal temperature, T_s. Above T_s the second derivative of the Gibbs free energy with respect to concentration is positive and the solid solution is metastable, this corresponds to nucleation and growth. Below T_s the second derivative is negative and the solid solution is unstable, this is the spinodal regime. Spinodal decomposition occurs as a result of weak long wavelength fluctuations in the alloy composition, which result in a lowering of the free energy. The amplitude of the fluctuation in composition grows as time proceeds, and eventually zones rich in the second phase element are formed. In the nucleation and growth regime, small fluctuations in concentration lead to an increase in free energy, and thus the system is stable to such fluctuations, however strong localised fluctuations (i.e. the creation of an embyro nucleus) leads to decomposition. A critical nucleus size has to be achieved before decomposition can proceed and the critical size is inversely proportional to the degree of supercooling under the single phase/two phase boundary. Thus, associated with the nucleation there is an incubation period during which no substantial phase separation occurs. The temporal evolution of the scattering associated with phase separation has been predicted using computer simulations [12,13,14] and universal scaling laws have been found to describe the decomposition. However, it is only in the latter stages of ageing that good agreement with experiments is generally found where the decomposition is better understood.

The initial phase separation need not necessarily lead to the equilibrium precipitate phase directly, but various intermediate precipitates, metastable precipitates and zones can be formed. In the discussion I will refer to all of these as precipitates.

Low Q neutron scattering

The importance of neutron scattering in the study of phase separation in alloys is that it provides structural information about precipitates ranging in scale from a few Angstroms up to defect regions with dimensions of the order of 0.5 µm in bulk specimens. It is a complementary technique to electron microscopy which provides very detailed information on precipitates in real space over tiny localised volumes in the specimen. Neutron scattering provides a diffraction pattern of the precipitates averaged over a volume of up to a few cubic centimeters. Of course, a diffraction pattern cannot be unequivocally translated into real space, but this is balanced by the ability to investigate the precipitate form and distribution throughout a large volume of a sample in a relatively short time (a few minutes on a high-flux neutron source).

Neutrons interact with matter in a variety of ways and because they are uncharged they can easily penetrate deep into a sample. The principal interaction with atoms is via the short-range, strong nuclear interactions which are independent of the atomic number of the nucleus and vary in an apparently random way from nucleus to nucleus. The short-range nature of the interaction ($\sim 10^{-15}$m) compared to the wavelength of the neutrons used in the study of condensed matter ($\sim 10^{-7}$m) means that the scattering from a single nucleus is spherically symmetric. The amplitude of the wave scattered by a single nucleus is denoted by b and is called the scattering length. The energy of the neutrons scattered by a crystal can either be the same as that of the incident neutron (elastic scattering) or can have gained or lost energy (inelastic scattering). In an ideal scattering experiment the momentum of the incident and scattered neutrons would be determined and in order to investigate the properties of a sample we need to determine the energy and momentum transfer and the orientation of the sample. In most SANS experiments the total scattering, elastic plus inelastic is measured, since the magnitude of the inelastic contribution is very small compared to the coherent elastic scattering in most alloy studies the inelastic contribution can be ignored. We will thus restrict the discussion to elastic scattering, for which the magnitude of the scattering vector, $\underline{Q}$, is $4\pi\sin\theta/\lambda$, which I will refer to as Q. The SANS range in materials studies is from the reciprocal lattice point (000) out to the first set of reciprocal lattice points ($0<Q<\pi/d$, where d is the interatomic spacing).

The coherent elastic differential cross-section per atom per steradian for n_p precipitates per unit volume in an alloy can be written [15]:

$$d\sigma/d\Omega = [(n_p\langle V_p\rangle^2/N)(\langle c_p\rangle-\langle c_m\rangle)^2].\langle S^2(Q)\rangle.I(Q) \qquad (1)$$

V_p is the volume of a precipitate, N is the number of atoms per unit volume in the alloy, c_p is the mean scattering length density in the precipitate and c_m that in the matrix, S(Q) is the form factor describing the scattering from a precipitate and I(Q) represents the interference scattering arising from the spatial distribution of precipitates. The $\langle..\rangle$ refer to averages over variations in precipitate size throughout the alloy. The first term contained within the square brackets determines the magnitude of the scattering, which is obviously zero if $c_p=c_m$. The last two terms $\langle S^2(Q)\rangle$ and I(Q) determine the shape of the SANS. For widely spaced (i.e. low concentrations) and randomly arranged precipitates the interference term I(Q) has a value of unity. However, for more dense precipitation the functional form of I(Q) depends on the exact way in which the precipitates are spatially distributed. Various models have been used [15] to describe the spatial arrangement and I(Q) then calculated. For n_p randomly distributed precipitates Messoloras [16] has shown that

$$I(Q)=1 - 3(\sin QD_p - QD_p\cos QD_p)/(QD_p)^3 = 1 - Z(QD_p) \qquad (2)$$

where D_p is the mean distance between precipitates. Now $\langle S^2(Q)\rangle$ decreases with increasing Q so that the differential cross-section becomes a maximum at the first maximum of the interference function, this is when $QD_p=5.763$. Thus a value for D_p can be obtained from the position of this maximum. For values of $QD_p>5.763$ the interference function $1-Z(QD_p)$ has a value close to unity and thus the Q dependence of the cross-section in this Q region is adequately described by $\langle S^2(Q)\rangle$ alone providing the precipitate volume fraction is not too high (<10%). This is an important result when it comes to fit experimental data using approximations for $\langle S^2(Q)\rangle$. The volume fraction of precipitates, f, is given by [16]:

$$f = n_p.V_p = (3/(4\pi D_p^3).V_p \qquad (3)$$

The Guinier approximation [17] is the most important approximation which is used in the interpretation of sans data. The form factor is approximated by

$$S(Q) = \exp(-Q^2R_g^2/6) \qquad (4)$$

The radius of gyration R_g is defined in analogy to classical mechanics by

$$V_pR_g^2 = \int_{V_p} r^2 dV \qquad (5)$$

where r represents the position of the volume element dV with respect to the centre of mass of the precipitate. For a spherical precipitate the radius $R_s = 1.29R_g$. The differential cross-section in the Guinier approximation is:

$$d\sigma/d\Omega = [(n_pV_p^2/N)(c_p-c_m)^2]\exp(-Q^2R_g^2/3) \qquad (6)$$

The first term in square brackets is referred to as the Guinier constant,G. The approximation is valid for values of QR_g up to values of 1.2 to 2 depending on the shape of the precipitates. A plot of $\log(d\sigma/d\Omega)$ against Q^2, the so-called Guinier plot, has a slope of $(-R_g^2/3)$ and an intercept at Q=0 of log(G). Thus from the slope of such a plot, if a linear region is found, the size of the precipitates can be found if their shape is known. The Guinier constant, G for an $A_{1-x}B_x$ alloy in which the volume per atom in the precipitate, v_p, and in the matrix, v_m, are the same can be written [15]

$$G = N.n_p.V_p^2.(m_p-m_m)^2(b_A-b_B)^2 \qquad (7)$$

where m_p is the concentration of B atoms in the precipitate, m_m the corresponding value in the matrix and b_A and b_B are the scattering lengths of the elements A and B. If the shape of the precipitate is known, V_p can be obtained from R_g and since

b_A, b_B and N are known the product $n_p(m_p-m_m)$ can be obtained from the experimentally determined value of G. If there is an interference peak a value for n_p can be obtained using (3) and hence a value for (m_p-m_m). In order to separate m_p and m_m we can use the fact that atoms are conserved during phase separation:

$$c_B = m_p n_p V_p + m_m(1-n_p V_p) \qquad (8)$$

where c_B is the total concentration of B atoms in the alloy. The values of m_p and m_m can also be obtained from the integrated intensity, without any assumptions having to be made about the shape of the precipitates [18]:

$$W = 4\pi\int (d\sigma/d\Omega)Q^2 dQ = 8N\pi^3(m_p-c_B)(c_B-m_m)(b_A-b_B)^2 \qquad (9)$$

The Porod approximation for $S^2(Q)$ applies for values of Q greater than the reciprocal of the largest dimension of the precipitate ($QR_g>2.5$ for a spherical precipitate). For randomly distributed and oriented precipitates with a well-defined surface the differential scattering cross-section is given by [15]

$$d\sigma/d\Omega = 2\pi G.S_p/(V_p{}^2.Q^4) + A = P.Q^{-4} + A \qquad (10)$$

where S_p is the surface area of the precipitate, P is the Porod constant and A is the contribution from incoherent scattering and other Q-independent scattering from the alloy. A plot of $Q^4(d\sigma/d\Omega)$ against Q^4 has an intercept at Q=0 of P, thus the ratio $S_p/V_p{}^2$ can be obtained. For a spherical precipitate $P=4.5G/R_s{}^4$.

Neutron studies of phase separation

The limits of the miscibility gap in a binary alloy can be obtained [18] from the integrated intensity (9). Expanding and rearranging (9) we can write

$$c^2_B + W/[8N\pi^3(b_A-b_B)^2] = U(c_B) = (m_p+m_m)c_B-m_p m_m \qquad (11)$$

Thus if we have at least two alloys (preferably 3) with different concentrations c_B and we plot $U(c_B)$ versus c_B we obtain (m_p+m_m) from the slope and $m_p m_m$ from the intercept at $c_B=0$ without knowledge of the shape of the precipitate. In the derivation of (9) $v_p=v_m$ is assumed but this can be corrected for [18].

For ternary alloys where phase separation into two phases occurs, the situation is more complicated. There are four concentrations to determine and in addition the tie line connecting the original alloy and the two equilibrium phases has to be determined since only alloys lying on a single tie line decompose into the same product phases. In order to obtain the limits of miscibility, the scattering length of one component in the alloy needs to be varied. This can be done by isotopic replacement, and silver isotopes have been used [19] to determine the limits of miscibility in Al-Ag-Zn. A combination of x-ray and neutron scattering which effectively varies the scattering length has been employed in a study of the miscibility gap in Al-Mg-Zn [20]. The accuracy of the limits of miscibility obtained depends on many aspects of the experiments, however, the most important source are the errors introduced in determining the integrated intensities, which involve extrapolation using the Porod approximation. A detailed discussion of the errors is given by Gerold et al [20] and Ghilarducci de Salva et al [19].

The phase boundary between the two phase region and the single phase region can be easily obtained by studying the SANS for a given alloy composition as a function of temperature. For example, figs 1 and 2 show the temperature dependence of the SANS from an Fe-Cr-Al alloy containing small amounts of Y [21] and SANS is seen for temperatures $\leq$480°C but not for temperatures $\geq$485°C. Thus the phase boundary lies between 480 and 485°C. The kinetics of growth and dissolution of precipitates in alloys can be investigated by following their growth as a function of time at a given temperature. In-situ experiments are possible [22,23,24] where the growth rate is sufficiently fast and dissolution has been followed "in situ" when the total time scale for complete precipitate dissolution is only a few minutes [22].

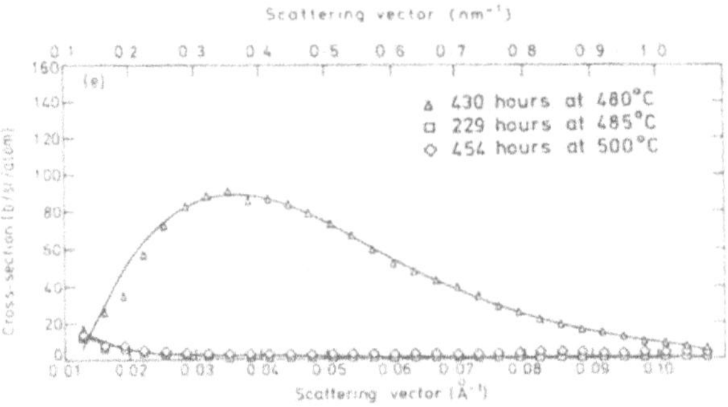

Fig. 1. The SANS from an Fe-Cr-Al alloy containing yttrium. The data clearly show that the phase boundary between the two phase region and the single phase region lies between 480°C and 485°C.

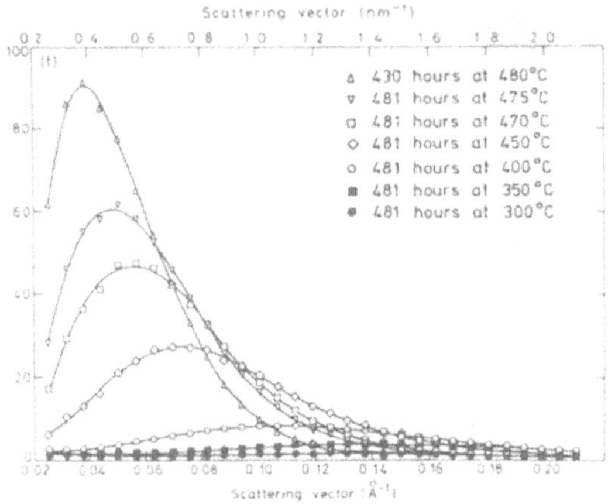

Fig. 2. The SANS from an Fe-Cr-Al alloy containing yttrium. The data clearly shows how the number of precipitates decreases with increasing ageing temperature (i.e. the peak in the sans moves to lower Q). The magnitude of the SANS also dramatically increases as the ageing temperature is increased.

The influence of trace elements on the decomposition can show marked effects. Messoloras et al [21] studied the effects of small additions (∿0.2at%) of yttrium on the precipitation of a Cr-rich phase in an Fe-15at%Cr-9.7at%Al alloy. No precipitation was found after 813 hours at 400°C in the Y-free alloy, whereas substantial precipitation was found at 300°C in the Y-containing alloy (fig.3). The radius of gyration of the Y-containing alloy showed growth from the earliest times measured (0.5 hours) [25] whereas an incubation period of at least 100hrs was observed in the Y-free alloy (figs 4 and 5). The precipitate size as a function of ageing time shows evidence for pre-existing nuclei in the Y-containing alloy. Messoloras et al [21] suggested that that these nuclei might be a complex of Y and Al. Hull [26] has investigated the influence of varying the Y concentration and also replacing it with Ce or La. For lower concentrations of Y (∿0.09at%) the rate of decomposition is reduced substantially and is similar to that of the Y-free alloy. Thus, lowering the Y content appears to remove the pre-existing nuclei, indicating that all the Y can then be held in solid solution. Ce and La effect the rate of decomposition in a similar way to Y.

The number of precipitates per unit volume can be obtained from the peak in the sans if some model of the spatial distribution of the precipitates is assumed. Thus, if the shape of the precipitates is known, the volume fraction $n_p V_p$ can be determined. Assuming a random distribution of precipitates in the matrix, the dependence of the volume fraction on ageing temperature for the spherical Cr-rich precipitates in the Fe-Cr-Al alloy can be obtained using (3) (fig.6).

The nickel superalloy Nimonic PE16 is a Ni-Cr-Fe based alloy strengthened by additions of Mo, Ti and Al. The principal precipitates are gamma prime precipitates ($Ni_3Al_xTi_{1-x}$) and carbides. Ali et al [27] have studied the kinetics of the gamma prime phase separation at a range of temperatures and compared their results with the growth theory of Lifshitz and Slyozov (LS) [28]. Since the concentration of

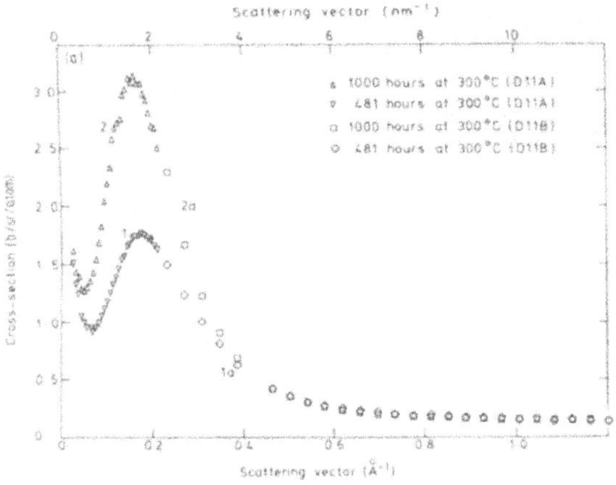

Fig.3. The SANS from an Fe-Cr-Al alloy aged at 300°C for various times. The alloy contains 0.19at% yttrium. The high Q data was obtained on the old high-angle detector bank on D11.

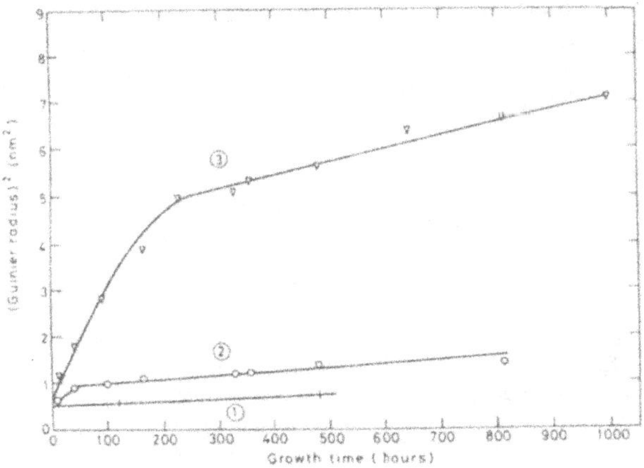

Fig. 4. The dependence of the radius of gyration (Guinier radius) on ageing time for an Fe-15.7at.%Cr-9.7at.%Al alloy containing 0.19at.% yttrium. The three curves correspond to ageing temperatures of 1:300°C 2:400°C and 3:475°C. Precipitate growth is apparent from the start and there is evidence for pre-existing nuclei, of about 0.8nm in radius of gyration, on which the precipitates grow.

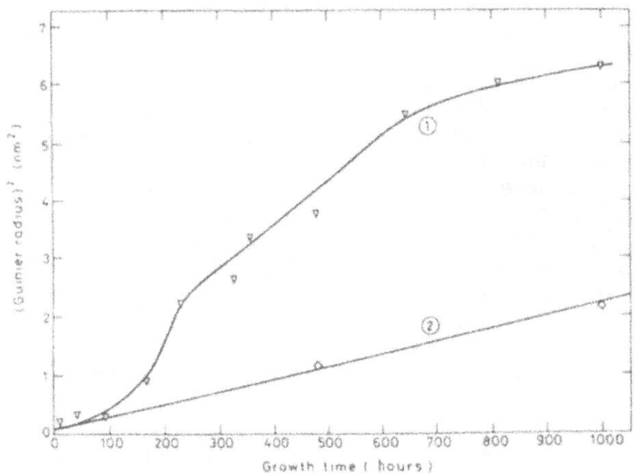

Fig. 5. The dependence of the radius of gyration (Guinier radius) on ageing time for an Fe-15.6at.%Cr-9.2at.%Al alloy without yttrium. Significant precipitate growth is not apparent until after 100hrs. The two curves correspond to ageing temperatures of 1:475°C and 2:450°C.

Al and Ti are low in PE16 the supersaturation is low and this is consistent with the assumptions of the LS theory. The LS theory predicts that the precipitate radius at the later stages of ageing is given by:

$$R_s^3(t)-R_s^3(t_o) = (4\alpha D/9T)(t-t_o) = K_1(t-t_o) \qquad (12)$$

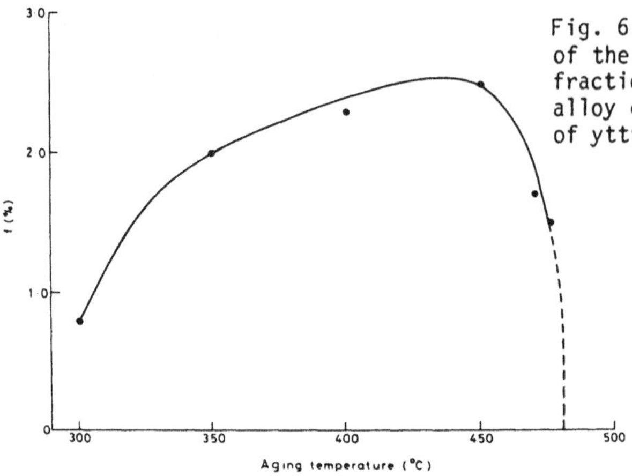

Fig. 6. The temperature dependence of the precipitate volume fraction, f, for an Fe-Cr-Al alloy containing small amounts of yttrium.

where R_s is the precipitate radius, t_0 is the time from which the gamma prime coalescence starts, D is the diffusion coefficient, T the temperature and α involves the interfacial energy and equilibrium concentration of solute in the matrix. The LS theory also predicts that the number of precipitates is given by:

$$n_p = \beta f_\infty / K_1 t = K_2/t \tag{13}$$

f_∞ is the equilibrium precipitate volume fraction and β is a numerical constant with a value of about 0.22. The temporal dependence of the Guinier constant is [27]

$$G = K_3.t = f_\infty (4\pi/3) K_1 (c_p - c_m)^2.t \tag{14}$$

The reduced data obtained from SANS measurements [27] are shown in fig.7 and a good correlation with slope 1 is observed. f_∞ was found to be between 0.07 and 0.10 depending on temperature. Assuming that the precipitates are of the form $Ni_3Al_xTi_{1-x}$ the value of $(c_p - c_m)$ determined from a combination of (12),(13) and (14) gives a value for x of 0.6 for the ageing temperatures 977K, 1021K and 1068K.

The use of single crystals permits more details about the shape and distribution of the precipitates to be obtained. The shape of the observed SANS is the Fourier transform of the shape function which describes the precipitate. A single ellipsoidal precipitate would have a SANS pattern with an elliptical shape the major axis of which would be perpendicular to that of the ellipsoidal precipitate. Single crystal studies on gamma prime precipitates in Ni-(12-14at%)Al have been made [29] and the sans showed no radial symmetry around the beam, therefore intensities were averaged into 10° sections around the <111>, <110> and <100> directions. Even for very short ageing times the gamma prime precipitates were found to align preferably along <100> directions. The form of the anisotropy of the sans patterns could only be understood in terms of a size distribution (initially of spheres) and an anisotropic distance distribution (where the average distance is constant but the width of the distribution depends on the crystal orientation). In the later stages model calculations of the sans based on elongated precipitates (aspect ratio $\sim$ 10) oriented in the <100> direction were found to reproduce the experimental data. Similar but weaker preferential alignment has been found in Al-6.8at%Zn single crystals where the precipitates are spherical [11].

For many years evidence to distinguish between spinodal decomposition and nucleation and growth using SANS has been sought. The early experiments [30,31,22] on Al-Zn alloys were consistent with nucleation and growth and no evidence for a spinodal decomposition was found. Many more experiments have been carried out on the Al-Zn system and Simon et al [32] have summarised the current position. In general, the conclusions are consistent with nucleation and growth followed by coarsening.

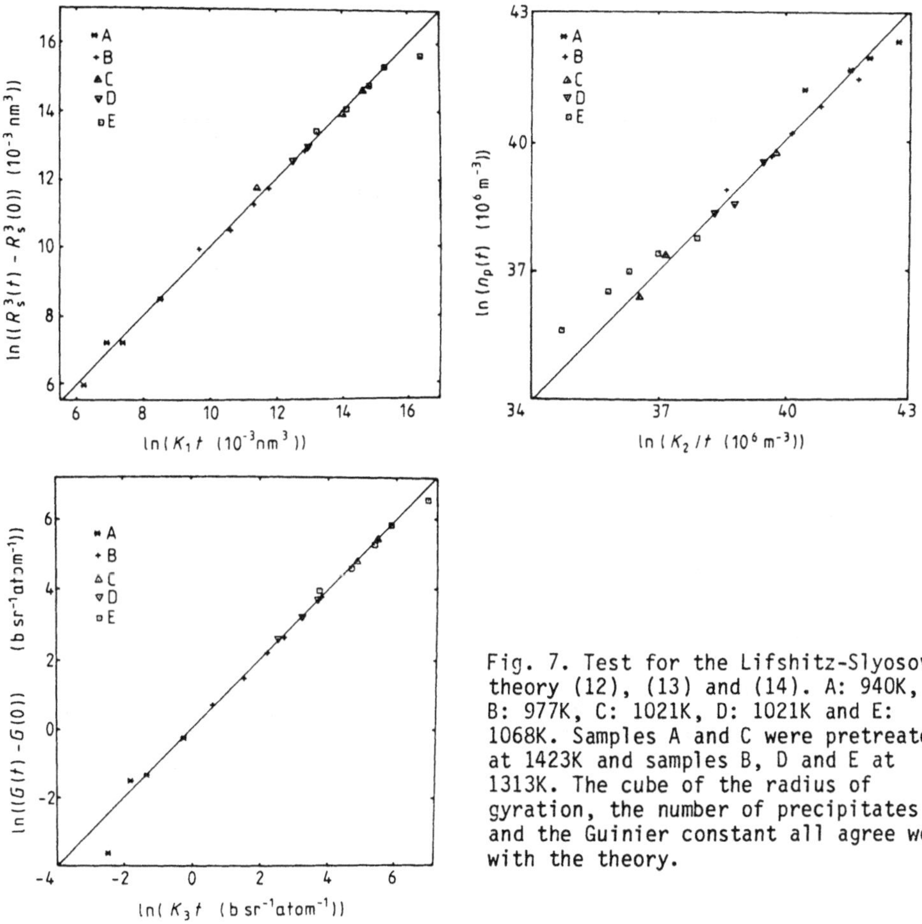

Fig. 7. Test for the Lifshitz-Slyosov theory (12), (13) and (14). A: 940K, B: 977K, C: 1021K, D: 1021K and E: 1068K. Samples A and C were pretreated at 1423K and samples B, D and E at 1313K. The cube of the radius of gyration, the number of precipitates and the Guinier constant all agree well with the theory.

Au-Pt alloys possess an almost symmetrical miscibility gap and are ideally suited to study the early stages of decomposition. An Au-60at%Pt alloy has been studied [33,34,11] and according to the linear theory of spinodal decomposition the peak intensity in the sans as a function of time should be a straight line. This was found to hold at longer ageing times,but not in the early stages where it should. Thus conclusive evidence for spinodal decomposition from SANS is illusive.

Lebowitz et al [12,13] and Fratzl et al [14] have developed a dynamical scaling technique from computer-simulated scattering functions. In the latter case,the scaling is achieved via the time dependence of the radius of gyration. The theoretically determined dynamical scaling applies to the later stages of decomposition and the time dependence is found to scale out,and the scattering function then remains self-similar during its evolution to equilibrium. Experimental neutron scattering studies on an AlMgZn alloy [35,36] were found to be consistent with the time-scaling behaviour after the first few minutes of decomposition and lasting over a time interval during which the precipitate volume fraction increased by a factor of 5 (5mins ageing to 4.5 years ageing). The time-scaling behaviour was found to be in good agreement with the computer simulations for two alloys of AlMgZn aged at three temperatures and subjected to different quenching procedures. In general,the results are consistent with the LS theory if a time-dependent diffusion constant is assumed. Similar experimental confirmation for the time-scaling behaviour was found in Al-Zn [23].

The precipitate size obtained from the Guinier radius is an average value weighted to the larger precipitates in the distribution, whereas the Porod distribution is weighted towards the smaller precipitates. For example, Messoloras et al [21] found for the spherical Cr-rich precipitates present in the early stages of the decomposition of an Fe-Cr-Al alloy that the precipitate radius obtained from Guinier plots were 1.36nm (9hr), 1.72nm (41hr), 2.1nm (91hr), 2.54nm (165hr) and 2.88nm (229hr) whereas from Porod plots for the corresponding ageing times the radii were 1.39nm, 1.49nm, 1.7nm, 2.05nm and 2.44nm. In the very early stages (i.e.<10 hrs) where there has been insufficient time for any significant size distribution to be established the results agree, whereas for longer times the results are consistent with the evolution of a size distribution. Baur and Gerold [37] have shown that for spherical precipitates R_g and R_p (the radius obtained from a Porod analysis) are equal to certain moment ratios of the size distribution

$$R_g^2 = 3\langle R_s^8\rangle/5\langle R_s^6\rangle \text{ and } R_p = \langle R_s^3\rangle/\langle R_s^2\rangle$$

Analysis of the size distribution of the precipitates in terms of size distribution functions have been made [38,39,40,41]. The problem with such determinations is that in the absence of additional information, a unique functional form for the precipitate size distribution is difficult to obtain. If the precipitation kinetics follow for instance the LS theory, then a LS distribution can be used, however, as discussed by Dusic [25], very small differences in the observed scattering can alter the form of the distribution, and thus careful measurements need to be made. However, in order to determine the size range the same accuracy is not necessary.

The sensitivity of the SANS technique in the study of the phase separation depends on the level of the background and parasitic (e.g. incoherent) scattering from the sample. In the ideal case of oxygen precipitation in dislocation, free silicon precipitation involving only a few ppm of oxygen has been successfully studied [42] and the kinetics of the phase separation determined. The improvements recently made on D11, the coming on line of the LOQ spectrometer at the S.N.S with its very wide simultaneously accessible Q range, and the proposed new instrument D22 on the second cold source at the I.L.L. should further stimulate this active field of study.

In this paper SANS has been shown to give information on the size, shape, distribution and composition of precipitates formed as a result of phase segregation in alloys. The temporal evolution of the phase separation can be followed and various kinetic constants can be obtained. Comparisons between the experimental observations and theoretical predictions of how the phase separation proceeds are continually becoming more quantitative, and promise much for the future of this field of study.

References

1. A.Guinier: C.R. Hebd. Seance Acad. Sci., Paris 204, 1115 (1937)
2. R.J.Stewart: Ph.D. Thesis, University of Reading, England (1969)
3. B.H.Meardon: Ph.D. Thesis, University of Reading, England (1970)
4. C.D.Clark and B.H. Meardon: Nature, Phys. Sci. 235, 18 (1972)
5. J.M.Raynal, J.Schelten and W.Schmatz: J. Appl. Cryst. 4, 511 (1971)
6. Ye.Z.Vintaykin, V.N.Dmitriivev and V.Yu.Kolontsov: Fiz. Metal. Metalloved. 29, 1257 (1970)
7. V.Gerold and G.Kostorz: J. Appl. Cryst. 11, 376 (1978)
8. G.Kostorz: Treatise on materials science and technology Ed. H.Herman, Volume 15, p227, Academic Press, New York (1979)
9. G.Kostorz: Physical Metallurgy 3rd. edition. Eds R.W.Cahn and P.Haasen p793, North Holland, Amsterdam (1983)
10. G. Kostorz: Microstructural Characterization of Materials by Non-Microscopical Techniques, 5th Riso International Symposium on Metallurgy and Materials Science, Riso National Laboratory, p65, (1984)
11. G.Kostorz: Physica 120B, 387 (1983).
12. J.L.Lebowitz, J.Marro and M.H.Kalos: Acta Met. 30, 297 (1982)
13. J.L.Lebowitz, J.Marro and M.H.Kalos: Comments Solid State Phys. 10, 201 (1983)

14. P.Fratzl, J.L.Lebowitz, J.Marro and M.H.Kalos: Acta Met. 31, 1849 (1983)
15. R.J.Stewart: Defects in Solids - Modern Techniques, Proceedings of the NATO Advanced Study Institute, Cetraro, Italy, Sept. 1985. Plenum Press, New York.
16. S.Messoloras: Ph.D. Thesis, University of Reading, England (1974)
17. A.Guinier and G.Fournet: Small-Angle Scattering of X-rays, Wiley, New York,(1955)
18. V.Gerold: J. Appl. Cryst. 10, 25 (1977)
19. A.Salva-Ghilarducci, J.P.Simon, P.Guyot and I.Ansara: Acta Met. 31, 1705 (1983)
20. V.Gerold, J.E.Epperson and G.Kostorz: J. Appl. Cryst. 10, 28 (1977)
21. S.Messoloras, B.C.Pike, R.J.Stewart and C.G.Windsor: Metal Science 18, 311 (1984)
22. D.R.Allen, S.Messoloras, R.J.Stewart and G.Kostorz: J. Appl. Cryst. 11, 578 (1978)
23. P.Guyot: Decomposition of Alloys: the early stages Eds P.Haasen, V.Gerold, R.Wagner and M.F.Ashby p41, Pergamon Press (1983)
24. J.G.Prince, J.G.Booth and R.J.Barton: J. Phys. F 13, 1333 (1983)
25. M.Dusic: Ph.D. Thesis, University of Reading, England (1982)
26. S.Hull: Ph.D. Thesis, University of Reading, England (1985)
27. K.Ali, S.Messoloras and R.J.Stewart: J. Phys F 15, 487 (1985)
28. I.M.Lifshitz and V.V.Slyozov: J. Phys. Chem. Solids 19, 35 (1961)
29. R.E.Beddoe, P.Haasen and G.Kostorz: Decomposition of Alloys: the early stages Eds P.Haasen, V.Gerold, R.Wagner and M.F.Ashby p233, Pergamon Press (1983)
30. D.Allen, J.E.Epperson, V.Gerold, G.Kostorz, S.Messoloras and R.J.Stewart: Proceedings of the Conference on Neutron Scattering, Ed R.M.Moon, vol 1 p102 NTIS, US Dept. of Commerce, Sringfield, VA, CONF-760601-P1 (1976)
31. G.Laslaz, G.Kostorz, M.Roth, P.Guyot and R.J.Stewart: Phys. Status Solidi A 41 , 577 (1977)
32. J.P.Simon, P.Guyot, A.Ghilarducci de Salva: Phil. Mag. A49, 151 (1984)
33. S.P.Singahl, H.Herman and G.Kostorz: J. Appl. Cryst. 11, 572 (1978)
34. S.P.Singahl: Thesis, State University of New York at Stony Brook (1981)
35. O.Blaschko and P.Fratzl: Phys. Rev. Lett. 51, 288 (1983)
36. O.Blaschko, P.Fratzl, G.Ernst, M.Bernole, F.Fujara: Phys. Rev. B30, 6498, (1984)
37. R.Baur and V.Gerold: Acta Metall. 12, 1449 (1964)
38. R.J.R.Miller, S.Messoloras, R.J.Stewart G.Kostorz: J. Appl. Cryst. 11, 583 (1978)
39. A.Boeuf, R.G.M.Caciuffo, R.Coppola, S.Melone, P.Puliti and F.Rustichelli: Nucl. Instr. Methods Phys. Res. A234, 562 (1985)
40. S.Abis, A.Boeuf, R.Caciuffo, S.Crice, P.Fiorini, M.Magnani, S.Melone, P.Puliti, F.Rustichelli, M. Stefanon: ILL technical report (1985)
41. A. Mathiot, S. Refiorentin and G. Robert: ILL report exp. no 07.03.191 (1982)
42. F.M.Livingston, S.Messoloras, R.C.Newman, B.C.Pike, R.J.Stewart, M.J.Binns, W.P. Brown and J.G. Wilkes: J. Phys. C 17, 6253 (1984)

SANS-Investigation of Early Stage Precipitation Kinetics in Cu-2.9 at.% Ti

H. Eckerlebe, R. Kampmann, and R. Wagner

GKSS-Forschungszentrum, Institut für Physik, D-2054 Geesthacht, F.R.G.

1. Introduction

According to joint analytical field ion microscopic (FIM) and transmission electron microscope (TEM) studies, the aging of supersaturated Cu-Ti alloys below the metastable β'-solvus line (Fig. 1) leads to the following phase separation sequence [1 - 3]:

α"-supersaturated solid solution ⟶ α'-matrix + coherent metastable β'-Cu_4Ti precipitates ⟶ α-matrix + incoherent stable Cu_3Ti-precipitates (only after extended aging).

FIM and TEM studies on Cu-2.9 at % Ti isothermally aged at 350 °C revealed the following microstructural features [1]: i) side bands in TEM electron diffraction patterns, ii) no measurable incubation period, iii) quasi-periodically aligned Cu_4Ti-precipitates, which in the later aging stages adopt the shape of rotational ellipsoids (aspect ratio u) with the major axis along the elastically soft <100> directions, iv) the absence of heterogeneous nucleation at lattice defects.

These microstructural features are expected if phase separation is initiated via spinodal decomposition [4]; however, their occurrence is not sufficient to distinguish a spinodal reaction from a first-order phase transition via the nucleation and growth of Cu_4Ti particles.

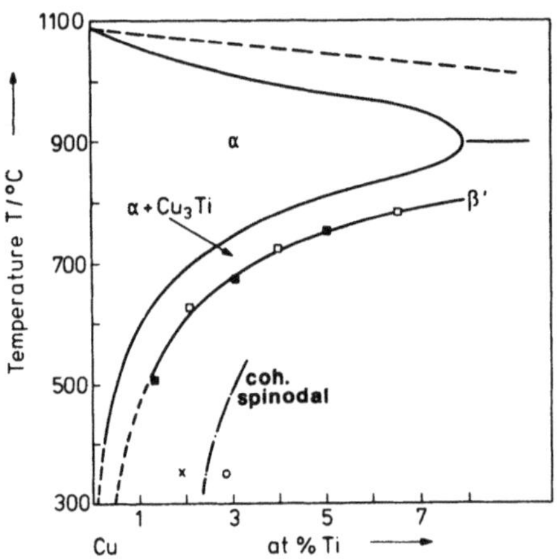

Fig. 1: Phase diagram of Cu-Ti. Also are shown the metastable β'-Cu_4Ti solvus line, the studied alloys, and the apparent spinodal line

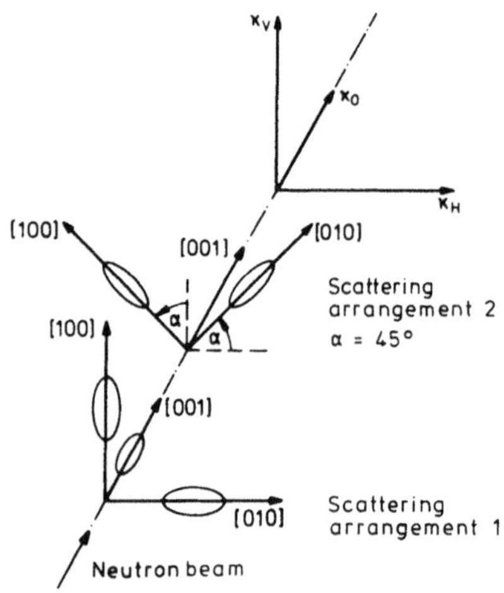

Fig. 2: Orientation of the rotation ellipsoidal particles within the single crystal with respect to the scattering vectors $\underline{\kappa}_H$ and $\underline{\kappa}_V$ in the two different scattering arrangements

The latter decomposition mechanism is thought to be operative in a less concentrated Cu-1.9 at % Ti alloy aged at 350 °C [2], where TEM and FIM showed i) no side bands, ii) randomly distributed particles, which according to FIM atom-probe analyses have the stoichiometrical composition Cu_4Ti already after aging for only 2.5 min., iii) the occurrence of some heterogeneous nucleation.

As a hypothesis, the differences in the microstructural features between the two alloys might be attributed to two different decomposition mechanisms in the early stages of phase separation; i.e. it is assumed that the coherent spinodal line at 350 °C lies between 1.9 at % and 2.9 at % Ti (Fig. 1).

The objective of this work was:

- to study the phase-separation mechanism in Cu-2.9 at % Ti which lies close to the apparent spinodal line,
- to study the influence of the quenching rate after homogenization on precipitation kinetics,
- to follow the time evolution of the mean cluster size $\bar{R}(t)$, the number density $n_\ell(t)$, the matrix supersaturation $\bar{c}(t)$, and the precipitate aspect ratio u(t),
- to compare the experimental results with recent theories of spinodal decomposition [6-8] and nucleation and growth [9].

Essentially, the theories and computer simulations on spinodal decomposition yield (though only qualitatively!) the time evolution of the structure factor $S(\underline{\kappa},t)$ ($\underline{\kappa}$, wave vector), i.e. the Fourier transform of the spatial concentration correlation function. As $S(\underline{\kappa},t)$ is proportional to the intensity $d\Sigma/d\Omega$ scattered with momentum transfer $\underline{\kappa}$, we employed SANS to follow $S(\underline{\kappa},t)$. Because of the positive and negative scattering length densities of Cu and Ti, respectively, SANS is particularly well suited to the study of two-phase Cu-Ti alloys.

2. Experimental

After solution treatment at various homogenization temperatures (T_H = 780, 870, 900, 960 °C) and quenching into brine, [001]-oriented Cu-2.9 at % Ti single crystals were aged at 350 °C for periods ranging between 1 and $3.6 \cdot 10^4$ min.

The SANS intensity was measured with two orthogonal linear position sensitive detectors in two different scattering arrangements (Fig. 2): - in arrangement 1 the scattering vectors $\underline{\kappa}_H$ (horizontal LPSD) and $\underline{\kappa}_V$ (vertical LPSD) were parallel to [010] and [100], respectively. - In arrangement 2, the specimen was rotated by 45° around [001], hence $\underline{\kappa}_H \| [\bar{1}10]$ and $\underline{\kappa}_V \| [110]$.

Measuring the scattering cross-section $d\Sigma/d\Omega$ in both arrangements 1 and 2, allowed us to detect anisotropic scattering from the rotational ellipsoidal particles aligned along <100> directions (Fig. 2).

3. SANS data evaluation

After suitable background corrections, the SANS data were evaluated in terms of the Guinier and Porod approximations, of the integrated intensity Q_0 (for t ≥ 25 min the Porod-regime was already accessible) and the characteristic length. From diffuse Laue scattering the matrix concentration $\bar{c}(t)$ was determined; with the knowledge of $\bar{c}(t)$ both the mean atomic volume and the scattering length density of the matrix could be obtained at any time during the decomposition reaction.

Because of the non-spherical morphology of the precipitates and their alignment along <100> in the later stages of decomposition, the SANS scattering laws had to be reformulated for rotational ellipsoids with aspect ratios u. $d\Sigma/d\Omega$ depends not only on κ and 2R (the short diameter of the ellipsoid) but also on the angle α between $\underline{\kappa}$ and its major axis (length $2R \cdot u$, $u \geq 1$), i.e. on $\kappa \cdot R \cdot \sqrt{u^2 \cdot \cos^2\alpha + \sin^2\alpha} \equiv \kappa \cdot R \cdot \Psi$. It is beyond the scope of this paper to present the lengthy, reformulated, scattering laws. These, and the procedure by which the various precipitation parameters $\bar{R}(t)$, $n_\ell(t)$, $\bar{c}(t)$, u(t) were inferred from the SANS data, have been discussed comprehensively by ECKERLEBE [10].

4. Results and discussion

4.1 Influence of the quenching rate on the decomposition kinetics

Figure 3 shows SANS curves from specimens which were solution treated at the given homogenization temperatures T_H; following the quench, each specimen was aged for 10 min/350 °C. The maximum in the scattering curves (max $d\Sigma/d\Omega \equiv d\Sigma^m/d\Omega$) is found to be the larger, and its position (κ_m) is found to be the smaller, the higher the homogenization temperature had been. This result indicates that phase separation in the early stages progresses the faster, the higher the homogenization temperature had been. Obviously, for the present Cu-Ti alloy, the number of quenched-in excess vacancies correlates with T_H and, correspondingly, accelerates the decomposition kinetics.

For comparison, Fig. 3 also shows the SANS curve from an as-quenched specimen homogenized at 780 °C. The increasing SANS intensity between $3 \gtrsim \kappa \gtrsim 1.5\ nm^{-1}$ indicates that some phase separation has occurred during the quench.

For future experiments it is suggested that the alloy is first solution treated at high temperatures, and subsequently equilibrated at a homogenization temperature slightly above the solvus temperature before it is quenched in order to minimize the effect of quenched-in vacancies.

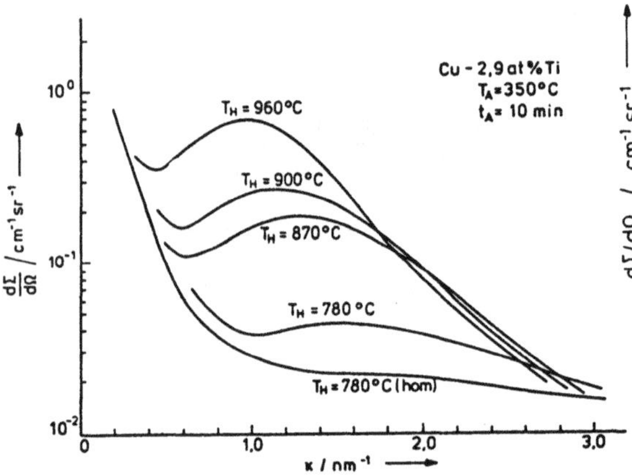

Fig. 3: SANS curves from specimens which had been quenched from the given homogenization temperatures T_H, and (apart from the curve on the bottom) subsequently aged for 10 min/350 °C

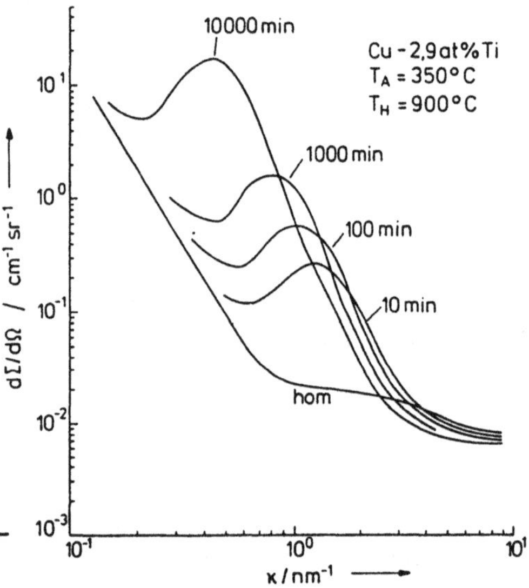

Fig. 4: SANS curves taken at various aging stages

4.2 Comparison of experimental results with theories on spinodal decomposition

Regardless of the chosen homogenization temperature, the SANS curves and their evolution with aging time showed the following features (Fig. 4):

- a maximum, the height ($d\Sigma^m/d\Omega$) of which increases with t, and the position κ_m of which decreases with t; i.e. the distance between Ti-rich clusters or the wavelength of solute fluctuations starts to increase instantaneously.
- no common cross-over point of the scattering curves from differently aged specimens.
- no wave vector, for which $d\Sigma/d\Omega$ (κ=const.) grows exponentially. At a fixed κ, $d\Sigma/d\Omega$ (κ=const.) rather grows slowly with t before it decays again (Fig. 5).

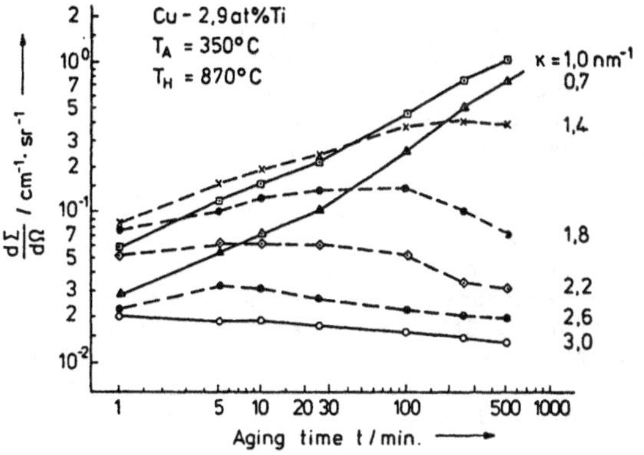

Fig. 5: Time evolution of $d\Sigma/d\Omega$ for different constant wave vectors κ

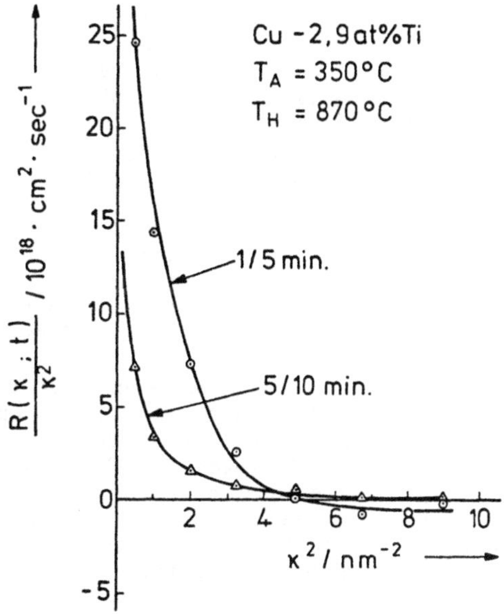

Fig. 6: Variation of $R(\kappa,t)/\kappa^2$ with κ^2 as determined for two different time intervals

- Plotting $R(\kappa)/\kappa^2$ as a function of κ^2, where the amplification factor $R(\kappa)$ for the growth of composition fluctuations is obtained from the scattering curves via [e.g. 7] $R(\kappa) \approx \frac{1}{2}\frac{\partial \ln(d\Sigma/d\Omega)}{\partial t}$ yields no straight lines (Fig. 6) at any time.

All of these features of the experimental scattering curves, as well as the shape of $R(\kappa)/\kappa^2 = f(\kappa^2)$ have been qualitatively predicted by results from more recent Monte Carlo computations and theories on spinodal decomposition [6 - 8].

Concerning the time evolution of the maximum of the structure factor $(\sim d\Sigma^m/d\Omega(t))$ and its position $(\kappa_m(t))$, the data from the various computational techniques are well represented by power laws, such as $d\Sigma^m/d\Omega \sim t^a$ and $\kappa_m \sim t^{-b}$ with the exponents a and b ranging from 0.69 to 0.74 and 0.17 to 0.25, respectively [e.g. 7, 8]. However, Fig. 7 shows that neither $d\Sigma^m/d\Omega(t)$ nor $\kappa_m(t)$ display power law behaviour. The shapes of both curves in Fig. 7 are qualitatively in good agreement with the generalized nucleation theory of BINDER et al. [6], which predicts a rather gradual transition from the nucleation and growth regime to the spinodal one if the supersaturation is increased.

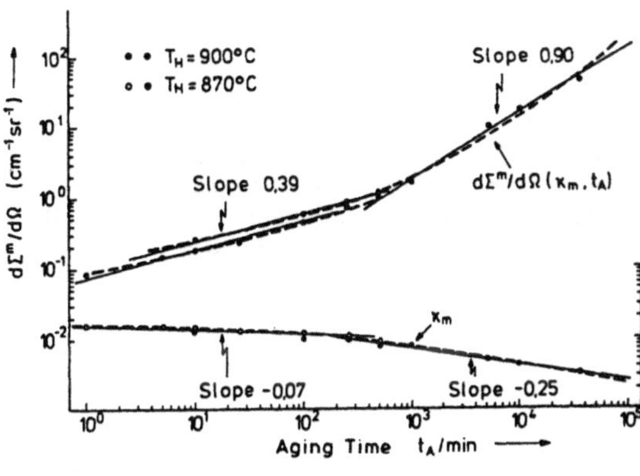

Fig. 7: Time evolution of the maximum of the structure factor $d\Sigma^m/d\Omega$ and of its position κ_m, as determined in scattering arrangement 1

From the apparent (qualitative) agreement between the experimental data and the theoretical results it might be concluded that phase separation in Cu-2.9 at % Ti is initiated via a spinodal reaction. However, FIM atom-probe results have shown that after aging for only 50 min/350 °C discrete Cu_4Ti particles with sharp interfaces have formed; i.e. the cluster formation process is terminated after aging times of less than 50 min regardless of whether initiation occurs by a spinodal or a nucleation and growth mechanism.

This experimental observation indicates that a comparison between experimental results and those from theoretical studies on spinodal decomposition ought to be confined to aging times of less than 50 min/350 °C. The good agreement between the experimental $d\Sigma^m/d\Omega(t)$ and $\kappa_m(t)$-curves and those from BINDER's et al. theory [6] beyond 50 min/350 °C must, therefore, be considered to be fortuitous.

From the present SANS-results it is not possible to decide unambiguously whether the decomposition reaction in Cu-2.9 at % Ti is initiated via a spinodal reaction. Since the microstructural development in Cu-2.9 Ti is different from that in Cu-1.9 Ti (section 1.), it will be necessary to follow the decomposition reaction in the less concentrated alloy also using SANS. This experiment, which is currently in progress, should indicate whether there are drastic differences in the time evolution of the structure factor between the two alloys, or whether there is a gradual transition from nucleation and growth to spinodal decomposition.

4.3 Interpretation of the SANS results in terms of nucleation and growth

As is shown in Figs. 3, 4 the quench-rate was not sufficient to suppress the formation of Ti-rich clusters during the quench. By assuming that these clusters already have the stoichiometrical composition Cu_4Ti in the as-quenched state, their radii and number density, as well as the supersaturation and the aspect ratio can be derived from the SANS data. Figure 8 shows that the solute concentration decreases from c_o = 2.9 at % Ti to 2.2 at % during the quench; with further aging the supersaturation decreases continuously through the formation of additional clusters and the growth of existing ones. After aging for ~ 100 min/350 °C the metastable β'-solvus line is reached at c_e ~ 0.22 at % Ti. The Ti concentration in the matrix was determined independently from diffuse Laue scattering, from the integrated intensity, and also from a Guinier-analysis, in the latter case by employing the continuity equation. The results agree well.

Figure 9 shows the variation of the aspect ratio with aging time; u was determined from the ratio of the integrated intensities, measured in the two scattering arrangements [10]. Already in the early stages of decomposition, the clusters have adopted an ellipsoidal morphology (u ~ 1.4); with increasing t they grow preferentially along the elastically soft <100> directions as is indicated by the increasing aspect ratio.

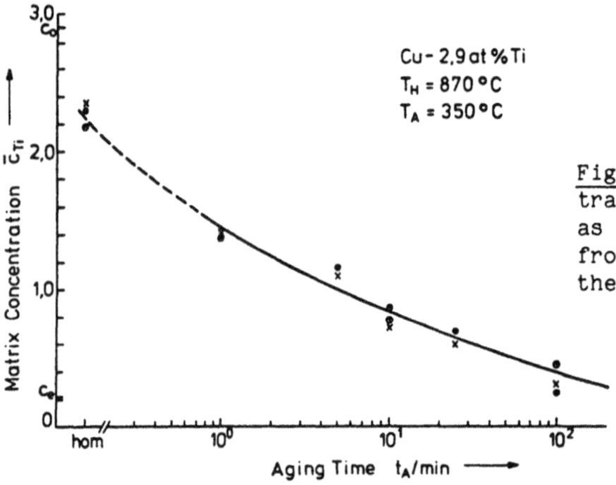

Fig. 8: Decrease of the solute concentration in the matrix with aging time, as determined from Laue-scattering (⊙), from the integrated intensity (x), and the Guinier approximation (•)

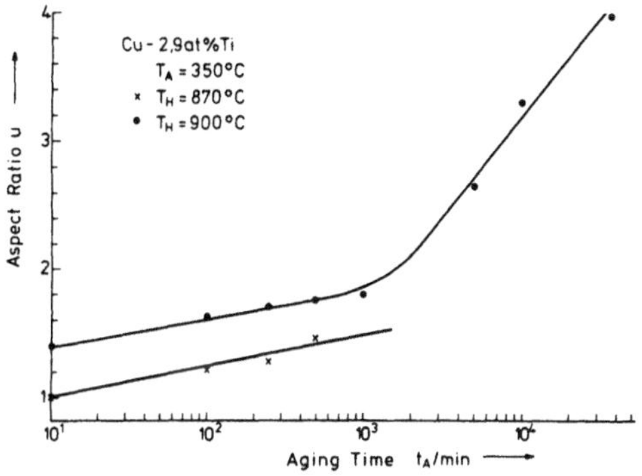

Fig. 9: Time dependence of the aspect ratio u for rotation ellipsoidal clusters

Fig. 10 shows the variation with aging time of the cluster number density, their mean radii $\bar{R} = \sqrt[3]{u \cdot \bar{R}_\ell^3}$ (R_ℓ determined from either Guinier analyses or from the characteristic length [10]), and the supersaturation. The experimental results (discrete symbols) are compared with predictions from a recent decomposition model [9] (full lines) which includes the time evolution of the precipitate size distribution, and treats nucleation, growth and coarsening of the clusters as competing processes. For the theoretical computation it was taken into account that at t=0 the alloy already contained ~ $2 \cdot 10^{19}$ clusters/cm³ with $\bar{R}$ ~ 0.7 nm; which can immediately grow by depleting the matrix of solute atoms. At t=0 the supersaturation is still large enough for nucleating new clusters with smaller radii at a nucleation rate J (Fig. 10a). Thus, during the first minutes of aging the cluster number density increases. At this stage, the alloy contains a sort of bimodal cluster distribution: the larger ones formed during quenching, and the smaller ones resulting from nucleation. Due to both nucleation of new clusters and growth of pre-existing ones, the supersaturation and, hence, the nucleation rate decrease rapidly; after aging for ~ 3 min nucleation is virtually terminated. After ~ 10 min the critical radius R*, which is correlated with the momentary supersatura-

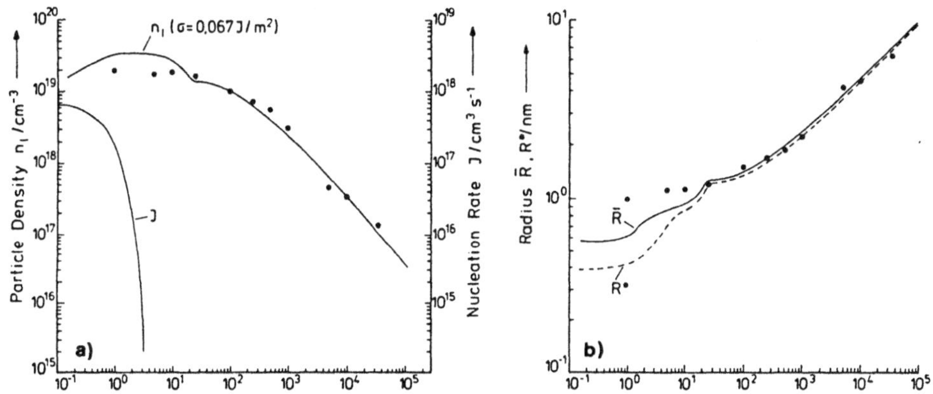

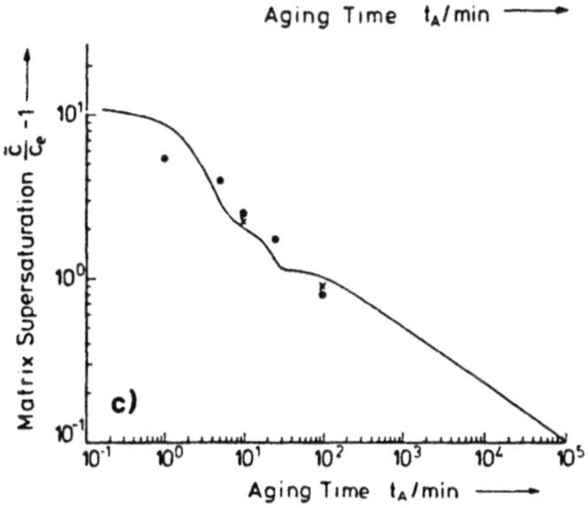

Fig. 10: Time evolution of a) the cluster number density and the nucleation rate J, b) of their mean radius and of the critical radius, and c) of the supersaturation. Experimental results: discrete symbols; computational results: full lines

tion, has reached the mean value $\bar{R}$ of the entire size distribution. At this point, R^* has grown beyond $\bar{R}'$ of the smaller freshly nucleated clusters; these now redissolve; this dissolution leads to a further decrease of n_ℓ. Now the size distribution is again governed by the larger precipitates which formed during quenching. After about 500 min, n_ℓ decreases in proportion to t as expected from the LSW theory [cf. 9].

With regard to the acccuracy of both the SANS experiment and, in particular, the SANS data evaluation, the agreement between the experimental kinetic data and those from the computations is rather good. In the early stages ($t \lesssim 10$ min) the experimental $\bar{R}$ is considerably larger than the theoretical one. This simply reflects the fact that the scattering power of a particle is biased with R^6; hence, for $t \lesssim 10$ min we essentially determine the radius of only the large particles within the bimodal distribution.

From these results, it is obvious that the decomposition reaction is rather fast, and that the SANS data can be well interpreted in terms of a nucleation and growth model.

References

1. K.E. Biehl and R. Wagner: Proc. Solid-Solid Phase Transformations, Eds. H.I. Aaronson, D.E. Laughlin, R.F. Sekerka and C.M. Wayman, Metall. Soc. of AIME, 1982, p. 185
2. L.v. Alvensleben and R. Wagner: Decomposition of Alloys: The early stages, 2nd Acta Scripta Metallurgica Conference, Eds. P. Haasen, V. Gerold, R. Wagner and M.F. Ashby, (Pergamon Press, 1984), p. 143
3. R. Wagner: Field Ion Microscopy in Materials Science, in Crystals-Growth, Properties and Applications, Vol. 6 (Springer Verlag, 1982)
4. J.E. Hilliard: Phase Transformations, (Metals Park, Ohio: Am. Soc. Met., 1970), p. 497
5. R. Wagner: Czech. J. Phys. B 31, 198 (1981)
6. K. Binder, C. Billotet and P. Mirold: Z. f. Physik B 30, 183 (1978)
7. J.S. Langer, M. Baron and H.D. Miller: Phys. Rev. A 11, 1417 (1975)
8. J. Marro, A.B. Bortz, M.H. Kalos and J.L. Lebowitz: Phys. Rev. B 12, 2000 (1975) and 15, 3014 (1977)
9. R. Kampmann and R. Wagner, see ref. 2, p. 91
10. H. Eckerlebe: SANS-Experimente zur Bestimmung der Entmischungskinetik in übersättigten Cu-2.9 at % Ti Einkristallen. Diploma-thesis, University of Hamburg, 1985.

Phase Transformations in Fe-Cu-Alloys – SANS-Experiments and Theory

R. Kampmann and R. Wagner

GKSS-Forschungszentrum, Institut für Physik, D-2054 Geesthacht, F.R.G.

1. Introduction

A precipitation reaction in a supersaturated metastable alloy starts with the formation of stable nuclei, then proceeds to a more or less marked growth regime and finally ends in a coarsening reaction. According to the classical nucleation theory, the nucleation rate J is mainly governed by the work of formation ΔF^* of a critical nucleus with radius R^* [1]:

$$J \propto \exp[-\Delta F^*/kT] \quad \text{and} \quad \Delta F^* = 4\pi\, R^{*2} \cdot \sigma/3 \qquad (1)$$

(σ: specific interfacial energy). A nucleated spherical particle of radius R and solute concentration c_P embedded in a supersaturated matrix of mean solute concentration $\bar{c}(t)$ will grow at a rate [2] (D: diffusion coefficient):

$$\frac{dR}{dt} = v(t) = \frac{\bar{c}(t) - c_R}{c_P - c_R} \frac{D}{R} \qquad (2)$$

The solute concentration c_R at the interphase boundary generally is calculated according to the Gibbs-Thomson equation [cf. 3].

By nucleation and growth of stable nuclei the matrix becomes depleted. This leads to an increase of the critical size R^* and consequently to a strong decrease of the nucleation rate (equ. 1). If, due to a high nucleation barrier ΔF^*, only a few precipitates are formed during the nucleation period, then there will be a marked growth regime during which the precipitates grow substantially by depleting the matrix according to equ. 2. In this case, a growth law for the mean size $\bar{R}$ of the form $\bar{R} \propto t^{1/2}$ might be observed within a short period - the growth rate, however, is not properly predicted by the theory of ZENER and HAM [3-5].

The coarsening reaction starts when the supersaturation of the matrix has been reduced so much that the critical radius R^* becomes larger than the radius of the smaller precipitates, which then start to redissolve. The LSW-theory describes the asymptotic (for $t \to \infty$) behaviour of the coarsening reaction, which is characterized by an $R \propto t^{1/3}$ growth law and a sharply peaked size distribution with $\bar{R} = R^*$ [6].

The course of a decomposition reaction and, hence, its mathematical formulation is rather complicated. Since such a formulation has not been available we have recently developed a decomposition model which includes the time evolution of the precipitate size distribution, and treats nucleation, growth and coarsening as competing processes [3, 4].

2. Experimental

2.1 Specimen preparation

The samples were solution treated (Fe-1.38 at% Cu: 48 h/920 °C; Fe 0.64 at% Cu: 48h/860 °C), quenched into water and then isothermally aged at 400 or 500 °C.

2.2 SANS experiment

Cu-precipitates and the surrounding α-Fe-matrix have a different Cu-content and magnetization. This results in a nuclear ($d\Sigma_{nucl}/d\Omega$) and a magnetic ($d\Sigma_{mag}/d\Omega$) scattering contribution. Both were determined by applying a strong magnetic field to the sample perpendicular to the incident beam, and by measuring the SANS intensity both parallel and perpendicular to the magnetization of the sample.

3. SANS analysis

3.1 Determination of the size distribution

TEM has established that small Cu-precipitates with $R \approx 3$ nm are spheres. During growth they become ellipsoidal - though the aspect ratio ε remains below 1.6 for $R < 10$ nm. Thus we can try to determine the size distributions from SANS data on the basis of (nearly) spherical particles.

From numerical simulations of the decomposition reaction we expect a LSW-distribution in the Fe-1.38 at% Cu alloy after aging for 200 h at 500 °C. Therefore, we tried to interpret the scattering curve of this specimen in terms of spherical, LSW-size-distributed particles. However, with these assumptions no satisfactory fit is obtained (e.g. 1; $\varepsilon = 1.0$ (LSW) curve). Furthermore, we interpreted the scattering curve with the assumption that the particles are LSW-distributed rotation ellipsoids with half-axes R and $R \cdot \varepsilon$. A reasonable fit is obtained for $\varepsilon = 2.72$; for $\varepsilon \lesssim 2.5$ significant discrepancies exist between the measured and the calculated cross-sections (Fig. 1). From TEM we know that $\varepsilon \lesssim 1.3$. So we have to conclude that the true size distribution does not agree with the theoretically expected one!

On grounds of this result the measured cross-sections were interpreted in terms of a logarithmic-normal size distribution $f_{log.norm.}$ of spherical precipitates, the standard deviation of which can be adjusted by a parameter C. In this way we obtained an excellent fit to the scattering curve from Fe-1.38 at% Cu, aged for 200h/500 °C (curve with C = 1.40 in Fig. 1). Similar good agreements were obtained for all other scattering curves from Fe-Cu samples.

Now we demonstrate for Fe-1.38 at% Cu aged for 1h/500 °C how accurately a size distribution can be determined by SANS: An excellent fit to the scattering curve was obtained by varying both the mean radius $\bar{R}$ and C independently (result: C = 1.29, Fig. 2; corresponding size distribution: Fig. 3). Further calculations were performed with different fixed values $C \neq 1.29$ and adjusting $\bar{R}$. For $C \lesssim 1.20$ and $C \gtrsim 1.40$ significant discrepancies between the measured and the theoretical scattering curves in the range of $0.6\ nm^{-1} < \kappa < 3\ nm^{-1}$ were obtained (Fig.2; corresponding size distributions: Fig. 3). Furthermore, a Guinier-like interpretation was performed with having fitted the scattering curve only in the range of $\kappa < 1.8\ nm^{-1}$ (C = 1.05; Fig. 2 and 3).

3.2 Determination of the volume fraction f and Cu-content ν_{Cu} of the precipitates

If the SANS intensity results from only one scattering contrast one cannot determine both f and ν_{Cu} from the scattering obtained from one sample. However, for Fe-Cu we can analyse the nuclear and the magnetic scattering contributions. Their ratio A depends on ν_{Cu} and on the magnetization of the precipitates. Following a study of KNELLER [7] we expect precipitates with $\nu_{Cu} \gtrsim 0.70$ to be paramagnetic and those with $\nu_{Cu} \lesssim 0.70$ to be ferromagnetic, the average magnetic moment per atom being roughly the same as for simple dilution. Thus, for ferromagnetic particles the ratio A is independent of ν_{Cu} (A = 10.0), in the case of paramagnetic precipitates the relation $A = 10.0/\nu_{Cu}^2$ holds.

The precipitated volume fraction f can be determined from the integrated magnetic scattering contribution Q_m; the relations $f \propto Q_m/\nu_{Cu}^2$ and $f \propto Q_m$ hold for ferromagnetic and paramagnetic precipitates, respectively. However, for all aging

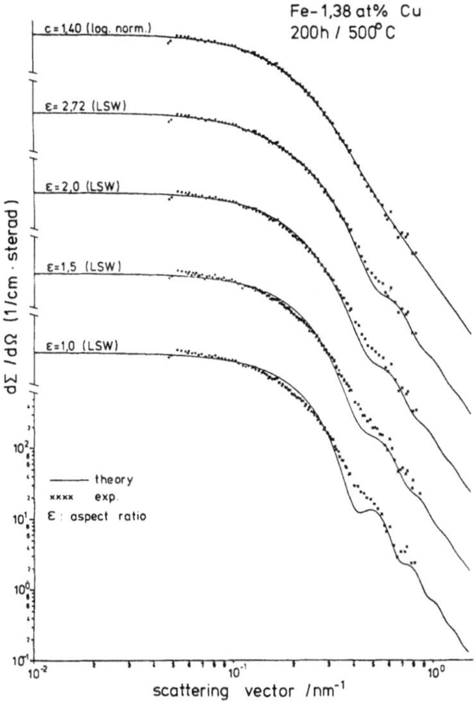

Fig. 1: Magnetic SANS intensity from Fe-1.38 at% Cu aged for 200h/500 °C. Theoretical curves are calculated assuming LSW-size distributions of ellipsoidal precipitates with various aspect ratios ε; one calculation is based on a log.-normal distribution of spherical precipitates with c = 1.40

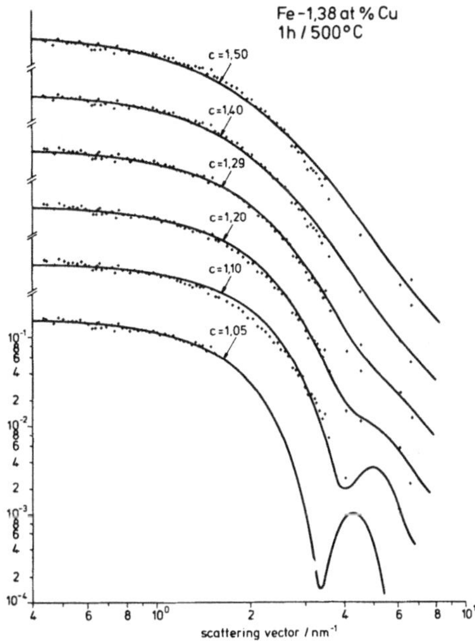

Fig. 2: Magnetic SANS intensity from Fe-1.38 at% Cu aged for 1h/500 °C. Theoretical curves are based on spherical precipitates and a log.-normal size distribution

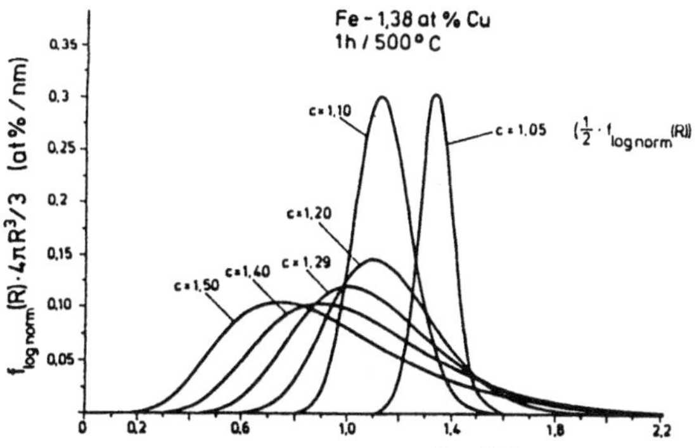

Fig. 3: Log.-normal size distributions which correspond to the theoretical scattering curves of Fig. 2

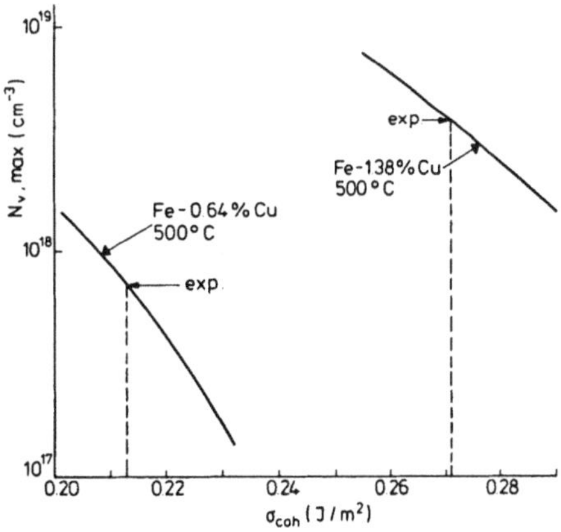

Fig. 4: Determination of interfacial energies from calculated (full lines) and experimentally determined maximum number densities $N_{v,\ max}$

stages for which Q_m and A could be determined with sufficient accuracy, we always calculated values for $f \cdot \nu_{Cu}$ which were larger than the nominal Cu-content of the alloy; thus, the precipitates are paramagnetic with $\nu_{Cu} \approx 1$.

4. Thermodynamic analysis of the Cu-Fe-System

Essentially from thermodynamic analyses, which were based on free energy calculations for the Fe-Cu phase diagram by KUBASCHEWSKI et al. [8], we calculated both the coherent and incoherent equilibrium solubilities, as well as the coherent and incoherent interfacial energies (σ_{coh}, σ_{inc}) for bcc and fcc Cu-precipitates, respectively; details are given in ref. 5.

5. Results and Discussion

5.1 Cu-content ν_{Cu} during the early decomposition stages

The nucleation barrier ΔF^* might be reduced and, hence, the nucleation enhanced if the stable precipitates originate from clusters of critical size, the composition of which was significantly different from the equilibrium one. In order to detect such a nucleation path, the determination of ν_{Cu} during the very early decomposition stage is of great importance.
On grounds of the thermodynamic analysis we have inferred that only very small precipitates ($R \lesssim 0.3$ nm) can energetically be favourable if $\nu_{Cu} \ll 1$. However, particles with $R \gtrsim 0.5$ nm, as observed in Fe-1.38 at% Cu after aging for only 0.5h/500 °C, are expected to have $\nu_{Cu} \sim 1$. As analyses of the SANS data have yielded the particles to be highly enriched with Cu at this aging stage, theoretical analysis and experimental results are in good agreement.

5.2 Coherent interfacial energy

The maximum number density $N_{v,max}$ of particles in a precipitation reaction depends sensitively on the interfacial energy σ (equ. 1), since $\Delta F^* \propto R^{*2}\, \sigma \propto \sigma^3$. This is the key for determination of σ by simulating the decomposition process numerically for various values of σ and comparing the calculated values of $N_{v,max}$ with the experimental ones [3,4]. In this way we obtained for Fe-1.38 at% Cu $\sigma_{coh} \approx 0.270$ J/m², whereas a significantly smaller value $\sigma_{coh} = 0.215$ J/m² was derived for Fe-0.64 at% Cu (Fig. 5). This is attributed to the fact that at the beginning of the decomposition reaction clusters of critical size in Fe-0.64 at% Cu contain about 30 atoms. In contrast, in Fe-1.38 at% Cu they only contain about 7 atoms. This implies a cubic morphology with a larger specific surface area than for spherical shaped particles.

5.3 Decomposition kinetics

By knowing the equilibrium solubilities and the interfacial energies we can employ our recently developed "Numerical Model" [4, 5] in order to describe the course of decomposition in various Fe-Cu alloys. Some of the theoretical and experimental results are shown in Fig. 5-7 for Fe-1.38 at% Cu aged at 500 °C. Nucleation is finished after about 15 min; at this stage, $\bar{R}$ is still smaller than 0.6 nm. This clearly demonstrates the difficulties one is faced with in studying nucleation processes in alloys. After the nucleation period the precipitates grow essentially by depleting the still highly supersaturated matrix (up to $\bar{R} \approx 0.8$ nm after 30 min aging). By inspection of the time evolution of $\bar{R}(t)$ and $N_v(t)$ (Fig. 5, 6), it becomes evident that theory predicts a significantly slower transition to the coarsening regime beyond t = 30 min than has been observed experimentally. This fact is reflected in a significant discrepancy between the measured and the computed standard deviation s of the size distributions (Fig. 7). As the numerical simulation describes correctly the time evaluation of the decomposition without limiting assumptions with respect to the mathematical treatment, we must conclude that some basic ideas concerning the microscopic mechanisms of particle growth must be modified. Indeed, we already have obtained better agreement between theoretical and experimental results by taking into account an interaction between the precipitates during their growth. This will be discussed in a forthcoming paper.

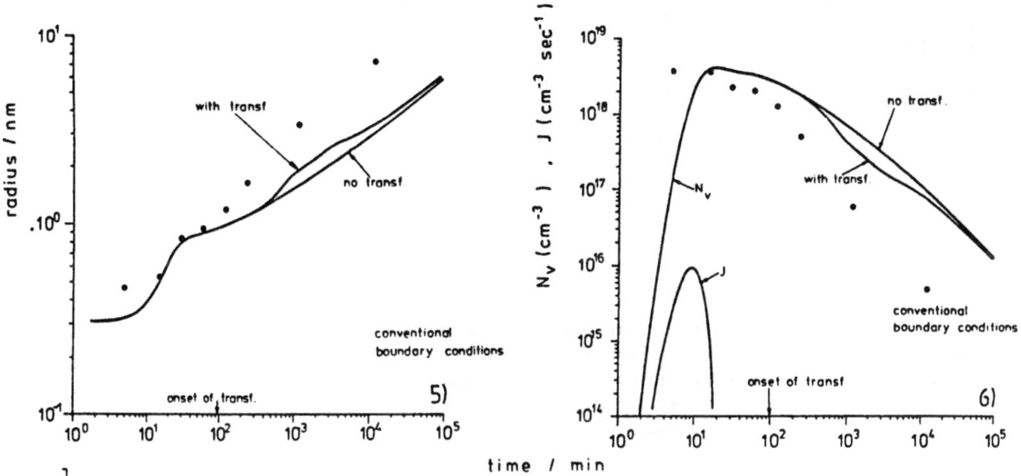

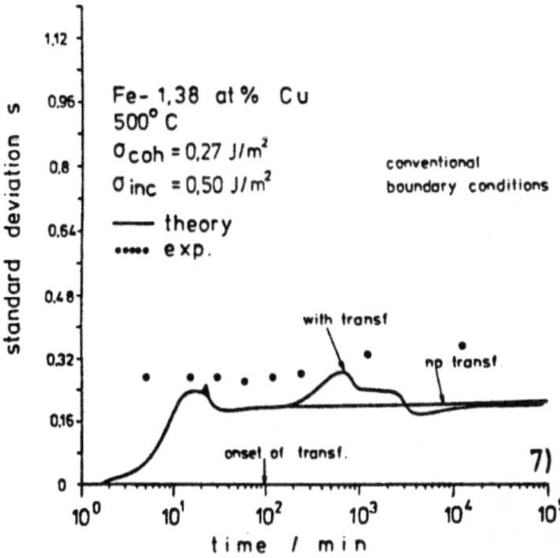

Fig. 5 to 7: Time evolution of $\bar{R}$ (Fig. 5), N_v and J (Fig. 6), and of the standard deviation of the size distribution (Fig. 7) during decomposition of Fe-1.38 at% Cu at 500 °C. Results from the Numerical Model: full lines. In one calculation the transformation of metastable bcc-particles into the fcc-equilibrium structure has been included (σ_{inc} = 0.500 J/m²)

References

1. R. Becker and W. Döring, Ann. Physik 24, 719 (1935)
2. C. Wert and C. Zener, J. Appl. Phys. 21, 5(1950)
3. R. Kampmann and R. Wagner, Decomposition of Alloys: The early stages, 2nd Acta Scripta Metallurgica Conference, Eds. P. Haasen, V. Gerold, R. Wagner and M.F. Ashby, (Pergamon Press, 1984), p. 91
4. R. Kampmann, Ph. D. thesis, University of Hamburg (1986)
5. C. Zener, J. Appl. Phys. 20, 950 (1949)
6. C. Wagner, Z. Elektrochem. 65, 581 (1961)
7. E.F. Kneller, J. Appl. Phys. 35, 2210 (1964)
8. O. Kubaschewsky, J.F. Smith and D.M. Bailey, Z. Metallkde 68, 495 (1977)

Kinetics of Order-Disorder Phenomena in Some Intermetallic Platinum-Based Alloys

M.C. Cadeville, C. Leroux, V. Pierron-Bohnes, and C.E. Dahmani

L.M.S.E.S. (UA CNRS n° 306), Université Louis Pasteur,
4, rue Blaise Pascal, F-67070 Strasbourg Cedex, France

We present a brief report on studies of the atomic mobility and of the order-disorder (O-D) transformation mechanism in some alloys belonging to Ni-Pt and Co-Pt systems. A prior section is devoted to kinetics phenomena inside single phase regions, solid solutions or ordered phases, whereas a second one concerns the region of the O-D transition which is believed to be of the first order kind in these systems.

The investigation we carried out using mainly "in situ" resistivity measurements and X-ray diffraction. We shall see that, in spite of its macroscopic and averaging character, the resistivity allows to get a detailed information on (i) the atomic mobility in both alloys and ordered compounds, and (ii) the role it plays in governing the O-D transformation. Such kinetics studies have to be considered like a preliminary step towards the understanding of the O-D phase transformation phenomena, before using more sophisticated techniques like TEM, XSAS, NSAS and so on.

1. ATOMIC MOBILITY INSIDE SINGLE-PHASE REGIONS

The atomic mobility has been investigated in two disordered alloys : $Ni_{.9}Pt_{.1}$ and Ni_3Pt and in five ordered compounds : Ni_3Pt, NiPt, CoPt, $Co_{.3}Pt_{.7}$ and $CoPt_3$. Our measurement principle relies on two facts : (i) the equilibrium LRO and SRO parameters are well-defined functions of T, (ii) the resistivity is a property quite sensitive to the distribution of atoms on the lattice sites (quantum effects). The methodology we use consists of 3 steps : (i) we start from a thermodynamic state at a given temperature T, characterized by X-rays, (ii) we create a small departure from this equilibrium state with the help of a temperature increment or decrement (ΔT) and then (iii) we measure the time necessary to restore the equilibrium state at $T + \Delta T$. The driving force is only the thermodynamic free energy variation associated with the temperature step. It is known to be very small (some meV) compared with the large potential barrier or activation energy (about some eV) that the atoms have to pass to attain the final equilibrium state. Details on experimental results will be found in ref. 1 and 2. We find by using small ΔT values (some degrees) that the atoms relax towards the equilibrium state through a simple exponential law (Fig. 1) as given by :

$$\Delta R_T(t) = \Delta R_T^{eq}(1 - \exp -(t/\tau_T)) \tag{1}$$

Such a behaviour was observed in all the investigated ordered phases but CoPt. Over the whole temperature range where the measurement of τ is experimentally accessible, we find that the kinetics of LRO is reversible, and the T-dependence of τ can be described by an Arrhenius law (Fig. 2) :

$$\tau_{(T)} = \tau_o \exp E_A/kT \tag{2}$$

The fitted values of τ_o and E_A are collected in Table 1, simultaneously with values obtained in the disordered phases, using a slightly different method [1,2]. It is to be noted that E_A is not very sensitive to the composition in Ni-Pt system, and is quite lower in Ni-Pt than in Co-Pt system. An important result is the similitude of E_A values in both ordered and disordered states of Ni_3Pt ($\Delta E_A^{O-D} \cong 0.1$ eV).

Table 1

	Ordered phases		Disordered phases	
Composition	τ_o(sec)	E_A(eV)	τ_o(sec)	E_A(eV)
$Ni_{.9}Pt_{.1}$			10^{-14}-10^{-15}	2.5-2.7
Ni_3Pt	10^{-13}-10^{-14}	2.7-2.8	10^{-13}-10^{-14}	2.6-2.7
NiPt	1.3×10^{-13}	2.48		
$Co_{.3}Pt_{.7}$	0.4×10^{-16}	3.28		
$CoPt_3$	0.6×10^{-15}	3.12		
CoPt	$\sim 10^{-15}$	$\sim$4.3-4.4		

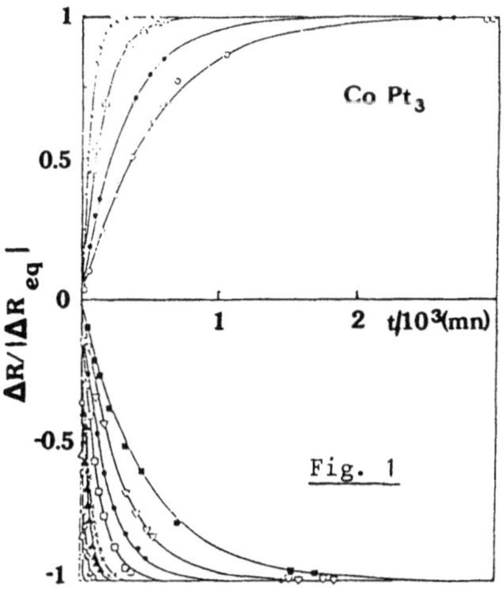

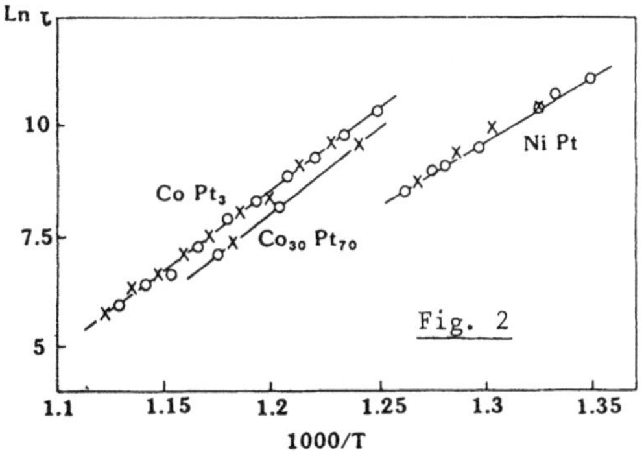

Fig. 1 : Isothermal time dependence of normalized ΔR measured at various temperatures on heating and cooling cycles. In full line, the fitted curves. Increasing T : o 814.4 K ; • 824.2 K ; Δ 833.6 K ; × 853.1 K ; decreasing T : Δ 885.9 K ; ▲ 848.5 K ; × 838.9 K ; □ 829,2 K ; • 819.7 K ; ∇ 810.2 K ; ■ 800.6 K

Fig. 2 : Arrhenius plots of the relaxation times measured at increasing (o) and decreasing (×) temperatures. In full lines the curves fitted with values of τ_o and E_A given in Table 1

The whole of these data and behaviours(a dominant relaxation time at each T, Arrhenius law, small ΔE_A^{O-D} values) is well described [1,2] by the unified model of Kikuchi et al [3] based on the Path Probability Method that is a generalization of the minimum free energy principle to time-dependent phenomena. This model, that treats on an equal foot both LRO and SRO kinetics, lies on the basic idea that LRO and SRO relaxations occur via a unique vacancy mechanism identical to that responsible for atomic and ionic diffusion. E_A represents rougly the sum of vacancy formation and migration energies, plus the ordering energy in the ordered phase.

In CoPt, a first investigation has not allowed to measure a reversible kinetics. One observes at increasing T some kind of disordering kinetics at only

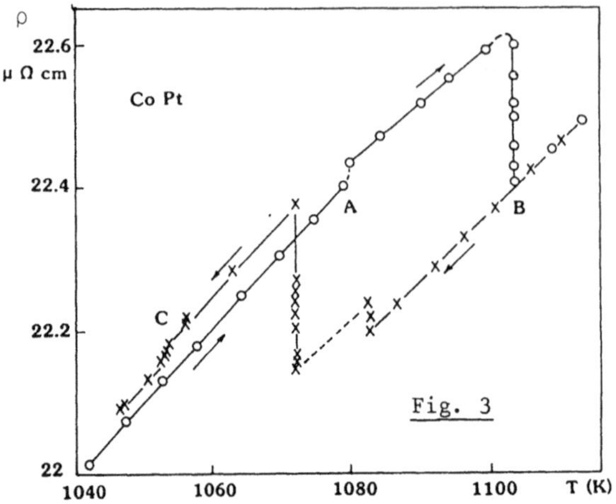

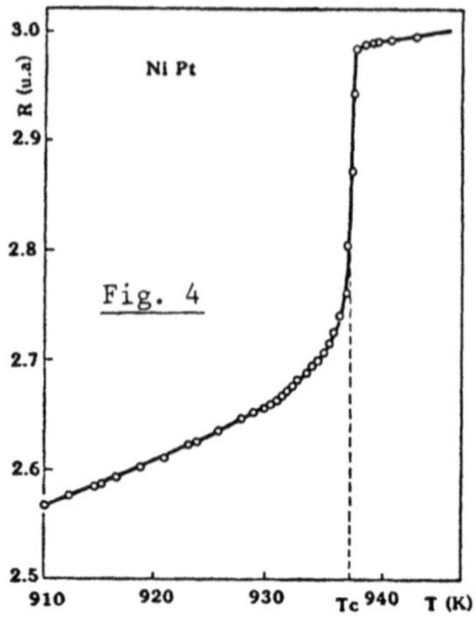

Fig. 3 : Dependence of the resistivity upon increasing (O) and decreasing (×) temperature in CoPt through the O–D transformation.

Fig. 4 : Dependence of the resistivity measured at equilibrium in NiPt upon increasing temperature through its O-D transformation.

1080 K (A point of Fig. 3) where ρ increases and correlatively the c/a ratio (or LRO parameter) deduced from "in situ" X-ray measurements [4] decreases. At decreasing T some ordering kinetics is observed around 1050 K (C point of Fig. 3). Both ordering and disordering kinetics are very sluggish. The measured relaxation times correspond to activation energies of 4.3-4.4 eV by using $\tau_o \cong 10^{-15}$ sec. One can consider that, in CoPt, under such experimental conditions, the atomic configuration is frozen in a not-well defined thermodynamic state over most of the investigated T-range. "In situ" at temperature measurements of LRO parameters are absolutely necessary to precise the order state.

2. RESISTIVITY and KINETICS at the O-D TRANSFORMATION

The determination of the quantities (τ_o and E_A) that govern the atomic mobility inside the ordered phases (but CoPt) allows to follow the T-dependence of the resistivity at equilibrium up to a transition point or line of the phase diagram. This investigation was carried out in the three Ni_3Pt, NiPt and $CoPt_3$ stoichiometric alloys and also in CoPt in its frozen metastable state. The compositions of NiPt and CoPt compounds are thought to correspond to the "critical point" of the phase diagram in the Gibb's sense. Both compounds present the same $L1_0 \rightarrow A_1$ phase transformation at respectively 935 and 1100 K. However, the observed results are quite different. NiPt which has a high atomic mobility ($\tau \sim 30$ s at T near T_c) displays an increase of ρ before T_c, associated with the decrease of the LRO parameter, followed by a rapid increase at T_c corresponding to the formation of the disordered A_1 phase (Fig. 4). The transformation is accompanied by a large increase of the relaxation time at T_c : 2200 mn is the characteristic time τ_c of the disordering. In CoPt where the atomic mobility is low at the vicinity of T_c ($\tau \sim 840$ mn), the phase transformation proceeds as a whole at 1100 K (B point on Fig. 3), quite rapidly ($\tau_c \sim 220$ mn), and the resistivity decreases at T_c. At decreasing temperatures, the transformation presents an hysteresis and is the more rapid, the larger the undercooling is. (In these experiments, T_c is defined within the temperature step which is about 5°).

By comparing these data with the atomic mobilities in the ordered phases, it is clear that the nature of the transformation is different in the two compounds : it is mainly governed by a diffusion process in NiPt where the atomic mobility is high at T_c, whereas it would be governed by some collective mode (martensitic-type transformation) in CoPt where the atomic mobility is quite low at T_c.

In Ni_3Pt and $CoPt_3$ we find that both the derivative of the resistivity and the relaxation times change over a temperature interval T_{c1}-T_{c2} (Fig. 5) and not at a fixed temperature like in NiPt or CoPt. Thus the stoichiometric Ni_3Pt and $CoPt_3$ compounds would not correspond to a critical point of the phase diagrams in the Gibbs sense. These values of T_{c1} and T_{c2} correspond to the limits of a two-phase region (Fig. 6) and are in agreement with previous determination of Ni-Pt and Co-Pt phase diagrams by Inden [4] and ourselves [2] and with X-rays studies of Berg and Cohen [5]. The increase of $d\rho/dT$ is mainly due to the formation of the disordered phase inside the ordered matrix. The large increase of τ is, like in NiPt and CoPt, mainly due to the extra energy necessary to make the first order phase transformation, which would be in that case governed by a diffusion process like in NiPt. The double peak structure of $\tau(T)$ observed in $CoPt_3$ could be explained by the various steps of the transformation, i.e. <u>step 1</u> : nucleation of the disordered domains inside the ordered matrix, <u>step 2</u> : disordered domain growth, <u>step 3</u> : domain coalescence. After the third step, τ rapidly decreases towards its value in the disordered alloy. Such interpretation implies that the extra energy necessary to the domain growth is smaller than the one necessary to create or suppress interphase boundaries. This is generally the case [5,6].

The double peak structure of τ is not observed in Ni_3Pt in which the relaxation times are 10^3 times larger than in $CoPt_3$ before the beginning of the transformation. The accessible range of τ being limited to some 10^3 min, the curve obtained for Ni_3Pt has to be considered like the lower part of the $CoPt_3$ curve. A similar double peak structure was already observed but not retained as significative by Hatta et al [7] in Ni_3Mn in which the O-D transformation also takes place over a two-phase region [8].

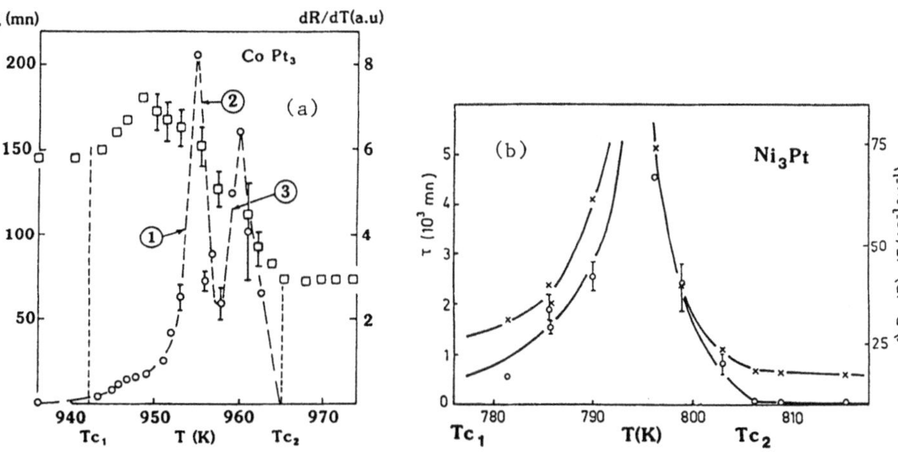

Fig. 5 :
T-dependence of $d\rho/dT$ (□, ×) and τ (o) in $CoPt_3$ (a) and Ni_3Pt (b) in the phase transformation range

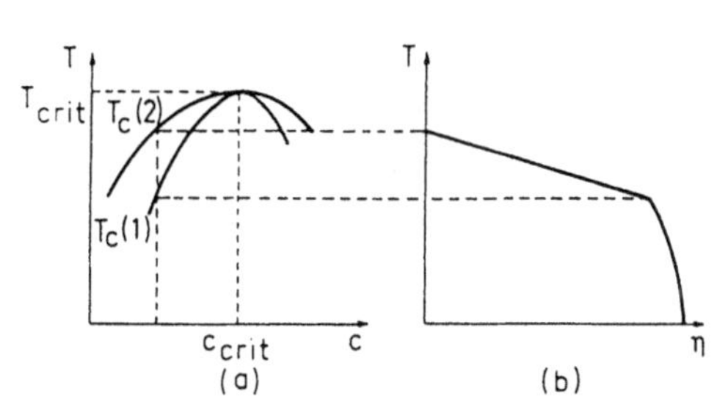

Fig. 6 :
(a) Schematic representation of a phase diagram near a first order O-D transition
(b) temperature variation of the long-range order η for a concentration different from the critical one

It is clear that further investigations and other kinds of experiments are necessary to go beyond these qualitative interpretations.

1. C.E. Dahmani, M.C. Cadeville and V. Pierron-Bohnes: Acta Metall. 33, 369-377 (1985).
2. C.E. Dahmani: Thesis of Louis Pasteur University, Strasbourg, France (1985).
3. H. Sato and R. Kikuchi: Acta Metall. 24, 797 (1976), and K. Gschwend, H. Sato and R. Kikuchi: J. Chem. Phys. 69, 5006 (1978).
4. G. Inden; Mat. Res. Soc. Symp. Proc. 19, 175 (1983), and private communication, Düsseldorf (1985).
5. H. Berg and J.B. Cohen: Metall. Trans. 3, 1979 (1972).
6. R. Kikuchi and J.W. Cahn: Acta Metall. 27, 1337 (1979).
7. I. Hatta and M. Shibuya: J. Phys. Soc. Japan, 45, 487 (1978).
8. J.C. Beers and L. Guttmann: Phys. Rev. B 9, 3941 (1974).

Studies of Heterogeneities of Micrometer-Range with Double Crystal Diffractometer

*D. Schwahn and M.H. Yoo**

Institut für Festkörperforschung, Kernforschungsanlage Jülich GmbH, Postfach 1913, D-5170 Jülich, F.R.G.

The resolution of the Double Crystal Diffractometer (DKD) at KFA is obtained at about 10^{-5} Å^{-1} by using channel-cut perfect Si single crystals. Procedures for the absolute intensity calibration and the slit-height deconvolution are demonstrated by an experiment utilizing Latex spheres of the known size distribution. Applicationsof High-Resolution Small-Angle Neutron Scattering using DKD in metal physics are discussed, with a specific example of creep-induced cavities in austenitic stainless steel.

Introduction

Even though large precipitates and cavities in micrometer (μm) range can be observed with conventional optical microscopes, the use of photon and/or neutron scattering methods is necessary for a quantitative analysis of precipitation and cavitation processes. The principle of scattering techniques involved in Double Crystal Diffractometer (DKD) and slit cameras for Small-Angle-Scattering (SAS) have been known for about the same time /1/. Whereas slit cameras for x-rays (SAXS) and neutrons (SANS) are widely used instruments today, till now the use of DKD instrument has been quite limited. Moreover, in nearly all applications, these DKD instruments have been utilized in a Q-range with reduced resolution where slit cameras are clearly superior to DKD systems. If one uses a DKD with its highest possible resolution, $\Delta Q \simeq 10^{-5}$ Å^{-1}, the superiority of the DKD to slit cameras can be attained, because beam cross-section and divergency are distinctly decoupled from the resolution /2/.

At the KFA-Jülich, we have built a prototype DKD instrument in the reactor MERLIN with resolution of about 10^{-5} Å^{-1}, with which SANS from μm-range particles can be resolved /3/. At the present time, we are in the process of installing this instrument at the new guide laboratory of the FRJ-2 DIDO Reactor where it will be used in complement to our two slit-camera SANS instruments.

The purpose of this paper is to demonstrate that a complete quantitative analysis of scattering particles in μm-size-range is possible by use of the DKD. Following a brief description of the experimental technique, we discuss two specific experimental measurements: (a) scattering from uniform Latex spheres of known size and volume fraction, and (b) detection of creep-induced cavities in an austenitic stainless steel. It is with the experiment with Latex spheres that the full potential of the DKD is realized by successful characterisation of absolute calibration, slit-height deconvolution, and particle-size determination. The second experiment is discussed here as an example for many similar problems, such as phase transformation, defect agglomeration, etc. in metal physics and physical metallurgy.

*Permanent adress: Metals and Ceramics Division, Oak Ridge National Laboratory, Oak Ridge, TN 37831, USA

Instrument*

A sketch of the DKD is given in Fig. 1a. Two perfect Si single crystals are mounted on an optical bench in a nondispersive arrangement. The rocking curve (Fig. 1b) is given by the convolution of the Darwin reflection curves of the two perfect crystals with a halfwidth of about 3 μrad. The sample is placed between the two crystals and the scattered neutronsare detected by detector 1 (Fig. 1a). To measure a neutron scattered into an angle Θ, the analyser crystal has to be rotated by the same angle Θ from its position parallel to the monochromator crystal. The background intensity from the Si-crystals is considerably reduced by using channel-cut crystals in which the neutrons are reflected several times (Fig. 1a). The high Q-resolution exists in one dimension only, therefore, as will be discussed below, slit corrections have to be applied.

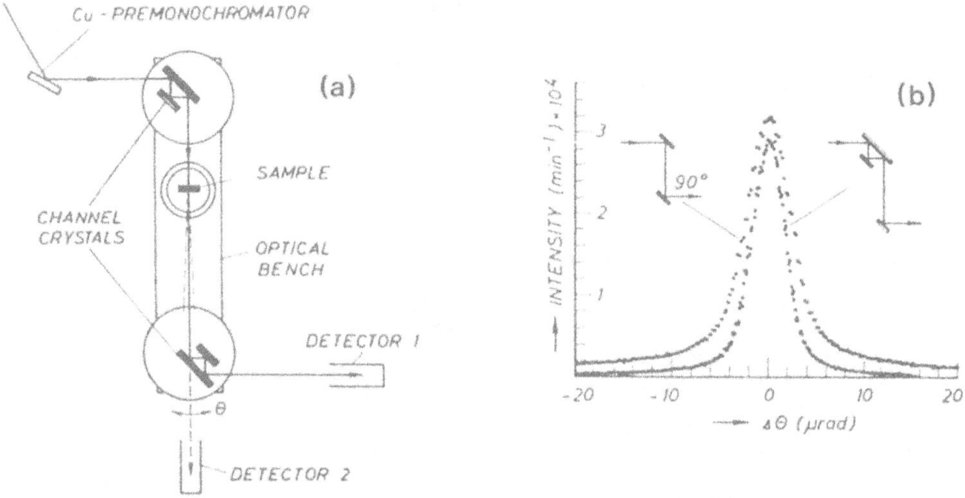

Fig. 1: (a) Sketch of the perfect crystal small-angle scattering camera and (b) measured rocking curves for double and four-fold reflections.

Uniform Latex Spheres

The measurements of uniform Latex spheres of known diameter and volume fraction /4/ discussed here will show how well DKD measurements give the complete information obtainable from SANS technique. As already mentioned DKD gives the slit-smeared cross-section, $\frac{d\tilde{\Sigma}}{d\Omega}$, which is related to $\frac{d\Sigma}{d\Omega}$ by

$$\frac{d\tilde{\Sigma}}{d\Omega}(Q) = \frac{1}{\Delta Q_V} \int \frac{d\Sigma}{d\Omega}(Q)\, dQ_V . \tag{1}$$

$\frac{d\Sigma}{d\Omega}$ is obtained in point-collimated instruments. Q_V is the scattering vector in the vertical slit direction. $d\tilde{\Sigma}/d\Omega$ in the absolute units is calculated in equ. (2) from the scattered intensity $\Delta I(Q)$ by use of the following equation:

$$\frac{d\tilde{\Sigma}}{d\Omega}(Q) = \frac{k_0 \quad \Delta I(Q)}{I_0\, F\, D\, T\, \Delta Q_V\, \Delta Q_H} \tag{2}$$

where k_0 is the incident wavevector, I_0 is the primary intensity, F. D is the area and thickness of the specimen, T is the transmission coefficient and ΔQ_V and ΔQ_H are the horizontal and vertical momentum resolution, respectively. No calibration measurement is necessary because the primary intensity (more exactly:

*A more complete description of the instrument can be found in /3/.

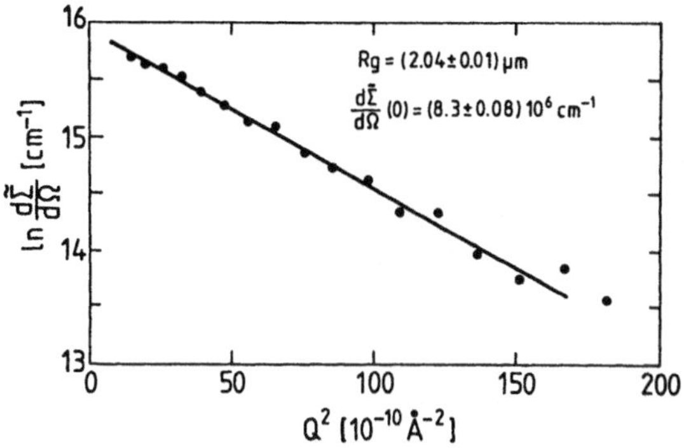

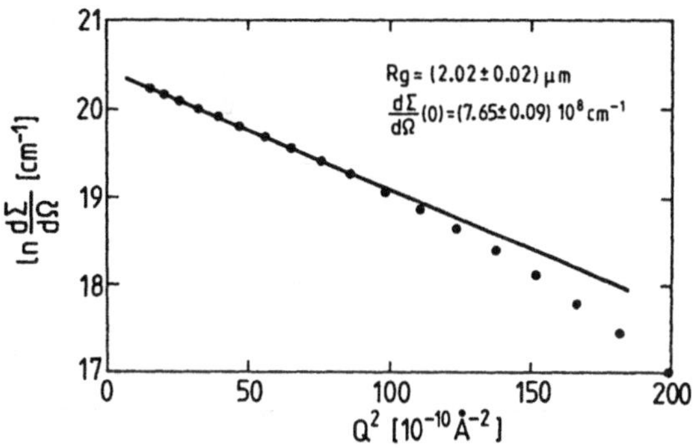

Fig. 2: Guinierplot of scattered intensity from Latex spheres with a mean diameter of (5.25 ± 0.01) μm

Fig. 3: Guinier plot of slit-height deconvoluted scattered intensity from Latex spheres

I_0 F T) is measured during the sample run at Q = 0. ΔQ_H is the halfwidth of the "empty" beam (ΔQ_H = 1.6 10^{-5} $Å^{-1}$) and, therefore, only ΔQ_V has to be evaluated (ΔQ_V = 1.4 10^{-2} $Å^{-1}$).

In Fig. 2 the background-corrected scattered intensity from Latex spheres is plotted in absolute units in a Guinierplot /5/. The "slit smeared" Guinier law is given by

$$\frac{d\tilde{\Sigma}}{d\Omega}(Q) = \frac{d\tilde{\Sigma}}{d\Omega}(0)\ \exp\left(-\frac{R_g^2\, Q^2}{3}\right). \tag{3}$$

Obviously the slope of the line shown in Fig. 2 is not affected by the smearing process. The forward scattering for point geometry, $d\Sigma/d\Omega(o)$, can be calculated from /6/

$$\frac{d\Sigma}{d\Omega}(0) = \frac{R_g\ \Delta Q_V}{\sqrt{(3\pi)}}\ \frac{d\tilde{\Sigma}}{d\Omega}(0), \tag{4}$$

$$\frac{d\Sigma}{d\Omega}(0) = p\ V\Delta\rho^2. \tag{5}$$

For a slit camera system with infinite slit heights and a beam of uniform intensity, the following analytical expression for deconvolution is applicable /1,5/:

$$\frac{d\Sigma}{d\Omega}(Q) = -\frac{\Delta Q_V}{\pi}\int_0^\infty \frac{\frac{d\tilde{\Sigma}'}{d\Omega}\left(\sqrt{(Q^2+x^2)}\right)}{\sqrt{(Q^2+x^2)}}\ dx \tag{6}$$

Both conditions are well fulfilled for DKD measurements since ΔQ_V is three orders of magnitude larger than ΔQ_H. The scattered intensity $d\Sigma/d\Omega$ from the Latex spheres as obtained from equ. (6) is plotted in Fig. 3.

With the known specimen parameters, i.e. the volume V, volume fraction p, and the known chemical composition and the atomic density of matrix and Latex particles, the scattering contrast $\Delta\rho^2$ and $d\Sigma/d\Omega(o)$ can be calculated from equ. (5). In Tab. 1 the diameter of the spheres (D = 2.6 Rg /5/) and $d\Sigma/d\Omega(o)$, all obtained from the known specimen parameters and the neutron scattering measurements are compared. Their agreement is excellent.

Cavities in Crept Stainless Steel

Creep-induced cavities in Fe-15% Cr-15% Ni austenitic stainless steel (SS) were investigated using the DKD. Following an annealing treatment for 2 h at 1273 K, the tensile sample were creep tested under constant load. The creep tests were performed at the initial applied stress of σ = 30 N/mm^2 and temperature of 1100 K

Table 1: Size and forward scattering cross-section from Latex spheres

	Calculated from Specimen Data	DKD-Experiment $d\tilde{\Sigma}/d\Omega$	Analytic Expression (equ. 6)
D [μ m]	5.25 $\pm$ 0.01	5.3 $\pm$ 0.03	5.25 $\pm$ 0.02
$\frac{d\Sigma}{d\Omega}$ (0) [10^8 cm^{-1}]	7.2	7.75 $\pm$ 0.07	7.65 $\pm$ 0.09

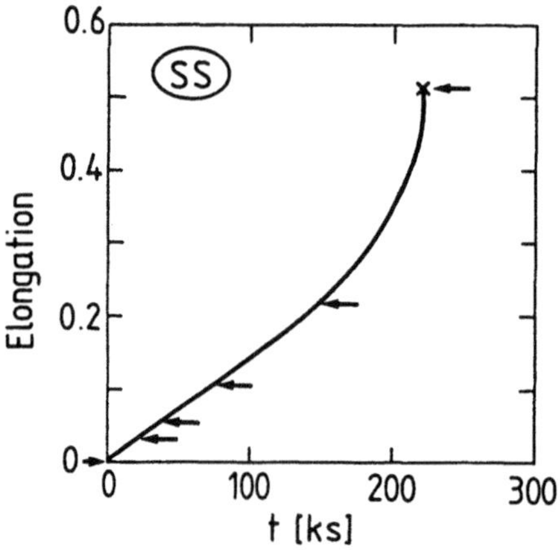

Fig. 4: Creep curve of SS /7/. The arrows indicate interrupted creep states at which DKD measurements were made.

under vacuum of about 10^{-5} mbar. The SS system is nearly free of precipitates according to the recent TEM analysis /7/. The time-dependent tensile elongation, ε , is shown in Fig. 4. The arrows indicate the interruptions of the creep test at which individual scattering measurements were made with the DKD.

Diffraction

Diffracted neutrons from cavities of a mean diameter less than 8 µm could be detected at a creep strain up to 20 %. The scattering data were evaluated in a similar manner as described above for the Latex particles. In Fig. 5 both the diameter and the forward scattering intensity are plotted against creep strain. A strong increase of cavity diameter from 2 to 9 µm is evident between 5 and 10 % strain. An optical microscope examination at ε = 10 % /7/ showed a mean cavity diameter of about 7 µm, which is in good agreement with the present neutron scattering results. The decrease of the mean diameter between 10 and 20 % strain is presumably due to the coarsening of cavities to a size range which contributes only to refraction as is discussed below. A volume fraction of cavities in the range of 10^{-4} and 10^{-3} is estimated from $d\Sigma/d\Omega(0)$ and D shown in Fig. 5. A more complete SANS study in a Q range between 10^{-5} and 10^{-1} $Å^{-1}$ will be made, and its results will be presented elsewhere.

Refraction

At the extremely small scattering angles under observation, refraction on the particles is observed. In Fig. 6 the beam broadening due to scattering centers normalized to 1 cm thick reference specimens is plotted against creep strain.

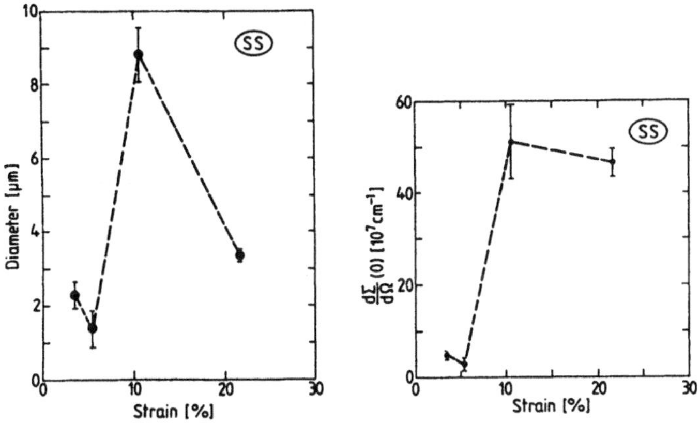

Fig. 5 : Diameter of cavities and forward scattering versus strain

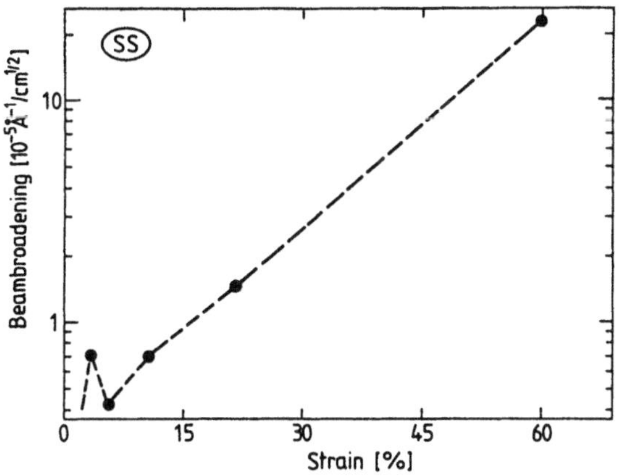

Fig. 6: Beam broadening versus strain due to cavities much larger than 8 μm diameter

For strains larger than 10 % a strong increase of the beam broadening is observed. The condition for refraction is given by /8/

$$\Phi >> \Delta\rho \; \lambda \; D, \tag{7}$$

where Φ is the phase difference between a neutron wave travelling through a particle (cavity) and around the matrix over the same distance, D, and λ (= 1.8 Å) is the wavelength. The observed broadening means that cavities with $\Delta\rho = 7.5 \times 10^{10} cm^{-2}$ and a mean diameter much larger than 8 μm are present in the SS specimen from the very early stage ($\varepsilon \simeq 3$ %) of the creep experiment. Since the beam broadening is independent of the size of the scattering particles /8/, the strong increase of the beam broadening shown in Fig. 6 directly implies a strong increase in the number of cavities with increasing strain beyond ε = 15 %.

Concluding Remarks

It is demonstrated that a quantitative analysis of scattering particles in micrometer size range can be performed using DKD. A high resolution of DKD, $\Delta Q \simeq 10^{-5}$ Å^{-1}, is obtained by using channel-cut perfect Si single crystals. Absolute intensity calibration is made with the primary beam in DKD system. That a reference specimen is not needed for calibration as in slit-cameras is clearly a distinct advantage of DKD. The excellent agreement between neutron scattering results and the known data of Latex particle size distribution indicates that applications of aproximate scattering law (as the Guinier Law) to slit-smeared data and of approximate analytic expression for slit-height deconvolution are fully justified. The example measurements of creep-induced cavities in stainless steel illustrate the potential applications of DKD in the study of cavitation and precipitation in metals and alloys.

Acknowledgement

We would like to thank Rita Thomas from the IFF-KFA Jülich for her computational work on the slit-height deconvolution problem.

Rerences

/1/ J.W.M. DuMond:
Phys. Rev. 72, 83 - 84 (1947)

/2/ J. Schelten and R.W. Hendricks:
J. Appl. Crys. 11, 297 - 324 (1978)

/3/ D. Schwahn, A. Miksovsky, H. Rauch, E. Seidl and G. Zugarek:
Nucl. Instr. and Methods in Physics Research A239, 229 - 234 (1985)

/4/ The specimen has been prepared by A. Rennie, ILL Grenoble

/5/ A. Guinier and G. Fournet: Small-Angle Scattering of x-rays (Wiley, New York, 1955)

/6/ V. Gerold in: Small-Angle x-ray Scattering, ed., H. Brumberger (Cordon and Breach, New York, 1967) p. 277

/7/ J.H. Schneibel, C.L. White, G.F. Petersen, and R.A. Padgett: in Proceedings Intern. Conf. on New Developments in Stainless Steel Technology, Sept. 1984 Detroit, Michigan (ASM), pp. 17 - 21

/8/ G.E. Bacon: Neutron Diffraction (Oxford Univ. Press, 1975)

Decomposition Kinetics in the Superalloy Nimonic PE16 and the System NiAlTi Investigated by TEM, FIM and SANS

W. Hein, H.P. Degischer[1], *D.J. Jensen*[2], *H. Strecker*[3], *W. Wagner, R.P. Wahi, and H. Wollenberger*

Hahn-Meitner-Institut für Kernforschung Berlin, Glienicker Str. 100, D-1000 Berlin 39

1 Introduction

Ni-based superalloys are strengthened by the $L1_2$-ordered γ'-$Ni_3(Al,Ti)$ phase. From TEM investigations it is known that the coherent γ'-precipitation shows the behaviour of classical Ostwald ripening. Such quantitative TEM analysis is restricted to mean diameters larger than about 10 nm; for smaller precipitates the size distribution extends by a considerable portion beyond the visibility limit. To obtain an insight into the precipitation at earlier stages, with the aim to cover the nucleation and growth regime prior to the onset of coarsening, we combined TEM with Small-Angle Neutron Scattering (SANS) and analytical Field Ion Microscopy (FIM). Nimonic alloy PE16 and a γ'-model alloy of the ternary NiAlTi system, Ni-5.0 at.% Al-5.8 at.% Ti were investigated. FIM as well as TEM yield informations on the precipitate morphology, size distributions and, by use of the atom probe device, the two phase compositions. The SANS measurements, analysed by model calculations and by dynamical scaling, allow to determine average precipitate size and interparticle distance. An interactive combination of all the results turns out to be essential for a well-founded evaluation of the undergoing processes.

2 Experimental

The NiAlTi specimens with the nominal composition Ni-5.0 at.%Al-5.8 at.%Ti were prepared by levitation melting in an induction furnace. The investigated Nimonic PE16 was commercial rod material with the composition: Ni-34 at.%Fe-17.6 at.%Cr-2.38 at.%Al-1.47 at.%Ti-1.9 at.%Mo-0.27 at.%C (X8 NiCrMoTiAl 43 16). SANS specimens of 1 mm thickness and typically 10 x 10 mm^2 area were sliced from the PE16 rods or cut from NiAlTi rolled foils, respectively. TEM specimens were punched from slices or foils of 0.2 mm thickness. For FIM specimens, wires of 0.2 mm diameter were drawn. Specimens of PE16 and NiAlTi were solution treated for 1 h at 1313 K and 1323 K, respectively, water quenched to room temperature and finally annealed at temperatures between 773 K and 1113 K. The final preparation of TEM and FIM specimens was done by electropolishing.

For the TEM investigations a conventional Philips EM 400 microscope was used. The Field Ion Microscope with Atom Probe used for the present investigation is described in details by Mertens [1]. The SANS experiments were performed at the RISØ SANS-facility, Denmark [2], with a neutron wavelength λ of 0.5 nm, covering a κ-range of $0.08 < \kappa < 1.65\ nm^{-1}$ ($\kappa = 4\pi\theta/\lambda$ with 2θ, the scattering angle). We note that the γ-matrix of NiAlTi is ferromagnetic. To avoid multiple refraction of the neutrons in magnetic domains with different orientation, the temperature during SANS measurements was adjusted to 423 K, i.e. well above the Curie temperature of about 370 K.

[1] now: Österreichisches Forschungszentrum Seibersdorf, A-1082 Wien, Austria

[2] RISØ National Laboratory, DK-4000 Roskilde, Denmark

[3] now: IBM Plant Mainz, D-6500 Mainz, Germany

3 Results and discussion

TEM superlattice dark field micrographs reveal a spherical shape of the γ'-precipitates in PE16 (Fig. 1a,b). No indication for preferred alignment or nonrandom distribution is found. These morphology characteristics remained unchanged over the entire investigated time and temperature range. In contrast, in NiAlTi there is a change in morphology from initially spherical precipitates to cubes with a pronounced alignment parallel to ⟨100⟩. The transition occurs for particle diameters of approximately 5 nm. The aligned cubes can be imaged in TEM (Fig. 1c) and FIM (Fig. 2), while the spheres are clearly resolved only in FIM (Fig. 2). This change in morphology observed in many Ni base alloys, is explained by the elastic strain at the particle-matrix interface and by elastic interactions between adjacent γ'-precipitates [3]. The missing morphology change in PE16 is due to the distinctly small lattice mismatch of the γ'-phase, which is $\Delta a/a \approx -10^{-4}$ in PE16 compared to $+10^{-2}$ in NiAlTi.

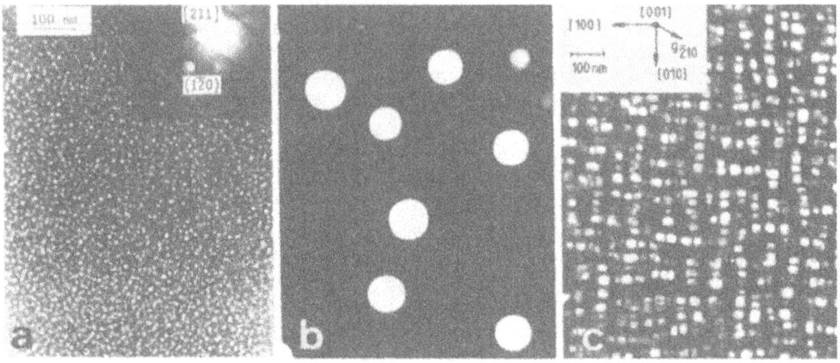

Fig. 1: TEM darkfield superlattice image of γ'-precipitation in Nimonic PE16 (a: 973 K/16 h; b: 1113 K/51 h) and in Ni-5.0 at.%Al-5.8 at.%Ti (923 K/48 h)

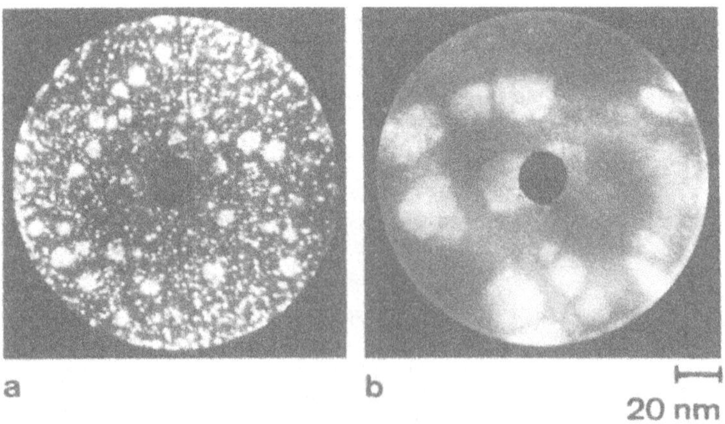

Fig. 2

FIM micrographs of γ'-precipitation in Ni-5.0 at.%Al-5.8 at.%Ti (a: 773 K/0.5 h; b: 923 K/48h)

The precipitate sizes can be determined from the TEM micrographs directly for a given magnification. In the FIM, brightly imaging particles tend to be overmagnified [4]. Precipitate sizes can be determined during continuous field evaporation by counting the evaporated atomic layers while a precipitate is visible. Results of precipitate size analysis are given in Fig. 3, with overlapping size distributions determined by FIM and TEM. We note that the limit for quantitative TEM ana-

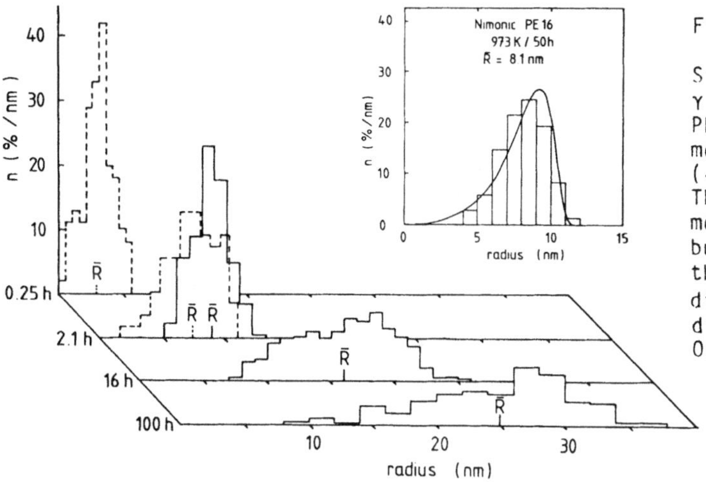

Fig. 3

Size distribution of γ'-precipitates in PE16 aged at 1073 K, measured with FIM (---) and TEM (——). The insert compares a measured size distribution with the theoretical size distribution of the diffusion-controlled Ostwald ripening

Table 1 Atom probe results of the two phase compositions

Nimonic PE16 973 K/16 h

	Ni	Fe	Cr	Al	Ti	Mo	Si	
γ	44.6±1.1	34.2±0.9	16.7±0.7	1.42±0.13	0.36±0.10	2.1±0.2	0.63±0.13	at.%
γ'	64.4±1.3	5.8±1.4	3.8±0.3	10.0±0.5	12.8±0.6	2.1±0.2	1.11±0.17	at.%

Ni-5.0 at.%Al-5.8 at.%Ti 873 K/31 h

	Ni	Al	Ti	
γ	91 ± 2	4.0±0.4	5.0±0.5	at.%
γ'	74.6±1.9	11.3±0.7	14.0±0.8	at.%

lysis is reached for diameters smaller than about 6 nm, while FIM still measures precipitates reliably for diameters around 1 nm.

The FIM atom probe allows the determination of the composition of both phases with a spatial resolution of about 2 nm (projected spot size). The measured composition of the γ'-phase and the γ-matrix are given in table 1. In NiAlTi the ratio Al/Ti of γ' is in good accord with the ratio Al/Ti of the homogenized alloy. This rule does not apply to PE16, where more Ti is found in the γ'-phase than Al. This result is in agreement with measurements by Mangen et al. [5], whereas microchemical analysis of γ'- extractions by Windsor et al. [6] reveal an excess of Al in γ'-precipitates.

Examples of SANS curves from γ'-precipitates in PE16 and NiAlTi are shown in Figs. 4 and 5, resp. The scattering contributions of γ'-precipitates show a maximum which arises from interparticle interference effects. Its position is correlated to the average precipitate distance. Carbides formed before ageing in Nimonic PE16 give rise to significant scattering at κ around and below 0.2 nm^{-1}. Their

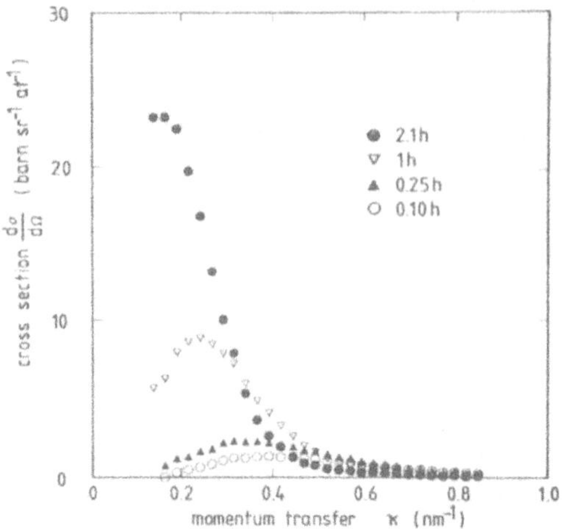

Fig. 4: Scattering cross section vs. momentum transfer of Nimonic PE16 aged at 1073 K

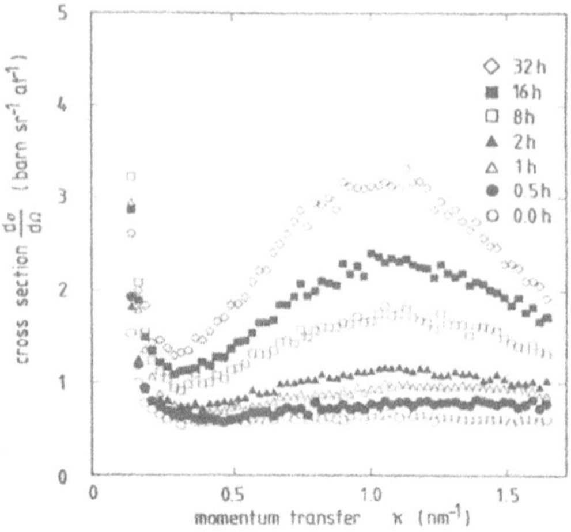

Fig. 5: Scattering cross section vs. momentum transfer of Ni-5.0 at.%Al--5.8 at.%Ti aged at 773 K

scattering contribution was eliminated by substracting the scattering of a homogenized reference sample. For NiAlTi early stages of decomposition were selected for Fig. 5. Here, the intensity maximum is growing with continuous ageing but its position remains almost unchanged. For later stages, the position shifts to smaller κ, like it does in the growth kinetics of PE16 shown in Fig. 4.

For a quantitative analysis of the neutron scattering curves, we calculated the structure function for spheres with a size distribution as determined by FIM and TEM. The interparticle interference was considered according to [7]. The calculated structure functions were fitted to the data. This fitting procedure combined with a dynamical scaling analysis according to Fratzl et al. [8] results in the relevant kinetics parameters "average size" and "average distance" as well as the volume fraction of the γ'-precipitates for all states investigated. The so-determined growth kinetics of precipitates are shown in a log-log plot in Fig. 6, (along with the radii measured directly with TEM and FIM). In PE16 over the entire investigated range the radii follow the t to 1/3 dependence expected for Ostwald-ripening according to [9]. The precipitate size distribution shows the corresponding shape from the very beginning (see Fig. 3). Further, the SANS analysis reveals constant volume fractions and a proportionality between average precipitate size and average interparticle distance. All this indicates that in PE16

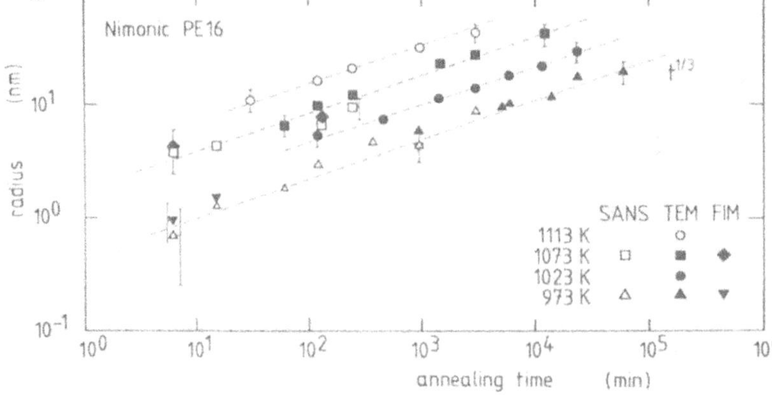

Fig. 6: Log-log plot of the average radii of the γ'-precipitates in Nimonic PE16 vs. annealing time for different annealing temperatures, combining SANS, TEM and FIM results

the Ostwald-ripening begins before the particles have reached a radius of 1 nm. This result is in accordance with calorimetric measurements, which give indications of very fast nucleation in this material [10]. Segregation seems completed on reaching the ageing temperature in specimens quenched as usual after the solution treatment. In contrast, in the alloy NiAlTi (Fig. 7) for small particles a regime is resolved with deviations from classical coarsening. The interparticle distance shows only a slight tendency to increase, approximately proportional to t to 1/18 in this regime, the growth rate is proportional to t to 1/6. The corresponding volume fraction increases with t to 1/3. We believe that in this regime two effects are competing: (i) those particles formed during quenching after the high-temperature solution treatment commence to coarsen, (ii) from the matrix which is still supersaturated at the ageing temperature, further precipitation of small particles occurs. Both effects together yield an increase of volume fraction with almost constant precipitate number density, and a precipitate growth which is retarded compared to classical growth theories.

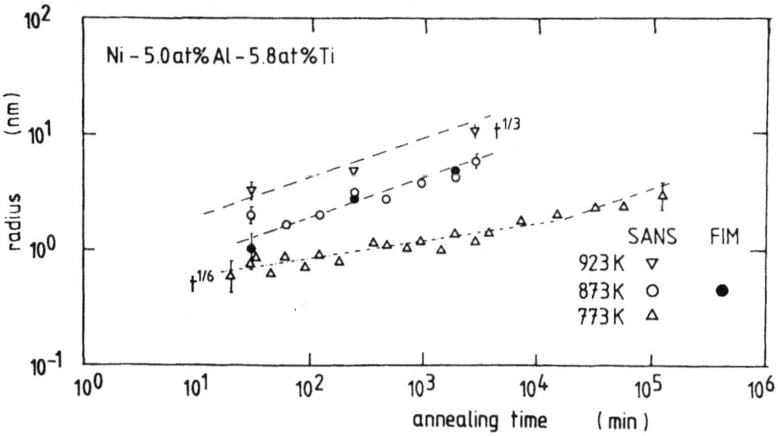

Fig. 7: As for Fig. 6, for Ni-5.0 at.%Al-5.8 at.%Ti

References

1 P. Mertens, U. Vidic and H. Becker, Journ. de Physique 45, 309 (1984)

2 J. Kjems, J. Linderholm, L.G. Jensen, B. Breiting, H. Bjerrum Møller and R. Bauer, RISØ Physics Department Annual Progress Report, RISØ-R-461 (1981)

3 A.J. Ardell and R.B. Nicholson, J. Phys. Chem. Solids 27, 1793 (1966)
Yu.D. Tyapkin, Ann. Rev. Mater. Sci. 7, 209 (1977)

4 R. Wagner, Field Ion Microscopy in Material Science, Vol. 6 of Crystals: Growth, Properties and Applications, Springer (1982)

5 W. Mangen, H. Schäfer and E. Nembach, to be published in Mater. Sci. and Engineering

6 C.G. Windsor, V.S. Rainey, P.K. Rose and V.M. Callen, J. Phys. F 14, 1771 (1984)

7 F. Zernike and J.A. Prins, Zeitschrift für Physik 41, 184 (1926)

8 P. Fratzl, J.L. Lebowitz, J. Marro and M.H. Kalos, Acta Metall. 31, 1849 (1983)

9 C. Wagner, Z. Elektrochemie 65, 581 (1961)

10 H.P. Degischer et al., to be published.

Comparative TEM, FIM and SANS Study of the Cu-2 at.% Co Decomposition

H.P. Degischer[1], *J. Piller*[2], *W. Wagner, and H. Wollenberger*

Hahn-Meitner-Institut für Kernforschung Berlin, Glienicker Str. 100, D-1000 Berlin 39

1 Introduction

Conventional Transmission Electron Microscopy (TEM), Field Ion Microscopy (FIM) and Small-Angle Neutron Scattering (SANS) are frequently applied methods for investigations of precipitation processes in alloys. They are in principle well suited for such investigations; however, in view of an increasing interest in distinct early decomposition stages, a classification of the methods with respect to resolution limit, accuracy and reliability in decomposition studies is urgently desired [1,2]. For this purpose, the three methods were applied together to investigate the alloy Cu-2 at.% Co. The system CuCo is known to form spherical precipitates of the Co-rich α-phase dispersed in a matrix of almost pure Cu [3]. The present combined study was just published in details elsewhere [4]. Here, we briefly review the results, focusing on different procedures of SANS evaluations using the complementary informations obtained by FIM and TEM. We show how the investigations can thus be extended to such early stages which could not be evaluated quantitatively up to now, if only one of the methods is applied.

2 Results

For the combined FIM, TEM and SANS study, two early decomposition stages were chosen, attained by 1 and 3 h annealing at 833 K. The SANS study was extended further to stages attained by short-time annealing at 833 K. The effective annealing times, considering the time temperature characteristic for heating, were 2 and 7 min. (2' and 7').

In the FIM the topology of the specimen tip surface is imaged via field-ionized gas atoms on a phosphor screen or an image intensifier. Depending on the image gas composition, the tip temperature and the elastic field strength at the tip surface, one obtains atomic resolution (Fig. 1a) yielding an image, which is originated by the surface steps of nearly circular atomic layers, or precipitate contrast (Fig. 1b) displaying the chemical heterogeneities. In this mode, the Co-rich precipitates, protruding from the surface due to higher resistance to field evaporation compared to the surrounding Cu matrix, are imaged as bright areas. Independent from the image magnification, the size of the precipitates can be determined with atomic resolution by counting the number of evaporated atomic layers while the precipitate is visible during continuous field evaporation. Figure 2a and b show so-determined distributions of precipitate radii for the 1 and 3 h annealing stages, along with the respective TEM results (Fig. 2c and d). The related numerical data are listed in Tab. 1.

For TEM imaging small elastic distortions around the precipitates cause a distinct diffraction contrast [5], yielding an image of two lobes of reduced intensity separated by a line of no contrast (coffee-bean-like contrast). Figure 3 gives examples of bright and dark field micrographs at $\underline{g}$ = (220). The length of

1 now: Österreichisches Forschungszentrum Seibersdorf, A-1082 Wien, Austria

2 now: Dornier System GmbH, D-7990 Friedrichshafen, Germany

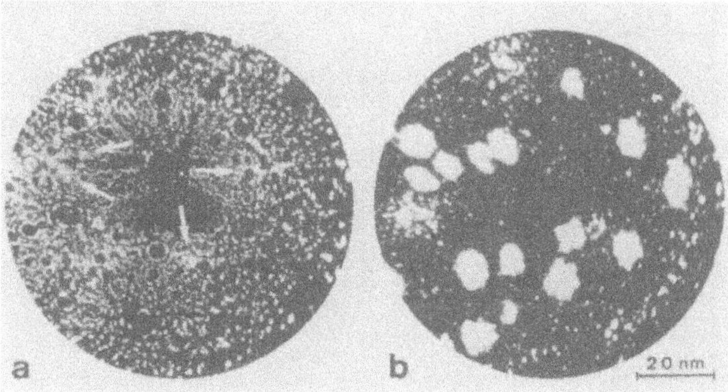

Fig. 1

Field-ion micrograph of Cu-2 at.% Co, annealed for 1 h at 833 K. Atomic resolution is obtained at a tip temperature of approx. 40 K (imaging gas: He). Here, the precipitates (some marked by arrows) are hardly visible. b) At higher tip temperature (approx. 75 K), precipitate contrast is achieved. The bright areas represent the intersections of the Co-rich precipitates with the tip surface (average R = 1,6 nm).

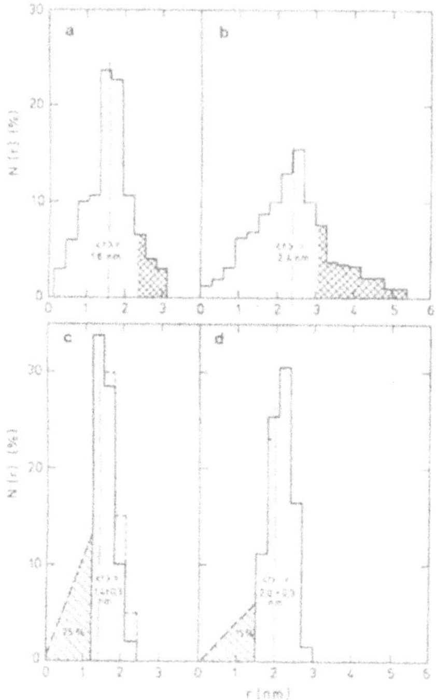

Fig. 2

Normalized particle size distributions for the two annealing states of Cu-2 at.% Co, as determined by FIM (a,b) and TEM (c,d). The left column (a and c) is for one hour and the right one (b and d) for three hours ageing time at 833 K. The hatched areas of the TEM distributions (25% resp. 15%) are particles beyond the limit of quantitative evaluation, which were classified according to the size distribution measured by FIM. The crossed areas of the FIM distributions are attributed to non-separated dimers and trimers.

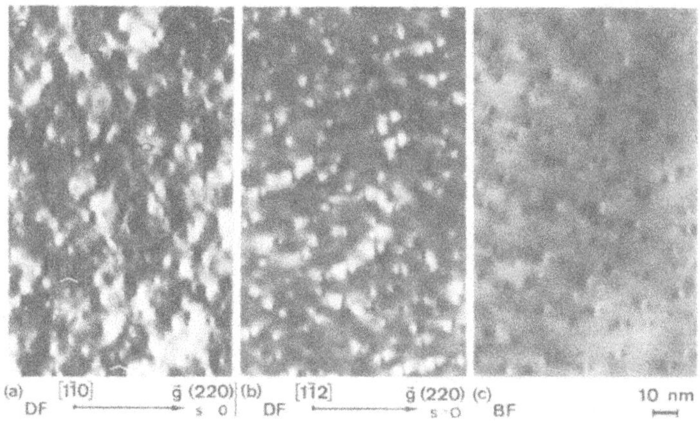

Fig. 3

(a) Dark field TEM micrographs of a foil aged 1 h at 833 K showing only a few symmetric strain field contrasts. (b) Dark and (c) bright field images of a foil aged 3 h at 833 K with a thickness of about 1.75 $\xi_g \approx$ 80 nm.

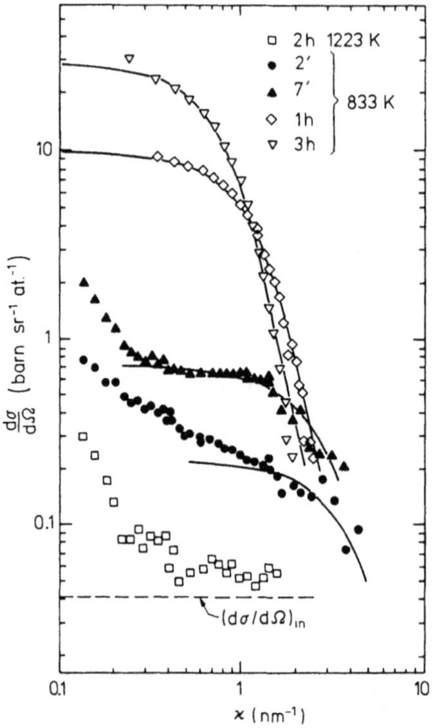

Fig. 4
Double-log plot of the neutron scattering cross-section vs. momentum transfer of Cu-2 at.% Co after quenching and annealing. The solid lines represent calculated SANS curves for a dispersion of spherical precipitates in a matrix, fitted to the 3 h ageing state and then scaled to the SANS data of the other ageing states. The results of the fit and the scaling are given in Tables 1 and 2.

Table 1: Particle sizes (average and maximum) and volume fractions as determined from FIM, TEM and SANS

	R (nm)				R_p (nm)	r_{max} (nm)	volume fraction f_p** (%)		
	FIM	TEM	SANS calculations with FIM size distr.	SANS calculations with TEM size distr.	SANS Guinier	TEM	FIM	TEM	SANS
1h 833K	1.5*	1.4	1.4*	1.55	1.87	2.4	2.4±1.2	> 1	2.2±0.8
3h 833K	2.1*	2.0	2.1*	2.2	2.67	3.0	2.4±1.2	> 1	2.2±0.8

* Tail of large particles (crossed regions in Fig. 2) omitted
** Magnetization measurements yield (1.80 ± 0.05)% [3]

	κ-scaling $R = R^{3h}\frac{\kappa^{3h}}{\kappa}$ (nm)	Intensity-scaling $R = R^{3h}\left(\frac{I}{I^{3h}}\right)^{1/3}$ (nm)
2' 833 K	0.46 ± 0.07	0.48 ± 0.04
7' 833 K	0.65 ± 0.05	0.71 ± 0.02

Table 2: Scaling results: average precipitate radii

the line of no contrast as well as the separation of the intensity minima in bright field or of the intensity maxima in dark field is directly correlated to the precipitate size. This correlation has been proved by intensity computations [4]. Thus, from images as shown in Fig. 3, the size of the visible precipitates can be measured, yielding the size distribution as shown in Fig. 2c and d. Precipitates smaller than about 3 nm in diameter could not be measured, but counted in number, and were accounted for according to the size distribution measured by FIM.

The neutron small-angle scattering curves of the alloy after homogenization and after the applied ageing treatments are shown in Fig. 4. The scattering intensity is given in absolute units of the differential cross-section per solid angle and atom ($d\sigma/d\Omega$) plotted versus the momentum transfer κ, defined in the usual way as $\kappa = 4\pi/\lambda \cdot \sin\theta$, with λ, the neutron wavelength and 2θ, the scattering angle.

The neutron scattering data have been evaluated in three different ways: i) by the classical Guinier evaluation, ii) by calculating the structure function for a dispersion of spheres with the size distributions determined by FIM and TEM, and iii) by applying a scaling procedure as suggested by Fratzl et al. [6]. The numerical result from these procedures are given in Tab. 1. In Tab. 2 the average precipitate radii of the 2'and 7' annealing stages are given as derived by scaling to the structure function of the 3h annealing stage.

3 Discussion

The particle size distributions (Fig. 2) show self explaining the resolution abilities and limits of FIM and TEM. For the small precipitates, only FIM gives quantitative results. For large precipitates marked differences in the distributions occur. The TEM results imply that the FIM data in this part of the size distribution must be erroneous. This is supported by the fact that no scaling could be achieved for the SANS curve computed with the FIM size distribution if the tail of large precipitates (crossed region in Fig. 2) is not omitted. Hence, these "large particles" in the present case are explained by unresolved dimers or trimers, aligned parallel to the tip axes. The correct precipitate size distributions as given in Figs. 2c and d were thus obtained by combining the results of all three methods. It represents the shape predicted by the model of diffusion controlled Ostwald ripening [7] with a characteristic steep drop at the maximum radius of about 1.5 R, where R is the average radius.

Comparing the results given in Tab. 1, FIM, TEM and the SANS scaling evaluation yield consistent values for the average particle radii within ± 5% omitting the tail of large particles of the FIM analysis. The particle radii found by the Guinier evaluation are about 25% higher than the average radii from FIM, TEM and the SANS calculation. This difference results from an inherent characteristic of the scattering law weighting the intensity contribution of a particle by its volume. Therefore the scattering curve accumulated from differently sized particles represents a volume-weighted particle size distribution. Regarding the volume fractions, the FIM and SANS measurements yield values somewhat above the literature result of $f_p = (1.8 \pm 0.05)\%$ determined by magnetization measurements [3]. TEM can give only an estimate of a lower limit, because in the considered size range TEM reveals only particles at certain depth levels in the foil. Therefore the counted particle number density has to be corrected by a factor of about 1.5. The uncertainties for f_p are remarkably high, in spite of the special carefulness spent in the data evaluation. This should be a warning for any kind of conclusion drawn from volume fraction results on the decomposition mechanism.

Precipitate sizes established by 2' and 7' annealing at 833K are definitely below the resolution limit of TEM. They still can be evaluated by FIM, however, uncertainties due to effects like unresolved dimers cannot be excluded.

The SANS measurements (Fig. 4) show that the scattering curves of these stages are clearly separated from the initial state of the solution-treated alloy. The scaling analysis can successfully be applied for the 7'-state, as shown in Fig. 4 by the solid curve. This curve represents the structure function calculated for the 3h annealing state and scaled to the 7' data. For the 2' annealing state, deviations from scaling occur for $\kappa < 1\ nm^{-1}$, but still scaling evaluation was applied as shown by the solid curve in Fig. 4. The scaling results are given in Tab. 2. The there given radii were determined i) from the κ-scaling with the scaling condition $\kappa \cdot R$ = const., and ii) from the intensity scaling, assuming $(d\sigma/d\Omega)/R^3$ = const. The latter relation is deduced from the scattering formula for constant volume fraction and scattering contrasts. This assumption holds only for the coarsening regime. Comparing the results given in Tab. 2, we find equal values for R from both procedures within the limits of uncertainty. This indicates that the final volume fraction and hence the coarsening regime is reached even after only 2' annealing time. The intensity scaling yields R-values somewhat higher than the κ-scaling. If this was true, it would imply that f_p of the 7' and 2' stages is even higher than f_p of the 3 h state, namely 30% for 7' and 15% for 2'. Again, the high uncertainty in the determination of f_p is obvious. Additional indications for the coarsening regime are obtained from the growth kinetics: as shown in Fig. 5, the precipitate radii follow the $t^{1/3}$-dependence of Ostwald ripening from the earliest state investigated. Hence, the appearance of nucleation between 1' and 5' as concluded from FIM observations [8] cannot be confirmed with our results. Our study reveals that the critical radius of nuclei in CuCo is smaller than 0.5 nm in radius (0.8 nm in ref. [8]). This limit is in accordance with the critical radius calculated to be 0.2 nm [8].

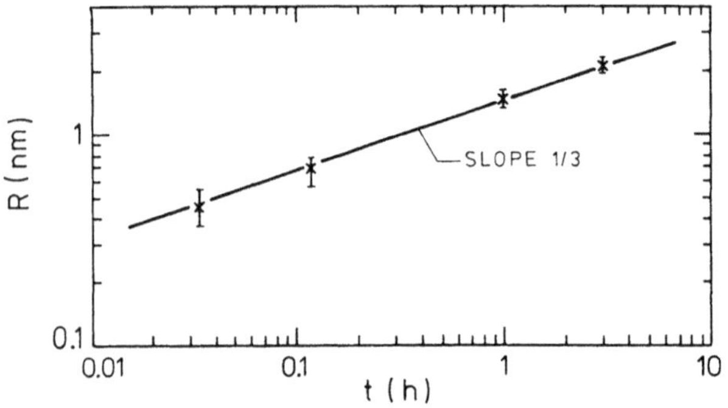

Fig. 5
Growth kinetics of the average precipitate radii as determined from the SANS data by calculation and scaling.

References

1 Decomposition of Alloys: The Early Stages, Proc. 2nd Acta-Scripta Conf., Sonnenberg, Eds. P. Haasen, V. Gerold, R. Wagner, M.F. Ashby, Pergamon Press, New York (1984)
2 G.M. Pound: Met. Trans. 16A, 487 (1985)
3 E. Nembach and M. Martin: Acta Metall. 28, 1069 (1980)
4 W. Wagner, J. Piller, H.P. Degischer and H. Wollenberger: Z. Metallkunde 76, no. 10 (1985)
5 M.F. Ashby and L.M. Brown: Phil. Mag. 8, 1083 and 1649 (1963)
6 P. Fratzl, J.L. Lebowitz, J. Marro and M.H. Kalos: Acta Metall. 31, 1849 (1983)
7 I.M. Lifshitz and V.V. Slyozov, Phys. Chem. Solids 19, 35 (1961)
8 H. Wendt and P. Haasen, Scripta Metall. 19, 1053 (1985).

Quantitative Measurement of Anomalous Small-Angle X-Ray Scattering (ASAXS) in Some Al-Zn-Ag Alloys

O. Lyon[1] *and J.P. Simon*[2]

[1]Laboratoire de Métallurgie Physique, bât. 413, Université Paris-Sud, F-91405 Orsay Cedex, France

[2]Inst. Nat. Polytechnique de Grenoble, Laboratoire de Thermodynamique et Physico-Chimie Métallurgiques, BP 75, F-38402 Saint Martin d'Hères, France

We have performed ASAXS experiments on AlZnAg alloys (and an AlZn alloy for comparison) using synchrotron radiation facilities at Stanford (SSRL) and at Orsay (LURE). By varying the X-ray energy near the Zn absorption edge (i.e. from 8.5 to 9.6 keV), we observed a decrease in the SAXS intensities due to the decrease of the Zn atomic scattering factor. The analysis of these data enabled us to determine the position of the different tielines on the AlZnAg metastable phase diagram. In the case of Al-14%Zn-4%Ag we clearly showed a rotation of the corresponding tieline when the unmixing process changes from Guinier-Preston (GP) zone formation to ε' formation.

In a crystalline alloy containing n elements, the n(n-1)/2 independent partial structure functions (PSFs) can be determined in principle from the same number of scattering experiments through the inversion of a system of linear equations. Independent scattering intensities may be obtained by isotopic substitution, in the case of neutron scattering, or by anomalous scattering, in the case of X-rays. In order to study clustering in ternary alloys, we performed small-angle neutron scattering (ISANS) of some AlZnAg alloys in the high-flux reactor of ILL (Grenoble) |1|. For a given chemical composition and metallurgical state, three different samples were prepared with either natural Ag, Ag_{107} or Ag_{109}. From these experiments the three partial structure functions S_{ij} (S_{Zn-Zn}, S_{Zn-Ag}, S_{Ag-Ag}) were extracted, one of which (S_{Ag-Ag}) seems to be negative. This unphysical behaviour seems to be due to the use of three different samples which, during the irreversible process of unmixing, may have reached different metallurgical states : small variations of the precipitate size can drastically modify the results |2|.

Another way to study ternary alloys is to take advantage of anomalous scattering. In the immediate neighbourhood of an absorption edge, the atomic scattering factor change can be written as $f = f_o + f'(E) + if''(E)$, where f' can be as large as -8 electrons in the range of photon energies E near K absorption edges. Thus, a whole experiment can be performed on the same sample and the possibilities of such a technique have already been explored |4|. Although ASAXS gives a rather ill-conditioned linear system for the PSFs' evaluation |2|, this drawback can be circumvented by making several measurements near absorption edges and performing a derivative analysis |2,3|. We illustrate the possibilities of ASAXS with results on unmixing of some AlZnAg alloys. After a brief presentation of the experimental conditions and procedure, we present the ASAXS results, develop the derivative analysis and finally show that these results allow the determination of the Ag/Zn ratio in the GP zones obtained in various ageing conditions.

1 EXPERIMENTAL CONDITIONS and PROCEDURE

Samples of Al-rich AlZnAg and AlZn alloys were kindly supplied by the Aluminium Péchiney Company. Their respective abbreviations and compositions (in at%) are the following : AZ22 : Al 22%Zn; AZ14S4 : Al 13.7%Zn 3.8%Ag;
AZ9S9 : Al 8.7%Zn 8.6%Ag; AZ4S14 : Al 3.9%Ag 12.8%Ag.

They were homogenized for one hour in an evacuated capsule, were water quenched and isothermally aged (100°C to 150°C). ASAXS measurements were made on line IV 2 at SSRL (Stanford) and on line 22 at LURE (Orsay). The designs of these two experimental sets are basically the same |Ref.3, Fig.1; 5| and consist of

(a) a double reflecting Si_{220} monochromator ;

(b) two monitors of the beam, one before and one after the sample (in order to measure the sample absorption and to accurately calibrate the monochromator position vs photon energy through EXAFS-like measurements) ;

(c) an evacuated tube and a position-sensitive detector.

The values of f' and f" had previously been measured |6|. EXAFS-like spectra were converted to f" using the optical theorem and f' was determined from the Kramers-Kronig dispersion relation. Finally, the detector efficiency at different photon energies E was calibrated by recording non-anomalous spectra (vitreous carbon).

2 EXPERIMENTAL RESULTS and ANALYSIS

As an example, the ASAXS spectra of AZ9S9, aged for 10 min at 125°C, are given in Fig.1.

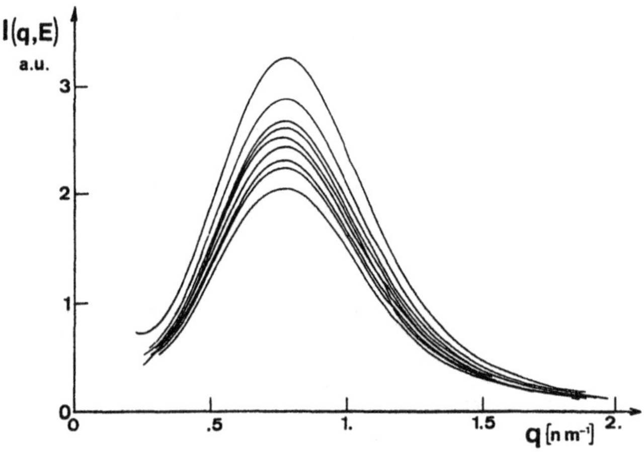

Fig.1. Variation of SAXS intensity for Al 9%Zn 9%Ag aged 10 min at 125°C ; the upper to lower curves were recorded at photon energies of 8500, 9000, 9400, 9500, 9560, 9625, 9640, 9650 and 9658 eV, respectively.

When the photon energy increases from 8500 to 9658eV, the ASAXS intensity monotonically decreases, as does the Zn atomic scattering factor. The amplitude of this variation decreases from AZ22, to AZ4S14 |Ref.3, Figs.2-5|, i.e. with the Zn content. More specifically, we saw no deformations of the curve shape for the different E.

For data analysis, let us recall the derivative method |3|. Write the intensity as a function of the PSFs,

$$I(q) = \Sigma_{ij} F_i F_j S_{ij}(q), \tag{1}$$

where $F_i = f_i - f_o$ is the difference of atomic scattering factors between the solute i (Zn or Ag) and the solvent o(Al), and the PSF $S_{ij}(q)$ is the Fourier transform of the corresponding pair correlation function. Since we are dealing with the variation of $f'_1(f'_{Zn})$ near the K_{Zn} edge only, F_1 may be written as

$$F_1(E) = \overline{F}_1 + dF_1(E) = (\overline{f_{Zn}} - \overline{f_{Al}}) + df'_{Zn}(E).$$

The other contrast, F_2 $(f_{Ag} - f_{Al})$, is constant in the studied E range. Then (1) may be rewritten as a Taylor expansion of $dF_1/\overline{F}_1$, which is exact since (1) is a quadratic form of $F_1(E)$.

$$\bar{I} = I(\bar{F}_1) = \Sigma\ K_{ij},$$

$$\bar{P} = \tfrac{1}{2}F_1.dI/dF_1(\bar{F}_1) = (K_{11} + K_{12}),$$

$$\bar{I}-\bar{P} = (K_{12} + K_{22}),$$

$$\bar{Q} = \tfrac{1}{2}F_1^{\,2}.d^2I/dF_1^{\,2}(\bar{F}_1) = K_{11},$$

$$\bar{I}(q) = \bar{I}(q) + 2\,\bar{P}(q).dF_1/\bar{F}_1 + \bar{Q}(q)\ (dF_1/\bar{F}_1)^2, \tag{2}$$

with $K_{ij} = \bar{F}_i\bar{F}_jS_{ij}$.

In theory, one should be able to determine $\bar{I}$, $\bar{P}$ and $\bar{Q}$, which ensures the best fit of I(E) as a quadratic function of $F_1(E)$. Unfortunately, as $(dF_1/\bar{F}_1)$ is small, of the order of 0.2 in AlZnAg, the second-order term is too small and the I vs F_1 plots appear as straight lines within the experimental uncertainty |Ref.3, Fig.6|. Then we can reasonably extract only two parameters, i.e. two linear combinations of the PSFs, for instance $\bar{P}$ and $\bar{I}-\bar{P}$, which are drawn in Fig.2 as a function of the reduced scattering vector q/q_{max} for the different alloys.

In the binary case where S12 and S22 are mathematically zero, Fig.2a confirms that the derivative method works and quantifies the precision obtained for the S_{ij}s (note that the K12 + K22 scale is amplified by a factor of 10). In the ternary alloys, $\bar{P}$ = K11 + K12 represents the influence of Zn concentration and hence decreases with Zn content, while $(\bar{I}-\bar{P})$ = K12 + K22 varies, of course, in an opposite way. Nevertheless these two functions have a common shape in q space : a

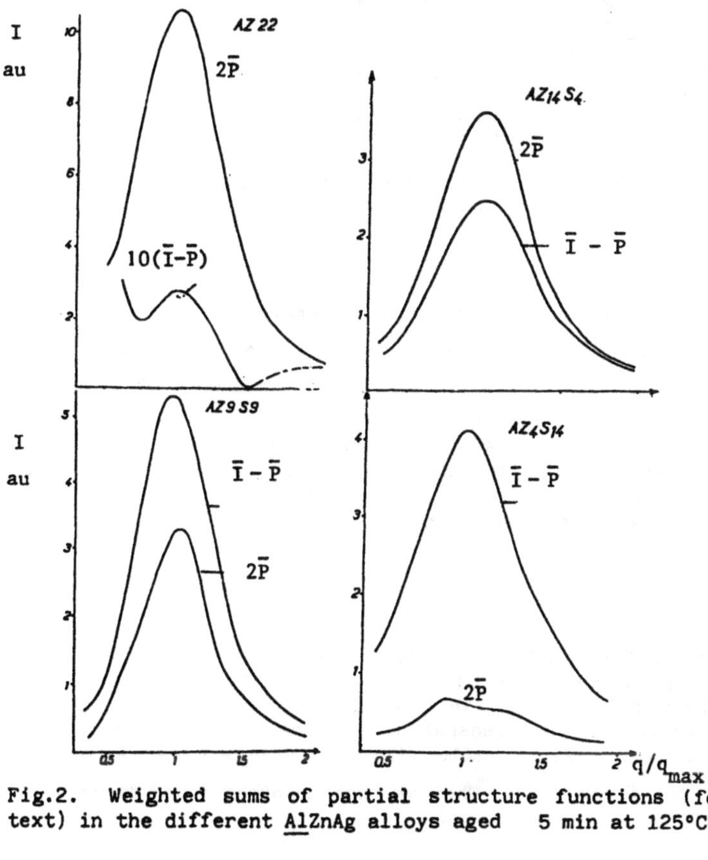

Fig.2. Weighted sums of partial structure functions (for the notations see the text) in the different AlZnAg alloys aged 5 min at 125°C

plot of their ratio vs q gives a plateau in the range $(0.5 < q/q_{max} < 1.5)$ as shown in |Ref.3, Fig.8|.

The homothety of the PSFs reminds us of a two-phase model consisting in a pseudobinary system of precipitates (p) embedded in a depleted matrix (m) of respective uniform compositions c_i^p, c_i^m. All three PSFs then have the same shape, namely

$$S_{ij}(q) = \Delta c_i \Delta c_j S_{pm}(q), \qquad \text{with } \Delta c_i = c_i^p - c_i^m. \tag{3}$$

Then the ratio of concentration differences (which would be the tie-line slope if equilibrium were reached) may be determined independently of q, simply from

$$\Delta c_2 / \Delta c_1 = (\overline{I} - \overline{P})/\overline{P} .$$

Another simple way of analysing these data is to rewrite (1) including the PSFs of (3):

$$I(q,E) = (\Delta c_1 F_1(E) + \Delta c_2 F_2)^2 S_{pm}(q).$$

Then $\sqrt{I}$ and $F_1(E)$ should be linearly correlated, from which the Ag to Zn ratio can be deduced. This procedure has been applied to the study of the unmixing kinetics in AZ9S9 and AZ14S4. Table I indicates the ageing conditions, the nature and size of the obtained precipitates and the Ag to Zn ratio (in all cases, a least squares fit over 6 to 8 values gives a residue better than 0.995).

TABLE I

Sample	Ageing and nature of the precipitation*		R_G [nm]**	$\Delta c_{Ag} / \Delta c_{Zn}$
AZ9S9	As quenched	GPZ	1.93	1.09 + 0.02
	3 min at 140°C	GPZ	2.52	1.13 + 0.02
	1 h at 140°C	GPZ	4.40	1.12 + 0.03
AZ14S4	3 min at 100°C	GPZ	2.38	0.47 + 0.02
	30 min at 100°C	GPZ	3.05	0.41 + 0.05
	3.5 h at 100°C	GPZ+ε'	5.78	0.33 / 0.27
	9.5 h at 100°C	ε'	7.07	0.25 + 0.02

* GPZ means Guinier-Preston zones (coherent, spherical particles); ε' are platelets of a hexagonal, semicoherent intermediate phase.
** Deduced from Guinier's plots.

The ratio $\Delta c_{Ag} / \Delta c_{Zn}$ is fairly constant during unmixing of AZ9S9, which indicates a rather balanced diffusion of Zn and Ag towards the GP zones. On the other hand, in AZ14S4, the $\Delta c_{Ag} / \Delta c_{Zn}$ ratio drastically changes when GP zones are overwhelmed by ε' precipitates. GP zones appear richer in Ag than the ε' phase. Finally the ratio measured for GP zones in AZ9S9 as well as in AZ14S4, is such that the two tie lines nearly converge towards the Al corner. This results is consistent with the metastable phase diagram, as thermodynamically calculated |Ref.1, Fig.1|, with a unique miscibility gap extending from <u>Al</u>Zn to <u>Al</u>Ag.

3 CONCLUSION

In <u>Al</u>ZnAg alloys, anomalous small angle scattering allowed us to show that phase separation occurs by precipitation of a single type of solute-rich GP zones embedded in a matrix depleted in Zn and Ag. We were also able to measure the slope of the tie line. This situation is not general : the analysis of recent ASAXS measurements on an unmixed $Cu_{42}Ni_{43}Fe_{15}$ sample allowed the determination of the three independent PSFs for the first time. They are not homothetic and their mean ratio is not consistent with a two phase system.

ACKNOWLEDGEMENTS : The authors entered upon this study during sabbatical stays in D. De Fontaine's group. Colleagues of this group as well as of LTPCM and of LURE are thanked for their help in this programme.

REFERENCES

1 A.Salva-Ghilarducci, J.P.Simon, P.Guyot, I.Ansara : Acta Metall. 31, 1705 (1983)
2 J.P.Simon, O.Lyon, D.De Fontaine : J.Appl.Crystallogr. 18,230 (1985)
3 O.Lyon, J.Hoyt, R.Pro, B.Davis, B.Clark, D.De Fontaine, J.P.Simon : J.Appl.Crystallogr. 18, 480 (1985)
4 P.Goudeau, A.Naudon, A.Fontaine, C.E.Williams : J. Physique Lett. 46, L255 (1985)
5 J.M.Dubuisson, J.M.Dauvergne, C.Depautex, P.Vachette, C.Williams : Nucl.Inst. and Meth. to be published Dec.85
6 J.P.Simon, J.Hoyt, O.Lyon, R.Pro, B.Davis,D.De Fontaine, W.K.Waburton : J.Appl.Crystallogr. 18, 181 (1985)

Part III

Hydrogen in Metals

Interstitial Point Defects (D,N,O) in Transition Metals

J. Peisl, H. Dosch, and A. von Schwerin

Sektion Physik der Ludwig-Maximilians-Universität München, D-8000 München 22, F. R. G.

Diffuse neutron or X-ray scattering close to the Bragg peaks (Huang scattering) and far away from Bragg peaks (Zwischenreflex-scattering) supplies detailed information on point defects and small agglomerates in crystals [1]. The defect structure, i.e. the lattice location, the displacements of the near neighbours and the strength and symmetry of the long ranged displacement field can be determined. In order to demonstrate the power of this technique, we report on recent experimental results. Interstitially dissolved N and O in Nb are located on octahedral sites and create rather large displacements in their vicinity. Their long ranged displacement field shows the symmetry of the defect site. The light interstitials H and D in Nb are located on tetrahedral sites and their long ranged displacement field shows the symmetry of the host lattice. A rather complicated defect model is necessary in order to explain the local defect structure.

1. Diffuse Scattering [2]

In a neutron or X-ray scattering experiment a well defined incoming wave with wave vector $\underline{k}_0$ is scattered by the assembly of lattice atoms (nuclei or electrons) and a scattered wave with wave vector $\underline{k}$ is detected under a scattering angle 2θ. For elastic scattering the wave length λ is not changed and the scattering vector is $\underline{K} = \underline{k}-\underline{k}_0 = (4\pi/\lambda)\sin\theta$. For an undistorted lattice the scattering intensity from the many atoms in a solid is zero except for the case that all atoms scatter in phase, i.e. whenever the scattering vector $\underline{K}$ is equal to a reciprocal lattice vector $\underline{G}$. For $\underline{K} = \underline{G}$ narrow Bragg peaks are observed. They contain all the scattering from the atoms periodically arranged in a crystal lattice. It leaves a wide range of $\underline{K}$ values with only little scattered intensity (incoherent and inelastic scattering in the neutron case, Compton and thermal diffuse scattering in the X-ray case) where scattering due to defects in the crystal lattice may be detected. Defects at lattice positions $\underline{r}_d$ may have a different scattering amplitude f_d and the lattice atoms are shifted from their regular sites $\underline{r}_m$ to new positions $\underline{r}'_m = \underline{r}_m + \underline{u}_m$. All defects contribute to the displacements $\underline{u}_m$ by their individual displacement field

$\underline{u}_m{}^d$. The scattering distribution changes accordingly: The Bragg peaks are shifted according to a change Δa of the lattice parameter a, due to an average, changed lattice $\underline{r}_m$. Deviations from this average lattice decrease the amount of constructive interference giving rise to the Bragg peaks. This attenuation is described by a static Debye-Waller factor exp(-2L). Between the Bragg peaks the scattered waves no longer interfere destructively to zero and a diffuse scattering intensity is observed.

The defect-induced scattering intensity for a low concentration of randomly distributed defects is obtained by properly subtracting the scattering intensity for a defect-free crystal.

$$I_d(\underline{K}) = c\,\exp(-2L)\exp(-2M)|f_d\exp(i\,\underline{K}\,\underline{r}_d) + i\,f\,\underline{K}\,\underline{u}(q)$$

$$+ f\,\Sigma_m{}^N \exp(i\,\underline{K}\,\underline{r}_m)\,\exp(i\,\underline{K}\,\underline{u}_m) - 1 - i\,\underline{K}\,\underline{u}_m)|^2 \qquad (1)$$

f is the scattering amplitude of the crystal atoms. Within this single defect approximation the diffuse scattering intensity is just c times the scattering intensity from a single defect. If the defects are correlated, the scattering amplitudes have to be summed properly and c has to be replaced by the Fourier transform of the concentration fluctuations.

The first term in (1) is the scattering amplitude of the defect, the so-called Laue scattering. The second term is the linear term for which $\underline{K}\,\underline{u}_m < 1$ is assumed and the scattering amplitude is given by the Fourier transform $\underline{u}(q)$ of the displacement field $\underline{u}_m{}^d$. Close to a Bragg peak $\underline{G}$ or for small $\underline{q}=\underline{K}-\underline{G}$, respectively, this term dominates and Huang diffuse scattering is observed. It stems mainly from the long ranged part of the displacement field. The third term in (1) is the scattering from the displacements of the near neighbours of the defect for which $\underline{K}\,u_m > 1$.

Close to a Bragg peak the scattering is dominated by the Huang term, the coherent scattering from the long ranged displacement field.The scattering intensity in this case (small q) is determined by the elastic constants C_{ij} of the crystal and the force dipole tensor P_{ij} [3].

$$I_{HDS} \sim c\,f^2\,(G^2/q^2)\,(\gamma_1\,\Pi_1 + \gamma_2\,\Pi_2 + \gamma_3\,\Pi_3) \qquad (2)$$

I_{HDS} shows a characteristic $1/q^2$ dependence. The γ's contain the elastic constants and depend on the $\underline{G}$ and $\underline{q}$ according to the measuring procedure. The π's are quadratic combinations of the force dipole tensor components. $\pi_1 = 1/3\ (\mathrm{Trace}\ P_{ij})^2$, $\pi_2 = 1/6\ (P_{ii} - P_{jj})^2$ and $\pi_3 = 2/3\ \Sigma_{i>j} P_{ij}^2$. In high symmetry lattice directions of $\underline{G}$ and $\underline{q}$ one or two of the γ's are zero. For these directions the π's can be determined most easily.

Far away from a Bragg peak (large q) the second term in (1) is small enough that also the contributions from terms one and three can be observed. This is the reason why the diffuse scattering from regions between Bragg peaks ("Zwischenreflex" scattering) gives information on the defect and its near neighbourhood [4]. The usual procedure is to compare the measured scattering distribution with model calculations. For this the displacements $\underline{u}^m$ are calculated by lattice statics using Kanzaki forces $\underline{f}^n$ in accordance with the experimental force dipole tensor

$$P_{ij} = \Sigma_n r_i^n f_j^n \tag{3}$$

and the lattice Green's functions G_{ij}^{m-n} obtained from experimental phonon dispersion curves

$$u^m_i = \Sigma_{n,j} G_{ij}^{m-n} f_j^n \tag{4}$$

2. Experimental Results

2.1 Nitrogen in Niobium

The force dipole tensor of N in Nb has been determined by Huang diffuse scattering of X-rays from Nb with interstitially dissolved Nitrogen [5].

$$P_{ij} = \begin{pmatrix} 12.6 & 0 & 0 \\ 0 & 5.0 & 0 \\ 0 & 0 & 5.0 \end{pmatrix} \mathrm{eV} \tag{5}$$

The coherent elastic diffuse scattering of neutrons on $NbN_{0.014}$ far away from Bragg peaks is shown in Fig. 1 together with model calculations [6]. It was assumed that N occupies either octrahedral or tetrahedral sites in the bcc lattice. We have calculated from (3) to (5) two sets of Kanzaki forces and the

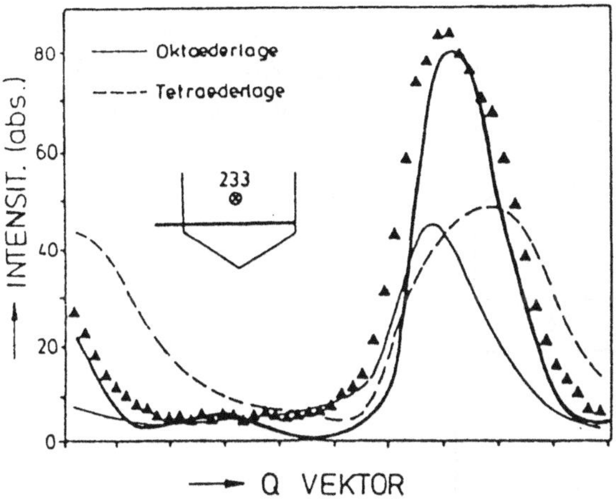

Fig. 1 Diffuse neutron scattering on $NbN_{0.014}$ single crystal. Experimental results compared with numerical calculations for the 2f model octahedral site (full line) and tetrahedral site (dashed line) and for the 3d model octahedral site (bold line)

corresponding displacements which are summarized in Table 1 and compared with calculations by TEWARY [7]. The simplest model that can produce P_{ij} according to (3) and (5) is radial forces f_1 and f_2 on the nearest and next nearest Nb atoms around the interstitial site ("2f model").A more extended force model with forces up to the third shell (" 3f model") was also considered. Comparison of the experimental data points with the calculated diffuse scattering cross section clearly shows that the 2f model can by no means reproduce the observed scattering intensities whereas the 3f model with N on

Table 1: Defect models for N in Nb: Kanzaki forces $f_{1,2,3}$ in eV/Å on first, second, third Nb shell. Local distortions $u_{1,2,3}$ in Å of first, second, third Nb neighbours.

	2f model		3f model	TEWARY [7]
	tetrahedral	octahedral site occupancy		
f_1	3,68	3,88	3,0	3,88
f_2	-0,48	0,94	-0,92	0,67
f_3	0	0	0,50	0
u_1		0,517	0,502	0,519
u_2		0,0048	-0,220	0,0031
u_3		0,0292	0,052	

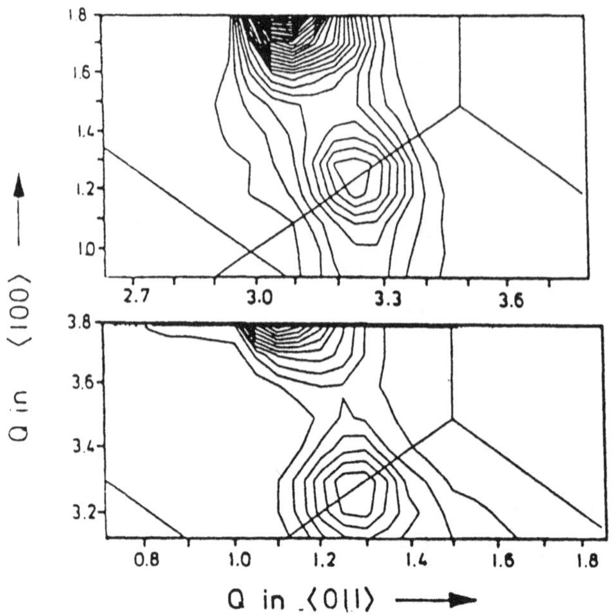

Fig. 2 Experimental diffuse neutron scattering distribution on $NbO_{0.02}$

octahedral sites gives rather good agreement. The 3f model explains also the results reported by ROWE and MAGERL [8].As also mentioned by these authors the diffuse scattering distribution pattern of N in Nb is quite similar to that of the ω-phase .

2.2 Oxygen in Niobium

A more extended diffuse neutron scattering distribution scan for O in Nb [9] is shown in Fig. 2 for two characteristic areas in reciprocal space. Like for N in Nb unusual diffuse scattering "peaks" are observed. Model calculation analogous to the ones described in the previous chapter gave the results shown in Fig. 3 and 4. Again a 3f model is needed in order to give good agreement between measurements and calculations. The 2f model cannot reproduce the salient feature observed experimentally. The diffuse scattering distribution observed here is indeed quite similar to the one observed from the ω-phase in $Zr_{0.8}Nb_{0.2}$ by KEATING and LA PLACA [10]. The relation between the ω-phase and the 3f model proposed here is not evident and a model beyond the Kanzaki formalism is needed. We suggest that impurities such as N and O favour the instability of the bcc lattice towards an ω-phase transition. They induce a local condensation of a new structure in the vicinity of the impurity which is described as a "local distortion" in the Kanzaki formalism. The volume mismatch of this local structure leads to long-ranged static displacements which can be detected by Huang diffuse scattering. In this picture the observed diffuse "peaks" appear as frozen in lattice modes in the presence of impurities.

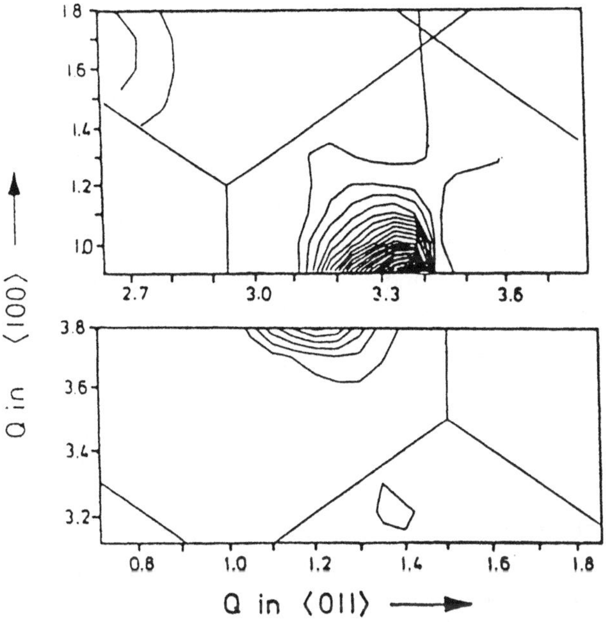

Fig. 3 Calculated diffuse neutron scattering distribution on $NbO_{0.02}$: 2-force model

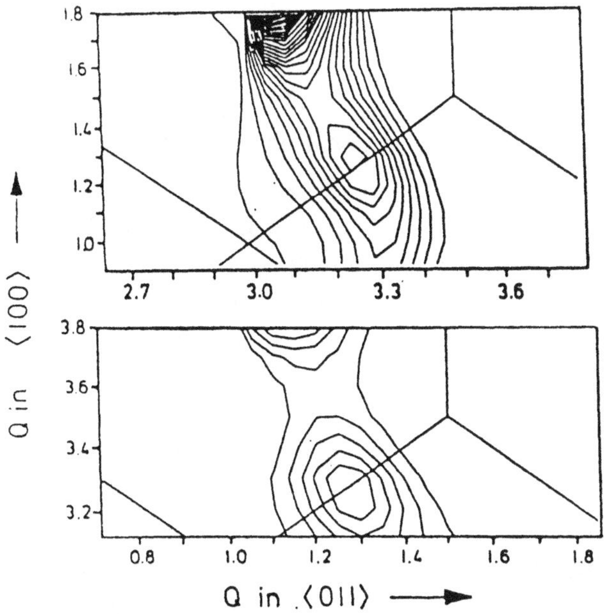

Fig. 4 Calculated diffuse neutron scattering distribution on $NbO_{0.02}$: 3-force model

2.3 Deuterium in Niobium

Huang diffuse scattering of neutrons from Nb with interstitially dissolved D has been observed by BURKEL et al. [11]. The force dipole tensor of D (H) in Nb has been determined by Huang diffuse scattering of X-rays by METZGER et al [12] and confirmed by BURKEL [13,2]. They obtained

$$P_{ij} = \delta_{ij}\, 3.3 \text{ eV} \qquad (6)$$

The most surprising result is that the force dipole tensor and the related long range displacement field have cubic symmetry. The local tetragonal symmetry of the defect site (H(D) occupies tetrahedral sites [14]) is obviously not transmitted to the long range displacement field. This is in contrast to the case of N and O in Nb, described in the previous chapters where the symmetry of the interstitial site (octahedral) is imposed on the long range displacement field. The force dipole tensor (6) with cubic symmetry for D on tetrahedral sites in Nb can be created by a 2f model with central forces on first and second neighbours, f_1 = 1eV/Å and f_2 = 0.23 eV/Å. If this parameterization of the displacement field by Kanzaki forces has a physical foundation, the interaction between the deuterium and the host lattice atoms creates the cubic displacement field. A conclusive answer to the physical origin can be expected from an experimental study of the local distortions in the close vicinity of the defect. Diffuse neutron scattering far away from Bragg peak is most suited to study these very local lattice distortions.

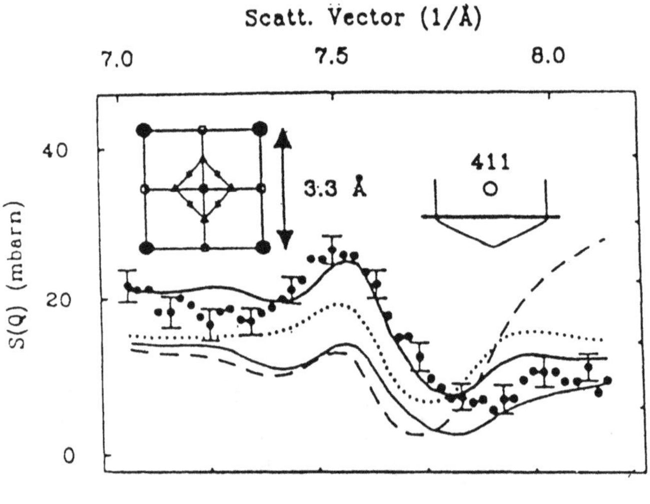

Fig. 5 Diffuse neutron scattering on $NbD_{0.017}$. Experimental results compared with numerical calculations for various defect models
——— tetrahedral site (Δ)
.... triangular site (□)
- - - - octahedral site (□)
▬▬▬ new model (see text)

The experimental results of the coherent elastic diffuse scattering distribution are given for a typical scan in Fig. 5 together with calculated intensity distributions for various defect models [15]. The data used in the model calculations are given in Table 2. The curves for D on tetrahedral (___), octahedral (----) and triangular (......) interstitial sites (see inset in Fig. 5) are obtained from a 2-force Kanzaki model in accordance with (6). The best agreement with the data points can be reached by the tetrahedral site occupancy, whereas the octahedral sites are clearly ruled out (see R-factor in Tab. 2). A relatively good agreement is also observed for the triangular sites. These sites

Table 2: Defect models for D in Nb: Various defect locations and Kanzaki forces $f_{1,2}$ on first, second Nb shell. R-factors R: $\Sigma(I_{obs} - I_{cal})/I_{obs}$

lattice site		tetrahedral	octahedral	triangular	new model
Kanzaki forces ($eVÅ^{-1}$)	f_1 f_2	1,0 0,23	1,02 0,72	1,0 0,5	1,0 0
R		2,08	3,24	2,75	0,76

have been discussed by FLYNN and STONEHAM [16] as being essential for the understanding of the microscopic diffusion of hydrogen in bcc metals. The best fit for the tetrahedral sites shows, however, a rather large quantitative intensity mismatch with the experimental data. This demonstrates that the 2-force model used so far is not yet correct.

A new model was proposed by DOSCH [15]. It is based on the idea that one has to include decaying lattice relaxations around previously occupied defect sites. For such effects to be relevant the inverse jumping frequency of the defect has to be comparable to the relaxation time of the lattice distortions. An estimation [15] based on a small polaron model shows that this is the case for D in Nb at room temperature. In order to take these relaxation time effects in the defect-induced lattice distortions into account the microscopic jumping path of the defect has to be considered. It has been shown [17] that the rapid motion of hydrogen in transition metals at high temperatures can be described by a 2-state model: In the so-called mobile state repeated jumps between tetrahedral sites occur. The number of tetrahedral sites involved is 3.2 at 293K and increases with temperature. In the immobile state tetrahedral sites are occupied.

This leads to the following new model to describe the lattice distortions: The lattice distortions of the defect on tetrahedral sites is described by Kanzaki forces on nearest neighbours only, $f_1 = 1.0\ eVÅ^{-1}$. Now the symmetry of the defect site is imposed on the displacement field. (These forces describe the immobile state.) In the mobile state the defect performs repeated jumps within a distorted region established around three adjacent tetrahedral sites. The lattice relaxations are (in a first approximation) assumed to be the same for the three defect sites involved and the resulting distortions are calculated by linear superposition of the three individual 1-force model fields. Thus in the mobile state the defect is located at one tetrahedral site within a distorted region extended over three sites. A tetrahedral site in the bcc lattice has four nearest neighbour tetrahedral sites. There are six possible configurations of three adjacent tetrahedral sites. With the three possible tetrahedral sites, one has to average over 18 possible the mobile state can be calculated without any free parameter . The finite flight phase of the defect in the mobile phase is taken into account by a finite occupation probability of triangular sites within the distorted 3-site region associated with the mobile state. (Eventual additional lattice relaxations due to the triangular site occupation have been neglected).

The total scattering cross section is obtained by summing up these contributions according to their, so far unknown, occupation probabilities p_{im} (immobile state), $p_{mb,t}$ (mobile state, tetrahedral site), $p_{mb,tg}$ (mobile state, triangular site). Because $p_{im} + p_{mb,t} + p_{mb,tg} = 1$ must be fulfilled, there remain two independent parameters for a least square fit to the experimental data. The final result is shown as the bold line in Fig. 5. The fit is best for p_{im} = 0.2, $p_{mb,t} = p_{mb,tg} = 0.4$, which is remarkably close to the value $p_{im} = 0.35$ obtained by LOTTNER et al. [17].

According to this new model the force dipole tensor $\langle P_{ij} \rangle$ averaged over all locally occurring distortions was calculated and Trace $\langle P_{ij} \rangle$ = 10eV is obtained in good agreement with literature values [12,2]. For the tetragonality of P_{ij} $t := (\langle P_{11} - P_{22} \rangle)/\mathrm{Trace}\,\langle P_{ij} \rangle = 0.05$ is obtained, very close to $t = 0$ for perfect cubic symmetry and $t = 0.02$ as observed by Huang diffuse X-ray scattering [12].

This work was supported by the Bundesministerium für Forschung und Technologie.

1 For a review see e.g.: H. Peisl, J. de Physique 37, C7-47 (1976).
P. Ehrhart, J. of Nucl. Mat. 69 & 70, 200 (1978)

2 J. Peisl, Lattice Distortion, Elastic Interaction and Phase Transitions of Hydrogen in Metals, in Festkörperprobleme (Advances in Solid State Physics), p. 45, P. Grosse (ed.), Vieweg, Braunschweig 1984

3 H. Trinkaus, phys. stat. sol. (b) 51, 307 (1972)

4 H.G. Haubold, J. Appl. Cryst. 8, 175 (1974)

5 U. Schubert, H. Metzger and J. Peisl, J. Phys.F 14, 2467 (1984)

6 H. Dosch and J. Peisl, Phys. Rev. B 32, 623 (1985)

7 V.K. Tewary, J. Phys.F 3, 1515 (1973)

8 J.R. Rowe and A. Magerl, Phys. Rev.B 21, 1706 (1980)

9 A. von Schwerin, Diplomarbeit LMU München 1985

10 D.T. Keating and S.J. La Placa, J. Phys. Chem. Solids 35, 879 (1974)

11 E. Burkel, H. Dosch and J. Peisl, Z. Phys. B, Condensed Matter 53, 33 (1983)

12 H. Metzger, J. Peisl and J. Wanagel, J. Phys. F: Metal Phys. 6, 2195 (1976)

13 E. Burkel, Doctoral Thesis, University of Munich 1982

14 H.D. Carstanjen and R. Sizmann, Physics Letters 40 A 93 (1972)

15 H. Dosch, Doctoral Thesis, University of Munich 1984

16 C.P. Flynn and A. M. Stoneham, Phys. Rev.B 1, 3966 (1970)

17 V. Lottner, J.W. Hans, A. Heim, K.W. Kehr, J. Phys. Chem. Sol. 40, 557 (1978)

Coherent Quasi-Elastic Scattering of Neutrons from Lattice Gas Systems

D.K. Ross[1], *D.A. Faux*[1], *M.W. McKergow*[1], *D.L.T. Wilson*[1], *and S.K. Sinha*[2]

[1]Department of Physics, University of Birmingham, Birmingham B15 2TT, UK
[2]CRSL, Exxon Research and Engineering Co. Ltd., Armadale, USA

Abstract The extension of existing theories for the incoherent quasi-elastic scattering of neutrons from lattice gas systems to include the case of coherent quasi-elastic scattering is presented. The effects of high concentrations of diffusing particles, particularily in the limits of weak and strong interactions are discussed using both analytic and Monte Carlo methods. The theory is applied to the use of the site-blocking model to calculate the ratio of the tracer to chemical diffusion coefficients for comparison with existing data on hydrogen in tantalum.

1) Introduction

In the present paper, we will describe expressions for coherent quasi-elastic scattering from atoms diffusing by simple hopping in a periodic potential generated by a host lattice . In the simple hopping model, diffusion is assumed to proceed by instantaneous stochastic jumps between lattice sites and where double occupancy of sites is forbidden. Here, it can be shown that all the terms in the scattering that involve the lattice nuclei will only contribute to the intensities of the Bragg reflections. Thus the diffuse scattering is determined by the diffusing atoms alone and hence is entirely quasi-elastic. Moreover, if the interstitial atoms cause lattice distortion, so long as the distortions due to different interstitials are additive, this effect only modulates the *intensity* of the quasi-elastic scattering. Examples of systems where coherent quasi-elastic scattering has been measured to date include deuterium in metals, superionic conductors and potassium intercalated into graphite. It is well known that up to now there have been no appropriate theories available to analyse such data, and the objective of the present note is to provide the necessary framework to deal with this problem.

The incoherent and coherent scattering functions are given by the Fourier Transforms of the self and total Van Hove correlation functions respectively [1]

$$S^{inc}(\underline{Q},\omega) = \frac{1}{2\pi}\int G_s(\underline{r},t)\exp(i(\underline{Q}.\underline{r}-\omega t))\,d\underline{r}dt \qquad (1)$$

$$S^{coh}(\underline{Q},\omega) = \frac{1}{2\pi}\int G(\underline{r},t)\exp(i(\underline{Q}.\underline{r}-\omega t))\,d\underline{r}dt$$

For energy transfers, $\hbar\omega$, that are small compared with kT, these correlation functions are real and can be defined as the probabilities that, if a given atom is at the origin at $t = 0$, then either the same ($G_s(\underline{r},t)$) or any atom ($G(\underline{r},t)$) is at the position $\underline{r}$ at time t.

From this definition, it can be seen that $G_s(\underline{r},t)$ is just the solution of the random walk problem for a given starting point, and that at low concentrations $G_s(\underline{r},t)$ and $G(\underline{r},t)$ are identical. Evaluation of $G_s(\underline{r},t)$ involves multiple convolutions, but its Fourier transform ($S^{inc}(\underline{Q},\omega)$) is easily obtained. By either solving the appropriate master equation [2] or expanding $G_s(\underline{r},t)$ in terms of the number of jumps at time t [3] we obtain the well-known Chudley-Elliott model.

$$S^{inc}(\underline{Q},\omega) = \frac{\frac{1}{\pi}(1/\tau(0)(1-F(\underline{Q}))}{((1/\tau(0)(1-F(\underline{Q})))^2+\omega^2} \qquad (2)$$

where $\tau(c)$ is the mean residence time for fraction, c, of sites occupied and $F(\underline{Q})$ is the Fourier Transform of the spatial distribution after one jump ie.

$$F(\underline{Q}) = \frac{1}{m} \sum_i \exp(i\, \underline{Q}.\underline{\ell}_i) \tag{3}$$

where $\underline{\ell}_i$ is one of m nearest neighbour jump vectors. Extension of this model to more complex jumps and non-Bravais lattices [4] is straightforward.

For large diffusion times, $G_s(\underline{r},t)$ becomes identical with the solution of Fick's Law (continuous diffusion) for a point source [5] and hence, at low Q, the incoherent scattering is Lorentzian with a half-width $D_T(c)\, Q^2$ [6]. Thus

$$D_T(0) = \ell^2/6\tau(0) = \langle r^2(t)\rangle/6t \tag{4}$$

where $D_T(c)$ is the tracer diffusion coefficient and $\langle r^2(t)\rangle$ is the mean square distance moved in time t

When we go to finite concentrations of the diffusing species, the incoherent and coherent scattering functions diverge. Our first important point is that, at low Q, incoherent scattering yields $D_T(c)$ the tracer diffusion coefficient, while coherent scattering yields $D_{chem}(c)$, the chemical diffusion coefficient, because both the total correlation function $G(\underline{r},t)$ and the chemical diffusion coefficient depend on actual concentration gradients.

We will now summarise the form of $S_{inc}(Q,\omega)$ at finite concentrations. For non-interacting [7] atoms, the tracer diffusion coefficient can be written

$$D_T(c) = f_T(c)\, \ell^2/\tau(c) = f_T(c)\, D_o(c) \tag{5}$$

where $\tau(c) = \tau(0)/(1-c)$ and $f_T(c)$, the tracer correlation factor describes the fact that, following an initial random jump, the next jump has an enhanced probability of being back to the original site. For a given lattice, the form of $f_T(c)$ can be most easily obtained by Monte Carlo simulation [8] although analytic derivations are possible. It is also straightforward to introduce interactions between diffusing atoms into a Monte Carlo simulation. The effects of finite concentrations on $S^{inc}(Q,\omega)$ were first given by ROSS and WILSON [9] who used Monte Carlo to calculate $R_{n'}(\underline{r})$, the spatial distribution after n' attempted jumps and hence obtained $S^{inc}(Q,\omega)$ using the fact that Poisson statistics still apply for attempted jumps. It is found that the resulting $S^{inc}(Q,\omega)$ coincides with the expected forms at low and high ω. At large ω (small time) it tends to the Chudley-Elliott model with the appropriate value of $\tau(c)$, while at low ω it tends to a narrower Lorentzian due to the fact that correlations reduce $\langle r^2(t)\rangle$ at large t. At low Q, the function is exactly Lorentzian and is narrowed as predicted by (5). These conclusions were subsequently confirmed by KUTNER et al [10] using a somewhat different Monte Carlo approach and later by TAHIR-KHELI and ELLIOTT [11] who used a Green's Function expansion. Again, however, only the Monte Carlo approach is easily extended to include the effect of interactions [12].

2. Coherent Quasi-Elastic Scattering

Turning to coherent case, we note that, in the absence of interactions, the scattering function takes the form first given by ROSS and WILSON [9].

$$S^{coh}(Q,\omega,c) = c(1-c)\, S^{inc}(Q,\omega,0) + \text{Bragg Terms} \tag{6}$$

where $S^{inc}(Q,\omega,0)$ is just the Chudley-Elliott model [2] with a residence time appropriate to zero concentration. The expression (6) has recently been extended by SINHA et al [13] to cover the case of interactions that are $\lesssim kT$. This approach yields the result

$$S^{coh}(Q,\omega,c) = \frac{S(Q)}{\pi} \frac{\Gamma'(Q)}{(\Gamma'(Q))^2 + \omega^2} \tag{7}$$

where $\Gamma'(Q) = \frac{(1-F(Q))}{\tau(0)} \left(1 - \frac{c(1-c)V(Q)}{kT}\right)$ with $V(Q) = \Sigma V_i \exp(iQ.r_i)$

Here V_i is the interaction energy of two diffusing atoms separated by distance r_i. Now, to the same level of approximation, it was shown by CLAPP and MOSS [14] that

$$S(Q) = I(T) \left[1 - \frac{c(1-c)V(Q)}{kT}\right]^{-1} \tag{8}$$

where $I(T) = c(1-c)\, B \left[\int_{BZ} \frac{1}{1 - \frac{c(1-c)V(Q)}{kT}}\right]^{-1}$

where BZ implies integration over the Brillouin Zone of volume B.

It will be seen that (7) implies that the scattering function is again Lorentzian with a width that varies like the Chudley-Elliott width divided by S(Q). Also, it can be shown that (7) fits with various known limits. For example, it clearly confirms both critical slowing down at the Mean Field transition temperature and equation (6) at high temperature. It also yields the correct Mean Field estimate of D_{chem}

$$D_{chem} = D_o(c) \left(1 + \sum_i V_i \frac{c(1-c)}{kT}\right) \tag{9}$$

There are two ways in which we can improve on this simple theory. The most straightforward is to introduce a temperature scaling factor to improve on Mean Field theory, for instance by comparison with Monte Carlo calculations [15]. The second effect is more subtle, being analogous to the effect of the tracer correlation on the incoherent scattering. First we must introduce the effect of correlations on $D_{chem}(c)$ itself which is defined by the equation.

$$J_x(c) = -D_{chem}(c)\, dc(x)/dx \tag{10}$$

where $J_x(c)$ is the net current in the x direction and c(x) is the concentration of all atoms at x. Now, net flow occurs because the concentration gradient sets up a chemical potential gradient, $d\mu/dx$, so that more generally

$$J_x(c) = -cMd\mu/dx \tag{11}$$

where M, the mobility, is the drift velocity in a unit potential gradient. Here M is usually defined from the Nernst-Einstein Equation

$$M = \ell^2/(6\tau(c)kT) \tag{12}$$

Now this equation is known to be inadequate. It is derived for the first jump due to the potential gradient and, for interacting atoms at finite concentrations, this may be followed by correlated jumps with components parallel to x which contribute to $J_x(c)$ and hence which give rise to a correction term in (12). This term can be calculated by Monte Carlo [16] and has been variously described as the intrinsic or physical correlation factor. However, as it arises in the Nernst-Einstein equation for the Mobility, in an exactly analogous way to the definition of the Tracer Correlation Factor in (5), it should more logically be called the Mobility Correlation Factor, $f_M(c)$. ie we should have written

$$M = f_M(c)\, \ell^2/(t\tau(c)kT) \tag{13}$$

There are two special forms for $f_M(c)$ to be considered. Firstly, if the diffusing atoms do not interact, apart from preventing double occupancy, one finds that $f_M(c) = 1$. This can be shown as follows. We can write

$$D_{chem}(c) = \frac{cf_M(c)\ell^2}{6\tau(c)kT} \frac{d\mu}{dc} = f_M(c)\ell^2/6\tau(0) = f_M(c)D_o(c) \tag{14}$$

because, for no interactions, $\frac{d\mu}{dc} = \frac{kT}{c(1-c)}$ and $\tau(c) = \frac{\tau(0)}{1-c}$. However, for this result to be consistent with (4) and (6) we must have $f_M(c) = 1$ for all c.

The opposite limiting behaviour occurs when repulsions produce strong short-range order. The first jump will now normally be from a favourable to an unfavourable site and hence the chance of jumping back is high and $f_M(c) \ll 1$. This effect has been demonstrated by MURCH and THORN [16] using Monte Carlo. It is now clear that the effect of $f_M(c)$ on the coherent scattering will be the same as that of $f_T(c)$ on the incoherent scattering. We would therefore expect that (7) would apply at large ω, but that the scattering would be narrowed at low ω as implied by $f_M(c)$ in (14).

3. Monte Carlo Simulations of S(Q,ω)

a) Low Q Case

At low Q, $S^{coh}(Q,\omega)$ depends only on $D_{chem}(c)$. Because potential barriers, and therefore $(1-c)(\tau(c))$ are concentration-dependent, it is more useful to try to model the ratio

$$D_{chem}(c)/D_T(c) = cf_M(c)(d\mu/dc)/(f_T(c)kT) \tag{15}$$

Here, the term $(c(d\mu/dc)/kT)$ is known as the "Thermodynamic Factor" and for a lattice gas model can be calculated analytically [17] or by Monte Carlo [18]. It can also be obtained experimentally from Gorsky Effect data, but a comparison depends on knowing the long-range interactions to put in the model. The remaining factor $(f_T(c)/f_M(c)$, the Haven's Ratio, can be obtained from Monte Carlo. We have performed calculations of site-blocking models for tetrahedral sites in b.c.c. systems out to third nearest neighbour distance (for details see [19]). In fig. 1 we show the variation of $f_T(c)$ and $f_M(c)$ for second-nearest neighbour blocking and also $f_T(c)$ for zeroth site blocking where $f_M(c) = 1$.

The ratio $f_T(c)/f_M(c)$, varies much less strongly than $f_T(c)$. It can be compared with experimental estimates given by POTZEL et al [20] for the TaH_x system (where here x = 6c). For x = 0.28 and 0.61, these authors found values of 1.2 ± 0.4 compared with our estimate of 0.939 at 0.28 and 0.873 at 0.61 for second nearest site blocking and 0.979 at 0.28 and 0.953 at 0.61 for zeroth site blocking.

b) General Case

Monte Carlo can also be used to evaluate $S^{coh}(Q,\omega)$ for the general case. Here, one simulates an interacting lattice gas at equilibrium, and calculates

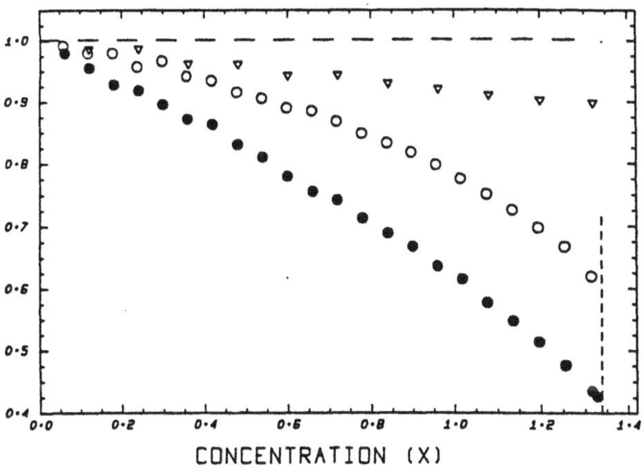

Fig. 1
Monte Carlo calculations of $f_T(x)$ and $f_M(x)$. For zeroth neighbour site blocking (no double occupancy) $f_M(x) = 1$ and $f_T(x)$ is shown (Δ). For second nearest neighbour site blocking $f_T(x)$ and $f_M(x)$ are (●) and (o) respectively.

$\alpha_i(t)$, the time-dependent short-range order parameters describing the occupation probability of sites at distance $\underline{r}_i$ from sites that are occupied at t = 0. From these parameters, the intermediate scattering function may be obtained using the equation

$$I(\underline{Q},t) = c(1-c) \sum_i \alpha_i(t) \, e^{i\underline{Q}\cdot\underline{r}_i} \tag{16}$$

Our discussion above implies that initially $I(\underline{Q},t)$ decays exponentially with a rate constant, $\Gamma'(\underline{Q})$ as in (7) but at longer time the decay slows down somewhat depending on the value of $f_M(c)$. Calculations along these lines are in progress.

4. Conclusion

We have presented new and general methods of obtaining $S^{coh}(\underline{Q},\omega)$ for lattice gas systems which can be used to analyse a range of data for deuterium in metals, superionic conductors and other lattice gas systems.

5. References

1 L. Van Hove, Phys. Rev. 95, 249 (1954)
2 C.T. Chudley and R.J. Elliott, Proc. Phys. Soc. 77, 353 (1961)
3 H.C. Torrey, Phys. Rev., 92, 962 (1953)
4 G. Blaesser and J. Peretti, in A. Seeger, et al. (eds) Proc. Int. Conf. on Vacancies and Interstitials in Metals. (KFA Julich 1968) 2, 836.
5 S. Chandrasehkar, Rev. Mod. Phys. 15, 1 (1943)
6 G.H. Vinyard, Phys. Rev. 110, 999 (1958)
7 A.D. Le Claire, Phys. Chem. 10, 26 (1970)
8 G.E. Murch, J. Nucl. Mater. 57, 239 (1975)
9 D.K. Ross and D.L.T. Wilson, Inelastic Neutron Scattering 1977 (IAEA Viennna 1978) 2, 383
10 R. Kutner, K. Binder and K.W. Kehr, Phys. Rev. B 23, 4931 (1981)
11 R.A. Tahir-Kheli and R.J. Elliott, J. Phys. C. 15, 5751 (1982)
12 I.S. Anderson, D.K. Ross and C.J. Carlile, "Neutron Inelastic Scattering 1977" (IAEA Vienna 1978) 2, 421.
13 S.K. Sinha and D.K. Ross, to be published
14 P.C. Clapp and S.C. Moss, Phys. Rev. 142, 148 (1966) ibid 171, 754 (1968) and 171, 764 (1968).
15 M.W.McKergow, private communication (1985)
16 G.E. Murch and R.J. Thorn, J. Phys. Chem. Solids 39, 1301 (1978)
17 G. Boureau, J. Phys. Chem. Solids 42, 743 (1981)
18 W.A. Oates, J.A. Lambert and P.T. Gallagher, Trans. Met. Soc. AIME 245, 47 (1969)
19 D. Faux and D.K. Ross, to be published
20 U. Potzel, J. Völkl, H. Wipf and A. Magerl, Phys. Stat. Sol (b) 123, 85 (1984)

Mechanism of Hydrogen Diffusion in Intermetallic Hydrides

R. Hempelmann[1] *and D. Richter*[2]

[1]Institut für Festkörperforschung, Kernforschungsanlage Jülich, Postfach 1913, D-5170 Jülich, F. R. G.
[2]Institut Laue-Langevin, 156 X, F-38042 Grenoble Cedex, France

Hydrogen diffusion in intermetallic compounds is quite complex on a microscopic scale. Intermetallics are composed of different metal atoms and correspondingly contain chemically different interstitial sites; for example, $TiMn_2$ contains tetrahedrons formed from four manganese atoms, from three manganese atoms and one titanium atom, and from two manganese atoms and two titanium atoms. The last mentioned is the most attractive for hydrogen because of the high hydrogen affinity of titanium. Owing to the generally rather complicated structures (Fig. 1 shows one unit cell of $Ti_{1.2}Mn_{1.8}D_3$ /1/ as an example) even chemically identical interstices may be of different sizes or be differently distorted, i.e. crystallographically non-equivalent. Thus intermetallic compounds generally contain a whole spectrum of interstitial sites which are energetically different for hydrogen. A microscopic description of the hydrogen diffusion in such a complex system, in particular on the basis of QNS data recorded from powder samples, must necessarily be restricted to the main features, which for intermetallic hydrides are as follows: (i) Chudley-Elliott behaviour of the hydrogen propagation over the energetically higher interstitial sites; (ii) trapping effects due to the existence of energetically deeper interstices; (iii) trap-blocking effects; (iv) correlation effects. Both (iii) and (iv) originate from the high hydrogen concentration. To a first approximation the energetically different interstitial sites in intermetallic compounds can be presented as a system of only two classes of interstitial sites: trap sites and free sites. In this sense an intermetallic compound provides a two-level system for hydrogen diffusion.

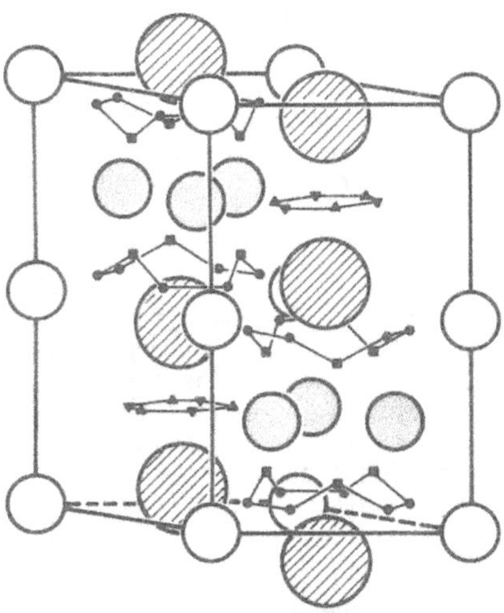

Fig. 1: Structure of $Ti_{1.2}Mn_{1.8}D_3$ /1/: ●, titanium; O, manganese and titanium; ● manganese; • D(1) (occupancy, 0.223); ■, D(2) (occupancy, 0.386); ▲, D(3) (occupancy, 0.220); ▼, D(4) (occupancy, 0.204).

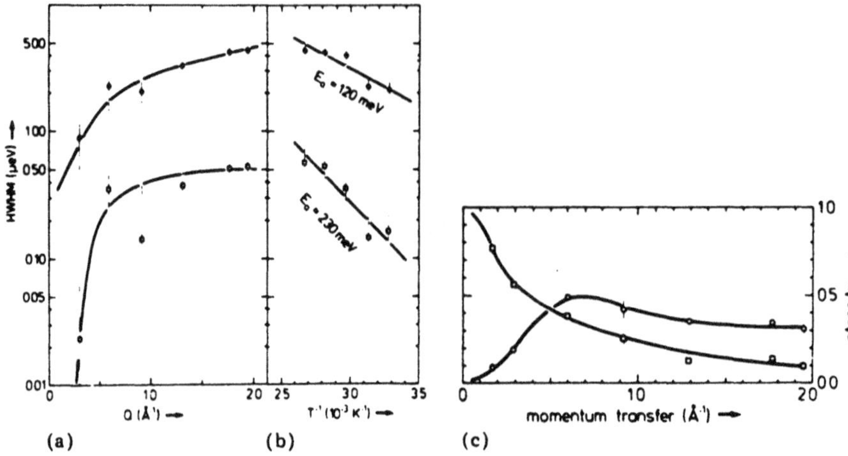

Fig. 2: (a) Q (T = 355 K) and (b) T (Q = 1.95 $\AA^{-1}$) dependence of the resolution-corrected linewidths obtained by fitting two lorentzian functions to the experimental QNS spectra of $Ti_{1.2}Mn_{1.8}H_3$; (c) Q dependence of the intensities of the two lorentzian functions /2/ (T = 374 K).

We performed a systematic QNS investigation of the hydrogen diffusion mechanism in $Ti_{1.2}Mn_{1.8}H_3$ and combined the features given above into a three-state model /2/. The motivation for this model arose from the model-independent results presented in Fig. 2: the experimental QNS spectra can be separated into two components by least-squares fits even at small momentum transfer. The Q dependence of the resolution-corrected linewidths has similarities to the two-state model /3/, and the fact that the narrow component has a substantially higher activation energy than the broad component unambiguously indicates the existence of energetically different interstitial sites. Figure 2 c shows the Q dependence of the intensities of both terms at T = 374 K relative to the total scattering intensity at 101 K; a substantial part of the intensity is missing at larger Q. This proves the presence of a further very broad component in the scattering function with a substantial weight ar large Q only, i.e. a rapid local motion. Two kinds of local motion are possible: (i) correlated hydrogen forth and back jumps and (ii) structural effects such as local hopping in an extended trap.

Our three-state model uses the microscopic parameters transport jump rate $^{-1}$, trapping rate $_1^{-1}$ and escape rate $_0^{-1}$. In view of the large amount of interstices in the unit cell, the diffusion in the free state is considered to be liquid-like; hence according to Chudley-Elliot /4/ $^{-1}$ enters the scattering function via

$$f(Q) = \frac{\hbar}{\tau} \left(1 - \frac{\sin Ql}{Ql}\right) \qquad (1)$$

An Arrhenius behaviour with the activation energies u and u_0 is assumed for τ^{-1} and τ_0^{-1}. The temperature dependence of the trapping rate, however, is complex. According to the kinetics of point defects /5/ τ_1^{-1} is related on the one hand to the transport jump rate (via D). On the other hand it is related to the number of trap sites in the elementary cell as well as to the hydrogen concentration, both via the concentration c_t^* of unoccupied trap sites which, owing to trap blocking effects, turns out to be rather dilute:

$$\tau_1^{-1} = 4 \quad R_t D c_t^* \qquad (2)$$

where R_t is the radius of the trap sites and $D = l^2/6\tau$ is the diffusion coefficient for free undelayed hydrogen diffusion. The effective trap concentration results from equalizing the chemical potential of the hydrogen atoms on free sites and on trap sites. The rapid local motion (jump rate, τ_b^{-1}) is for simplicity assumed to extend over a dumbbell of length d; its scattering function consists of an originally elastic contribution of weight $F = 0.5 + 0.5 \times \sin(Qd)/Qd$ (which, however, is broadened owing to the long-range diffusion) and a lorentzian peak of weight $1 - F$ with a Q-independent linewith $\Lambda_3 = \hbar \tau_b^{-1}$.

For correlated hydrogen jumps the local motion is associated with the free state as in eqn. (3), whereas for hopping in a trap it is related to the trapped state. Thus the total incoherent scattering function can be written as

$$S_{inc}(Q,\omega) = R_1 F \frac{\Lambda_1/\pi}{\Lambda_1^2+(\hbar\omega)^2} + (1-R_1)\frac{\Lambda_2/\pi}{\Lambda_2^2+(\hbar\omega)^2} + R_1(1-F)\frac{\Lambda_3/\pi}{\Lambda_3^2+(\hbar\omega)^2} \tag{3}$$

This scattering function was multiplied by the Debye-Waller factor calculated from the Debye temperature of $Ti_{1.2}Mn_{1.8}H_3$ and the frequencies of the localized hydrogen modes /6/, and was simultaneously fitted to 48 QNS spectra at eight temperatures and six Q values. Multiple-scattering corrections, which are essential for this procedure, were included. Some quantitative results are shown graphically in Fig. 3. The local jump rate τ_b^{-1} is ten times higher than the transport jump rate, but has almost the same activation enregy. Although not a proof, this supports the suggestion that the rapid local motion is related to correlation effects. The energetic difference between free sites and trap sites is given by the difference between the activation energies for the transport jump rate and the escape rate and is 90 meV. The effective hydrogen diffusion coefficient D_{eff} is reduced relative to the free-diffusion coefficient D:

$$D_{eff} = \frac{\tau_1}{\tau_1 + \tau_0} D \tag{4}$$

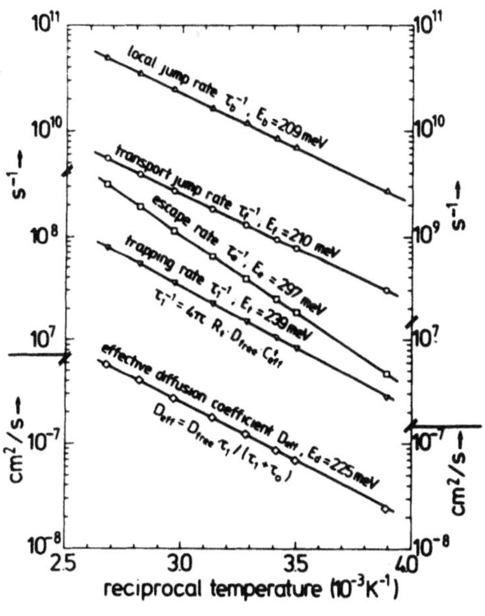

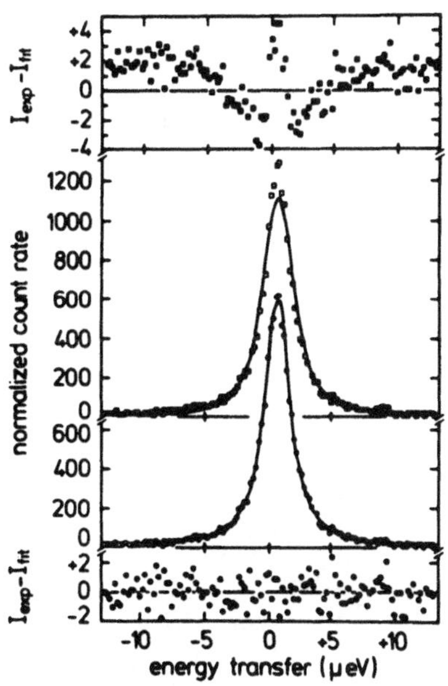

Fig. 3: Arrhenius diagrams of the local jump rate, the transport jump rate, the escape rate, the trapping rate and the effective diffusion coefficient of hydrogen in $Ti_{1.2}Mn_{1.8}H_3$.

Fig. 4: QNS spectrum of $LaNi_5H_6$ fitted with (a) one and (b) two lorentzian functions separation; the resulting error distributions are shown at the top and the bottom of the figure respectively ($Q = 1.70$ Å^{-1}; T = 303 K).

where τ_1 is the time for free diffusion within the time unit $\tau_1 + \tau_0$. $LaNi_5H_6$ exhibits a similar diffusion behaviour. Figure 4 shows a QNS spectrum fitted with one and two lorentzian functions. The two-lorentzian fit is significantly better, as is apparent from the error distribution which is purely statistical in this case. Thus the quasi-elastic spectra within the energy window $\pm$ 11 μeV are composed of narrow and broad components which appear to have different temperature dependences /7/. The reverse concentration dependence of the hydrogen diffusion coefficient also indicates trapping effects: the trap sites are preferentially occupied, and thus the ratio of empty free sites to empty trap sites increases with increasing hydrogen content and the hydrogen diffusion becomes faster, as can also be seen by inserting eqn. (2) in eqn. (4). As in $Ti_{1.2}Mn_{1.8}H_3$, part of the quasi-elastic scattering intensity is missing at large Q. This again indicates the existence of a broad component in the scattering function and thus a rapid localized motion /7/.

We conclude that, although the structures of $Ti_{1.2}Mn_{1.8}H_3$ and $LaNi_5H_6$ are completely different, the hydrogen mobility in both these intermetallic hydrides can be described in terms of three motional states: (i) the free state where the hydrogen atoms propagate over the energetically higher sites; (ii) the trapped state where the hydrogen atoms are located in structural traps; (iii) a rapid local hydrogen motion which may be due to correlation effects in hydrogen diffusion. Thus the three-state model may have general validity for hydrogen diffusion in intermetallic hydrides. It should also be noted that the diffusion in superionic conductors appears to proceed similarly according to QNS investigations of Ag^+ diffusion in AgI /8/.

Attempts have been made to relate the local hydrogen motion in $LaNi_5H_6$ to the structure of this hydride /9,10/. However, these interpretations were based on structure models /11 and 12, resp./ which turned out to be unrealistic /13,14/. The structure as determined by Thompson et al. /14/ is shown in Fig. 5. The main features are : (i) three types of H sites (La_2Ni_4 octahedra, La_2Ni_2 tetrahedra Ni_4 tetrahedra), (ii) displacement along c-axis of the intermediate Ni layer, and (iii) long-range correlations in the D occupancies resulting in a doubling of the unit cell. Because of the correlations in the time-averaged occupancies, strongly correlated hydrogen jumps can be expected to occur in the diffusion process, and perhaps a single particle picture of the diffusion process is insufficient.

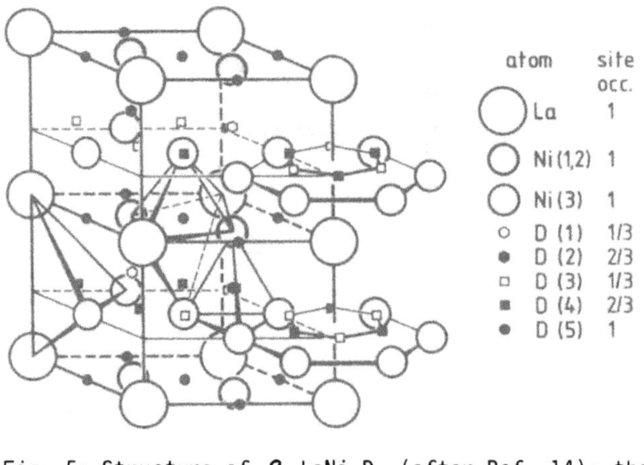

Fig. 5: Structure of β-$LaNi_5D_7$ (after Ref. 14); the D sublattice consists of La_2Ni_4 octahedra (circles) La_2Ni_2 tetrahedra (squares) and Ni_4 tetrahedra (hexagons); the two latters have alternately high (full symbols) and low (open symbols) occupancies in both half cells.

References

/1/ D. Fruchart, J.L. Soubeyroux and R. Hempelmann, J. Less-Common Met. 99, 307 (1984)

/2/ R. Hempelmann, D. Richter and A. Heidemann, J. Less-Common Met. 88 343 (1982)

/3/ D. Richter and T. Springer, Phys. Rev. B 18, 126 (1978)

/4/ C.T. Chudley and R.J. Elliott, Proc. Phys. Soc. London 77, 353 (1961)

/5/ T.R. Waite, Phys. Rev. 107, 471 (1957)

/6/ G. Sicking, e. Mag omedbekov and R. Hempelmann, Ber. Bunsenges. Phys. Chem. 85, 686 (1981)

/7/ D. Richter, R. Hempelmann and L.A. Vinhas, J. Less-Common Met. 88, 353 (1982)

/8/ G. Eckold, K. Funke, J. Kalus and R.E. Lechner, J. Phys. Chem. Solids 37, 1097 (1976)

/9/ D. Noreus, L.G. Olsson and P.E. Werner, J. Phys. F 13, 715 (1983)

/10/ C. Lartigue, A. Percheron-Guegan, J.C. Achard, M. Bee and A.J. Dianoux, J. Less Common Met. 101, 391 (1984)

/11/ P. Fischer, W. Halg, L. Schlapbach and Th. v. Waldkirch, Helv. Phys. Acta 51, 4 (1978)

/12/ A. Percheron-Guegan, C. Lartigue, J.C. Achard, P. Germi and F. Tasset, J. Less-Common Met. 74, 1 (1980)

/13/ C. Lartigue, A. Percheron-Guegan, J.C. Achard and J.L. Soubeyroux, J. Less-Common Met. 113, 127 (1985)

/14/ P. Thompson, J.J. Reilly, L.M. Corliss, J.M. Hastings and R. Hempelmann J. Phys. F 15, .. (1985)

Hydrogen Diffusion in Amorphous Liquid-Quenched $Pd_xSi_{1-x}H_y$ Measured by QNS

*G. Driesen, R. Hempelmann, and D. Richter**

Institut für Festkörperforschung, Kernforschungsanlage Jülich, Postfach 1913, D-5170 Jülich, F. R. G.
*Institut Laue-Langevin, 156 X, F-38042 Grenoble Cedex, France

By means of quasielastic neutron scattering we show that the H-diffusion mechanism in $Pd_{80}Si_{20}H_3$ is characterized by two well-separated regimes of jump rates, in clear contradiction to present ideas on diffusion in amorphous substances. A semiquantitative interpretation of the data is possible in terms of a diffusion-trapping model. Experiments on $Pd_{85}Si_{15}H_{7.5}$ reveal a similar diffusion mechanism as for $Pd_{80}Si_{20}H_3$.

Introduction

The study of H-diffusion in metallic glasses rapidly gains interest, not only because of possible applications, but also from a more fundamental point of view, as such systems are disordered with respect to topology and H-site energies. Snoek-effect measurements /1/ revealed a broad though distinctly skewed distribution of jump rates. Electrochemical diffusion and H-solubility data on Pd_xSi_{1-x} /2,3/ are interpreted assuming a single Gaussian distribution of the H-site energies in connection with Fermi-Dirac occupation statistics (this takes into account mutual blocking effects). This Gaussian model is also supported theoretically /4/; using a model H-metal potential, the distribution of relaxation times (i.e. activation energies) is related to a distribution of distances between metal atoms and interstitial holes and metal atoms and saddle points.
Our QNS-investigation of the hydrogen diffusion in amorphous Pd_xSi_{1-x} led to the surprising observation of two well-separated regimes of jump processes, obviously in disagreement with the anticipated continuous distribution of jump rates.

Experiments

In this paper we report quasielastic neutron scattering investigations on H-diffusion in $Pd_{80}Si_{20}H_x$ and $Pd_{85}Si_{15}H_x$. Both samples were prepared by melt-spinning technique. An examination by X-ray and neutron diffraction did not show crystalline fractions of importance. The samples were held in closed Al pressure cells containing fixed amounts of hydrogen gas. According to the pressure composition isotherms, which were measured in advance, we obtained a composition of $Pd_{80}Si_{20}H_{3.1}$ at 373 K for a charging pressure of 47.7 bars H_2. Cooling the sample to 328 K changed the H-content slightly to $Pd_{80}Si_{20}H_{3.3}$.

For $Pd_{85}Si_{15}$ we used a H-content of 7.5 % $H/Pd_{85}Si_{15}$. Since broad distributions of H-jump rates were anticipated, we performed scattering experiments at different spectrometers thereby varying the energy resolution and the energy range. We employed the backscattering spectrometer IN10 (ILL Grenoble) offering a resolution δE = 1 μeV and a dynamic range $\Delta E = \pm$ 12 μeV and the time-of-flight (TOF) spectrometers IN5 at the ILL and MIBEMOL at Saclay ($\delta E \sim$45 μeV, $\Delta E \gg \delta E$).

Results

Typical IN5 and IN10 results are given in Fig. 1. The scattering contributions of the H_2-gas and of the host metal including the sample holder have been measured separately and are already substracted /5/. From these spectra it is obvious that two strikingly different regimes of jump rates exist: Slow jump rates which cause broadening of the IN10 spectrum (Fig. 1b) with linewidths in the sub-μeV range. They appear as an "elastic" contribution in the TOF- spectrum (Fig. 1a). Its quasielastic broadening is indicative for fast jump rates, which are even several times faster than in crystalline Pd. The two classes of jump rates differ by two orders in magnitude. They are well separated from each other, in contradiction to an intensity profile which is an envelope of a continuous distribution of jump rates as required by the Gaussian model. Similarely MIBEMOL and IN10 experiments on $Pd_{85}Si_{15}H_{7.5}$ again point in the direction of well-separated regimes of jump rates.
The lower part of Fig. 2 presents the temperature and Q dependence of the resolution-corrected linewidths obtained for $Pd_{80}Si_{20}H_3$. Thereby the IN10 data (narrow component) were fitted with one Lorentzian while the IN5 spectra were described by a superposition of an elastic and a quasielastic component. At small Q the width of the narrow component increases with Q^2 indicating that it results from long-range diffusion. The diffusion coefficients at 332 K and 373 K are of the order of $4 \cdot 10^{-7}$, $6 \cdot 10^{-7}$ cm^2/s, respectively. Astonishingly, we observe only a weak temperature dependence for the broad linewidths.
In the upper part of Fig. 2 we show the Q and T dependence of the weight of the narrow component. The vanishing weight of the broad component at small Q indicates that the fast jump process occurs across spatially restricted regions. With increasing temperature the fast jumps gain in weight; apparently the fast moving state is reached by thermal activation.

Discussion

a) The Gaussian Model

In order to test the idea of a Gaussian site energy model in more detail we performed a quantitative comparison with our data. Thereby we utilized the fact that at large momemtum transfer Q, QNS basically observes the single jump rates while at small Q's it measures long-range diffusion. According to the Gaussian model /2,3/, the

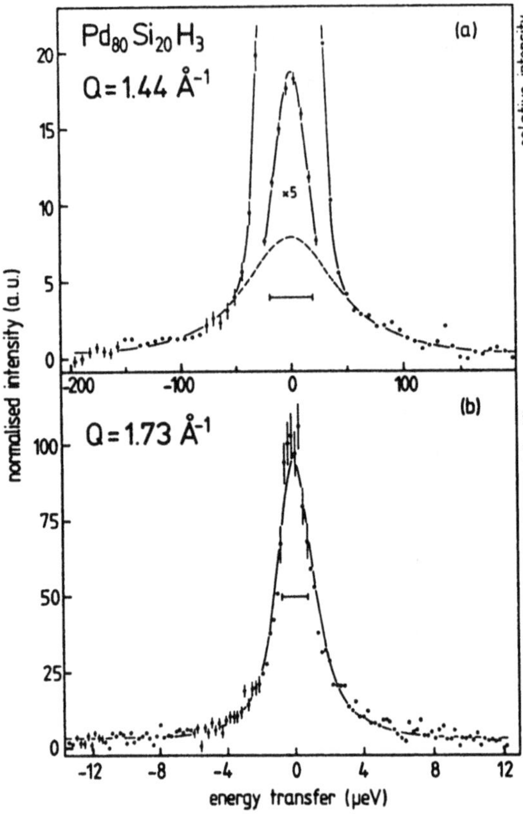

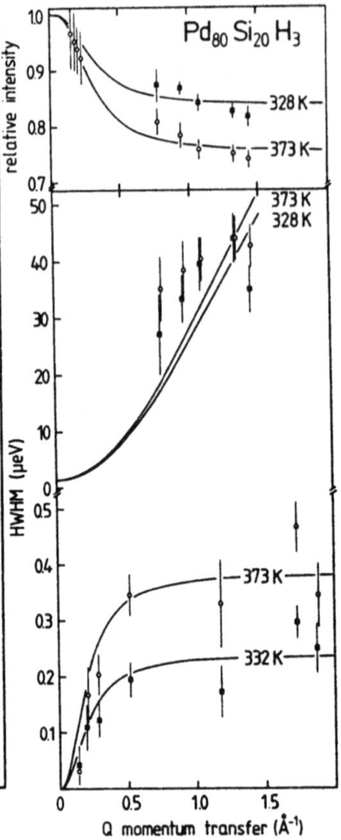

Fig. 1: QNS-spectra of H in $Pd_{80}Si_{20}$ at 373 K: (a) TOF-result (IN5), solid (left) line; fit with a superposition of an elastic and a quasielastic scattering contribution (dashed line). (b) Backscattering result (IN10). The bars indicate the respective resolution widths.

Fig. 2: Upper part; weight of the elastic component in the IN5 spectrum as (right) a function of temperature and momentum transfer. Lower part; T and Q dependence of the quasielastic linewidths (HWHM) observed at IN5 and IN10. The solid lines represent a fit with the two-state model on these data.

site energy distribution is given by:

$$n(E) = \frac{1}{\sigma\sqrt{\pi}} \; e^{-(E/\sigma)^2} \tag{1}$$

with σ the width of the Gaussian. The site energies E are defined with respect to the centre of the Gaussian (E = 0) and E < 0 corresponds to the energetically lowest sites. Site blocking imposes Fermi-Dirac site occupation statistics:

$$p(E) = \frac{1}{1 + e^{(E-\mu)/k_BT}} \tag{2}$$

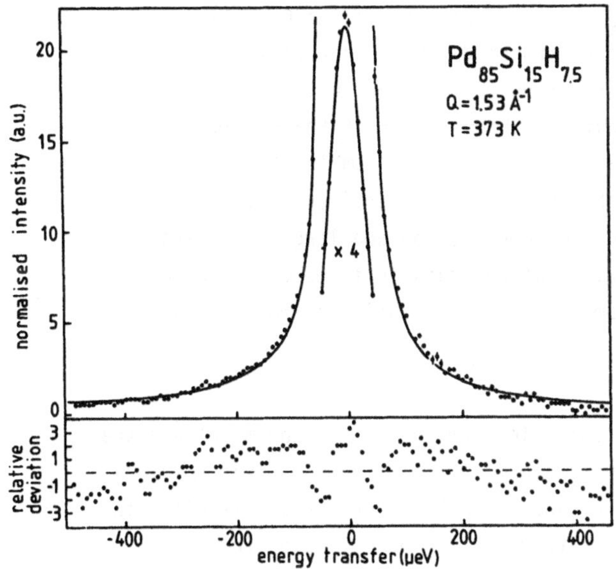

Fig. 3: QNS-spectrum of H in $Pd_{85}Si_{15}$: TOF-result (MIBEMOL), the solid line represents a fit with Eq. 3 (Gaussian model). The weighted difference between experiment and theory is shown in the lower part.

with k_B the Boltzmann constant and μ the chemical potential. All saddle point energies are supposed to be at the same level. Then, assuming that at large Q QNS measures directly the jump rates in the system, the scattering function at large Q can be written as

$$S(Q,\omega) = \frac{1}{c} \int_{-\infty}^{\infty} dE \;\; n(E)\; p(E) \;\; \frac{1}{\pi} \;\; \frac{\Lambda\, e^{E/k_BT}}{(\Lambda\, e^{E/k_BT})^2 + (\hbar\omega)^2} \tag{3}$$

with $\Lambda = \hbar/\tau$ the linewidth corresponding to the jumprate $1/\tau$ from site E = 0 and c the H-concentration. Using 33 evenly distributed site energies between $-2\sigma \leqslant E \leqslant 2\sigma$ we fitted Eq. 3 on the $Pd_{85}Si_{15}H_{7.5}$ data (Fig. 3). Clearly, the obtained fit shows systematic deviations from the experimental data. The thus determined parameters of the Gaussian model are: $R\sigma$ = 10.6 KJ/mol in close agreement with Kirchheim and $\hbar/\tau$ = 390 μeV. With these numbers we calculated the model prediction for the chemical diffusion coefficient. According to Kirchheim /2,3/ we have:

$$D_{ch} = D_o \; \frac{d}{dc} \; e^{\mu/k_BT} \tag{4}$$

where Do is the tracer diffusion coefficient in a hypothetical one-level system with site energy E = 0, $D_o = d^2/6\tau$. Assuming an individual jump distance d of 2 Å the chemical diffusion coefficient becomes D_{ch} = 12.8 10^{-6} cm^2/s. The chemical and the tracer diffusion coefficient, D_{ch} and D, are related by the normalized chemical factor $(d\mu/dc)/(d\mu/dc)_{c\to o}$. In the framework of the Gaussian model this factor amounts 1.7 . The corresponding tracer diffusion coefficient is $D = 7.5\ 10^{-6}\ cm^2/s$. This value is in strong disagreement with the observed tracer diffusion coefficient at IN10: The Q^2 behaviour of the resolution-corrected linewidths for small Q revealed $D = 9.4\ 10^{-7}\ cm^2/s$. Thus, in addition to the poor fit of the large Q-data, the Gaussian model applied to the QNS-results leads to a significant inconsistency between the low and large Q-results. In the next chapter we show that a description of our data in terms of diffusion and trapping overcomes the above problem and allows a consistent interpretation.

b) The Two-State Trapping Model

The starting point for an interpretation in terms of a trapping model are the experimental results for the spectral weight of the broad component (Fig. 2). Obviously, protons are activated from energetically more favourable sites and thereafter are very mobile. The mobility range can be estimated from the Q-dependence of the weight function. We consider it as the sum of the elastic incoherent structure factor (EISF) for the fast motion, which eventually decays to zero at large Q and a Q independent fraction arising from those protons which do not participate in the fast jumps. Thereby the EISF may be visualised as the Fourier transformed of the region visited by a proton in the fast jumping mode. Interpreting it in terms of the EISF for diffusion in a sphere we obtain a diameter of the order of 10 Å. Thus, the following picture emerges. The protons are generally bound in deep levels which we call traps. With an escape rate $1/\tau_o$ they can enter the mobile state, where they perform fast jumps until they are trapped again. The trapping rate $1/\tau_1$, the coefficient of fast diffusion D_o and the spatial extent of the motion s are related by the Kubo equation $\langle s^2\rangle = 6\ D_o\ \tau_1$.

The motional behaviour outlined above is similar to that observed for hydrogen diffusion in Nb doped with N /6/ and in the intermetallic compound $Ti_{1.2}Mn_{1.8}H_3$ /7/ and was described quantitatively in terms of a two-state trapping model. The solid lines in Fig. 2 represent a fit with this two-state trapping model on the observed linewidths and weights /8/. Obviously a microscopic description of H-diffusion in amorphous $Pd_{80}Si_{20}$ is possible. The diffusion coefficients provided by the model $D = 7.5\ 10^{-7}$ and $12.3\ 10^{-7}\ cm^2/s$ at 332 K, 373 K respectively agree with our estimates from the Q^2-law within a factor of two. In comparison with the literature we get good agreement with Gorsky relaxation experiments /9/, which extrapolated to 332 K and 373 K give $D = 4.8\ 10^{-7}$ and $12.6\ 10^{-7}\ cm^2/s$, whereas the electrochemical diffusion coefficient at 332 K $D = 4.1\ 10^{-8}\ cm^2/s$ /3/ is one order of magnitude smaller /10/. Although the fast jumps between the traps are faster than in α-Pd, the long-range diffusion is smaller due to the trapping events.

Microscopically our model can be regarded as evidence for the existence of two kinds of interstitial sites, possibly comparable with octahedral and tetrahedral sites with different site energies. However, due to the disorder we rather expect energy distributions. The smooth distribution of relaxation times observed with reorientation relaxation is probably related to that of the trap sites, since these experiments are performed at low temperatures /1/ where most of the protons are trapped.

In summary, we performed a high-resolution QNS study in order to unravel microscopic details of the H-diffusion process in amorphous $Pd_{80}Si_{20}H_3$. Surprisingly, we discovered a distinct fraction of protons jumping with rates one order of magnitude faster than in pure Pd, whereas the macroscopic diffusion coefficient is below that of Pd. The data are understood in the framework of a two-state trapping model. Thus H-diffusion in $Pd_{80}Si_{20}$ is a two-state process, where periods of fast diffusion are interrupted frequently by trapping events. A model based on a single Gaussian distribution of H-site energies, which was assumed so far, does not describe our Pd_xSi_{1-x} data.

We wish to express our thanks to Prof. T. Springer for many stimulating discussions, to Dr. G. Coddens, Dr. I. Anderson, Dr. A. Magerl for help during the experiments at MIBEMOL, IN5 and IN10 and to Dr. R. Kirchheim for the preparation of the samples.

References

/1/ B.S. Berry and W.C. Pritchet, Scripta Met. 15, 637 (1981) and Phys. Rev. B 24, 2299 (1981)

/2/ R. Kirchheim, F. Sommer and G. Schluckebier, Acta Met. 30, 1059 (1982)

/3/ U. Stolz, R. Kirchheim, J.E. Sadoc and M. Laridjani, J. Less-Common. Met. 103, 81 (1984)

/4/ P.M. Richards, Phys. Rev. B 27 2059 (1983)

/5/ For a detailed description of the data evaluation we refer to G. Driesen, R. Hempelmann, D. Richter to be published

/6/ D. Richter and T. Springer, Phys. Rev. B 18 126 (1978)

/7/ R. Hempelmann, D. Richter and A. Heidemann, J. Less-Common Met. 88 343 (1982)

/8/ D. Richter, G. Driesen, R. Hempelmann, I.S. Anderson and R. Kirchheim, to be published

/9/ B.S. Berry and W.C. Pritchet, in G.E. Murch, H.K. Birnbaum and J.R. Cost (Eds.), Proc. "Non-traditional Methods in Diffusion", The Metallurgical Society of AIME, Philadelphia 1983, p. 83

/10/ The chemical factor at C_H = 3.0 % can be considered to be close to one.

Part IV

Diffusion in Alloys and Hydrogen in Metals

Quasielastic Neutron Scattering Study of Fast Diffusion of Co in β-Zr

W. Petry[1], *G. Vogl*[2], *T. Flottmann*[2], *and A. Heidemann*[1]

[1]Institut Laue-Langevin, 156 X, F-38042 Grenoble Cedex, France
[2]Freie Universität Berlin und Hahn-Meitner-Institut, Glienicker Str. 100, D-1000 Berlin 39

Introduction

As a common feature, group IVb elements - Zr, Ti, Hf - exhibit a martensitic phase transition from the closed packed hcp structure (stable at low temperature) to the more open bcc structure. Diffusion properties in these metals are dominated by this phase transition. For instance the Zr self-diffusion increases by five orders of magnitude at the phase transition at about 840°C. A similar behaviour is known for most of the metal impurities in Zr of which the diffusivities are in the same order of magnitude as for Zr self-diffusion.

However, a group of transition metals deviates from this general behaviour with diffusivities orders of magnitude larger than the Zr self-diffusion (see Fig. 1). Among these impurities are the group VIII elements Fe and Co [1]. The maximal solubility of Fe in β-Zr reaches 8 at % at 934°C and Co disolves in β-Zr with maximal 3.4 at % at 981°C [2]. These figures are extraordinarily high for fast diffusers. Diffusivities in Ti and Hf hosts exhibit similar features.

Due to the macroscopic character of tracer methods - they measure a net mass transport - they fail to explain the mechanisms of such an anomalous fast diffusion. It is the virtue of microscopic methods like quasielastic neutron scattering (QNS) or quasielastic Mössbauer spectroscopy to detect diffusion mechanisms on an atomic scale [3]. Combining both information - mass transport constants and diffusion jump vectors and rates - it is possible to elaborate a detailed picture of hitherto unknown diffusion mechanisms.

To do so we performed a series of QNS studies on polycrystalline β-Zr 2 at% Co samples[+]. The system has been studied earlier by the tracer technique [5]. For QNS ZrCo has two advantages: i) The solubility limit in the β-phase is in the order of several at%; ii) Co has a large cross-section for incoherent neutron scattering (σ_{inc} = 4.8 (3) barn) and Zr has a low one (σ_{inc} = 0.04 (1) barn). Both facts are essential to get reasonable counting statistics, e.g. it would be impossible to study by means of QNS one of the well-known fast diffusers in lead, with solubility limits always less than several 100 ppm. On the other hand, one has to be careful when comparing fast diffusion in β-Zr with that in Pb. The different solubilities may be an indication of different mechanisms. Compared to the best energy resolution which can be achieved for incoherent neutron scattering ($\geq$ 0.3 µeV fwhm for the backscattering instrument IN10 at the ILL) the diffusivity has to be larger than $\sim 5 \times 10^{-12}$ m^2/s to give rise to a measurable broadening of the elastic peak. For ZrCo this is always fulfilled above the phase transition.

Results

First results of our measurements have been reported, recently [6]. The experimental findings are as follows:

i) Different to what one may have expected, it was found that all 2 at% Co participate in the fast diffusion, i.e. they perform more than one diffusion jump within the time resolution of the instrument ($\leq 10^{-9}$ s). This was established by the fact

[+]Independently fast diffusion of Fe in β-Zr has been studied by Mössbauer spectroscopy [4].

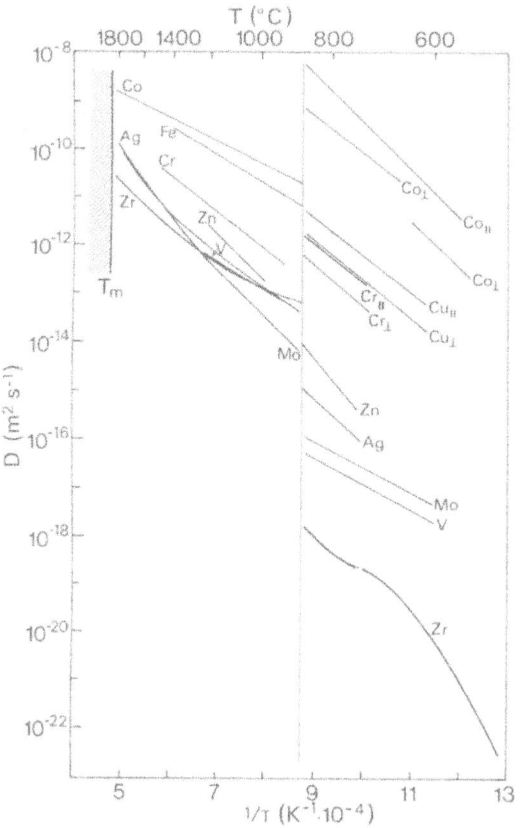

Fig. 1: Diffusivities in Zr. Data taken from "Diffusion and Defect Data" F.H. Wohlbier and D.J. Fisher (ed.), Trans. Tech. Publ. (Switzerland) and references therein.

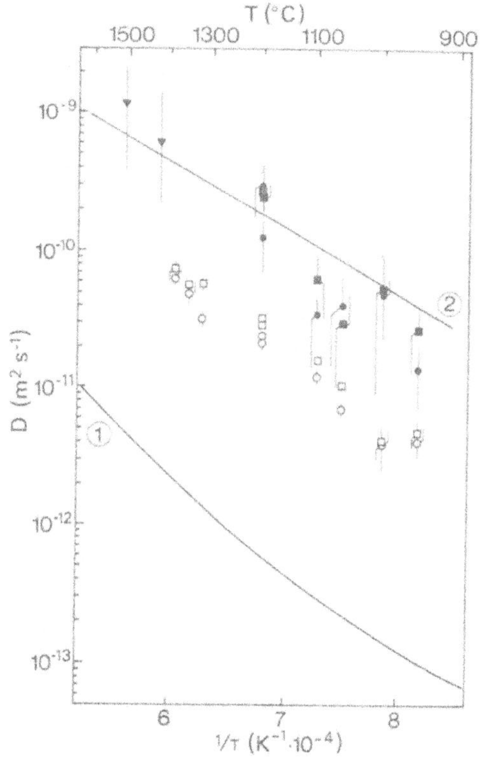

Fig. 2: Diffusion coefficient D_{Co} in Zr - 2at% Co as deduced from QNS data. Open symbols: assuming one quasielastic component in the QNS spectra; full symbols: assuming a broad and a superbroad quasielastic component in the QNS spectra. For comparison D_{Zn} ① and D_{Co} ② measured by tracer technique [5,8] are shown.

that all the intensity attributable to scattering from the Co atoms exhibits a quasielastic broadening.
ii) This quasielastic intensity can be fitted by a single Lorentzian yielding an Arrhenius-like increase of the quasielastic broadening Γ (fwhm) with temperature. For the scattering vector $Q = 1.9$ Å^{-1} a fit with an Arrhenius law yields $\Gamma = 3.3 \cdot 10^4 \exp(-1.20\ \mathrm{eV}/k_BT)$ µeV.
iii) Γ varies with Q, showing a maximum between 1.5 and 2.0 Å^{-1}.
iv) The quasielastic scattered intensity decreases with increasing temperature far beyond what is expected from a Debye Waller factor with a Debye temperature $\theta_D = 220$ K.

For an isotropic approximation the diffusional broadening is described by

$$\Gamma = \frac{2\hbar}{\tau}\left(1 - \frac{\sin Qd}{Qd}\right)f \qquad (1)$$

with the jump frequency $1/\tau$ and the jump distance d. f is a factor smaller than one which takes into account the correlation between subsequent jumps, i.e. that a jump might be immediately followed by its backward jump. Comparison of equ. (1) with the measured maximum of Γ between 1.5 and 2.0 $Å^{-1}$ yields a jump distance of the order of 2.5 to 3.3 Å, a figure which is compatible to the nearest neighbour distance in β-Zr d = 3.1 Å.

Combining equ. (1) with the Einstein-Smoluchowski relation

$$D = d^2 f/(6\tau) \tag{2}$$

we can now calculate the diffusivity D_{Co} as measured by QNS, assuming a nearest neighbour jump length and further assuming that the correlation factors f entering into equ. (1) and (2) are the same [7]. The results for D_{Co} for two different scattering vectors are shown in Fig. 2. For comparison Fig. 2 shows also the diffusivities of Co in Zr [5] and Zr self-diffusion [8] measured by the tracer technique. As expected, the QNS results exhibit the fast diffusion of Co in β-Zr in comparison to the Zr self-diffusion. However, the diffusivities are nearly an order of magnitude smaller than measured by tracer techniques.

From this contradiction, and the fact that the measured quasielastic intensity is far below what is expected, we have to conclude that a further very broad quasielastic component exists. This second component is too broad to be measured with the small energy window of IN10 of ± 14 μeV.

Two attempts have been undertaken to prove the existence of this very broad quasielastic component: i) Measurements on polycrystalline ZrCo samples with a time-of-flight spectrometer with a larger energy range indicated a quasielastic contribution in the order of magnitude of the expected very broad component. ii) All IN10 spectra could be fitted with an improved quality with a set of two broadened components. Summarising, we can state that about 25% of the total quasielastic intensity can be attributed to a second very broad component, whose width is about 20-40 times larger than that of the narrower component.

Models for the fast diffusion of Co in β-Zr

Additional to the experimental findings mentioned so far, an appropriate model for fast diffusion of Co in β-Zr has to account for two further experimental facts: The enhancement of β-Zr self-diffusion through alloying 2 at% Co is $D_{Zr}(Zr\text{-}2at\%Co)/D_{Zr}(Zr) = 2.22$ at 933°C [5]. The dehancement of the Co diffusivity is small $D_{Co}(Zr\text{-}2at\%Co)/D_{Co}(Zr) = 0.82$ at 933°C [5].

We can definitely exclude several possibilities to explain fast diffusion of Co in β-Zr. Enhanced diffusion along dislocations or grain boundaries can be excluded because it is not conceivable how 2 % of all atoms could at the same time ($\leq 10^{-9}$ s) be associated with dislocations. Fast diffusion of a Co-oxygen or nitrogen complex has to be excluded because samples prepared with different oxygen content showed the same diffusivity. Diffusion via vacancies alone is incompatible with the measured low enhancement of β-Zr self diffusion. Even in the extreme case of tight bound Co-vacancy pairs an enhancement $D_{Zr}(Zr\text{-}2at\%Co)/D_{Zr}(Zr) \simeq 15$ is calculated [9,10]. What remains are models with two different sites for the Co atoms and where at least parts of the Co atoms diffuse via interstitial sites. The postulation of two different sites is a consequence of the observed two different quasielastic components in the QNS spectra of β-ZrCo.

We first assume that at any moment 75% of the Co atoms are dissolved on substitutional sites and 25% on interstitial sites. The broad quasielastic component is related to the slower diffusion via vacancies and the very broad component to the fast moving interstitials. The effective diffusivity D_{Co} is a weighted sum of two independent diffusion constants. Again a lower limit of the enhancement of Zr self-diffusion can be calculated yielding $D_{Zr}(Zr\text{-}2at\%Co)/D_{Zr}(Zr) \simeq 2$) [10]. Although this model can account for the enhancement of the Zr diffusivity and for the two broadened components in the QNS spectra, we have to realize that the estimation for the enhancement represents the extreme case of tight binding, a situation which has never before been found experimentally [9].

As an alternative we discuss secondly the case of fast exchange (more frequently than $10^9\ s^{-1}$) between the two sites. Different to the preceding model diffusion on substitutional sites is of the order of Zr self-diffusion and too slow to be observed. Now, the broad component is due to the fast release jumps of the Co atoms from substitutional sites ot interstitial sites. The very broad component is due to the extremely fast motion of Co atoms in the interstitial sublattice [11,12]. The fast exchange between substitutional and interstitial sites implies a tremendously high vacancy concentration (of the order of 0.5%) which would lead to a strong enhancement of Zr self-diffusion in contradiction to the experimental findings. This makes the model inapplicable to our results.

What remains is to explain fast diffusion of Co in Zr entirely via interstitial sites. The two components are then due to jumps on the two sublattices. Since it is difficult to imagine that the Co atoms exist on two different independent interstitial sublattices, we shall not pursue that case further.

The alternative case which we consider finally, looks much more promising. Here the exchange between the interstitial sites is fast compared to the time resolution of QNS. We call the site where diffusion is slower a "trap". An exchange jump (release) can lift the Co atom into the other site where it moves faster. After a few jumps retrapping occurs. The mathematical model can follow the model of Schroeder et al. [11] and Richter et al. [12] for interstitial diffusion hindered by a trap.

The essential features of this model are:

i) For small $Q \ll 2\pi/a$ the scattering law averages over the different jump vectors and rates and can be described by one broadened Lorentzian, the width of which measures the net mass transport.

$$\Gamma \simeq 2\hbar D_{Co} Q^2 \tag{3}$$

ii) For large Q, i.e. in the order of the Brillouin zone boundary, the scattering law decomposes into two broadened components. The width of the narrower component Γ_1 yields the release rate K_r; whereas the broader component Γ_2 is a direct measure of the free (unhindered) interstitial diffusion D_{Co}^{int}.

$$\Gamma_1 \simeq 2\hbar K_r \tag{4}$$

$$\Gamma_2 \simeq \frac{12\hbar}{d^2} D_{Co}^{int} \cdot \left(1 - \frac{\sin Qd}{Qd}\right) \tag{5}$$

iii) At the Brillouin zone boundary the relative weights of the components are a direct measure of the two site occupations, i.e. at any time $C_1 \simeq 75\%$ of all Co atoms are trapped, whereas $1 - C_1 \simeq 25\%$ diffuse very fast through the interstitial lattice.

iv) The effective diffusivity D_{Co} is equal to the Co diffusivity in the fast moving interstitial state D_{Co}^{int} times a trapping term $(1-C_1)$

$$D_{Co} = D_{Co}^{int}(1-C_1). \tag{6}$$

Results for D_{Co} determined according to equ. (6) and assuming a jump length of the order of the nearest neighbour distance are shown in Fig. 2. Obviously the diffusivities determined by QNS are of the same order as the one determined by tracer technique.

Summary

It has been shown by means of QNS that all Co atoms, i.e. 2 at%, are fast diffusing atoms in β-Zr. There are strong indications that the Co atoms occupy two different sites. The most prospective model explains the fast diffusion of Co by fast interstitial diffusion hindered by traps.

This work was partly funded by the Bundesministerium für Forschung und Technologie (03-B72A37-1).

References

1. R. Tendler, E. Santos, J. Abriato and C.F. Varotto in "Thermodynamics of Nuclear Metals", Vol. 2, IAEA-SM-190/75, Vienna (1975) p. 71
2. M. Hansen and K.L. Anderko, "Constitution of Binary Alloys", McGraw-Hill, New York, (1958) and supplements
3. G. Vogl and W. Petry in "Festkörperprobleme (Advances in Solid State Physics), Vol. XXV, P. Grosse (ed.), Vieweg, Braunschweig (1985) p. 655
4. R. Stehr, Diplomarbeit, Freie Universität Berlin (1983) unpublished; Y. Yoshida, W. Petry, R. Stehr, K.-H. Steinmetz, W. Miekeley and G. Vogl, to be published
5. G.V. Kidson and G.J. Young, Phil. Mag. 20, 1047 (1969); G.V. Kidson and J.S. Kirkaldy, Phil. Mag. 20, 1057 (1969)
6. G. Vogl, W. Miekeley, A. Heidemann and W. Petry, Phys. Rev. Lett. 53, 934 (1984)
7. D. Wolf, Phil. Mag. 47, 147 (1983)
8. Ch. Herzig and H. Eckseler, Z. Metallkunde 70, 215 (1979)
9. A.D. LeClaire, J. Nucl. Mat. 69 & 70, 70 (1978)
10. W. Petry, G. Vogl, A. Heidemann, J.H. Steinmetz, to be published
11. K. Schroeder, Z. Physik B 25, 91 (1976)
12. K.W. Kehr and D. Richter, Solid State Comm. 20, 477 (1976); D.Richter and T. Springer, Phys. Rev. B 18, 126 (1978)

Systematic Investigation of Self-Correlation Functions in Cubic Lattices; Mono- and Divacancy Mechanisms

K. Ruebenbauer and B. Sepioł

Institute of Nuclear Physics, ul, Radzikowskiego 152,
PL-31-342 Krakow, Poland

The radial parts of the spatial self-correlation functions for a long-range diffusion of an isolated impurity via the monovacancy mechanism have been calculated for the monoatomic SC, BCC and FCC lattices within the encounter approximation, varying relevant parameters of the five-frequency model. An interaction of the impurity with the strongly bound divacancy has been studied for the FCC lattice within the encounter approximation and a high temperature limit. Simple expressions allowing to calculate the average number of impurity jumps per encounter Z and a scalar part of the self-correlation function W_o in terms of the five-frequency model parameters have been found.

Introduction

A diffusional part of the self-correlation function $G_s(\bar{k},t)$ takes on the following form in the encounter approximation for the long-range diffusion via monovacancies [1,2]

$$G_s(\bar{k},t) = \exp\{-(\omega_D t)\,[Z^{-1}\gamma(\bar{k})]\}, \tag{1}$$

where $\bar{k}$ stands for the wave-vector of the emitted radiation (Mössbauer effect) or scattering vector (neutrons), t stands for time, ω_D^{-1} is the average time interval between elementary jumps of the impurity, the latter occurring between the nearest neighbour lattice sites, Z is the average number of impurity elementary jumps per encounter and [1,3]

$$\gamma(\bar{k}) = 1 - \frac{1}{3}\sum_{n=0}^{\infty} W_n \alpha_n(\bar{k}) \quad \text{with} \quad \sum_{n=0}^{\infty} W_n = 1 \tag{2}$$

for lattices being locally symmetrical upon the total reflection [3]. Here W_n stands for the probability to reach the n-th coordinate shell within the encounter and hence can be called a radial part of the self-correlation function. The angular part $\alpha_n(\bar{k})$ depends solely upon the wave-vector and lattice parameters and it has been discussed in detail elsewhere for the monoatomic cubic lattices [3,4]. It has to be noted, that $\alpha_o(\bar{k}) = 3 = \text{const}$ and therefore W_o contributes to the isotropic line broadening, the latter being equal to $2\omega_D Z^{-1}(1-W_o)$

[2,3]. Hence, it is essential to estimate at least Z and W_o in order to compare bulk diffusion data with the microscopic data obtained upon polycrystalline samples exhibiting not too excessive broadening.

Relevant Parameters of the Five-frequency Model

The radial part, i.e. Z and W_n depend only upon the following ratios of frequencies [3,5,6] : $w_e=w_1/w_o'$, $w_c=w_2/w_o'$, $w_r=w_3/w_o$ for the FCC lattice; $w_e=w_1/w_2$, $w_r=w_3/w_o$ for SC and BCC lattices. The situation is illustrated in Fig. 1 and more details have been published elsewhere [3,4].

It has to be noted that in the high temperature limit [3,4] $w_e = (M_m/M_i)^{1/2}$, where M_m and M_i stand for the mass of the host atom and impurity, respectively, while the remaining frequency ratios equal one.

Divacancy Mechanism

As far as the long-range bulk diffusion via strongly bound vacancy complexes is considered, the only possible mechanism for monoatomic

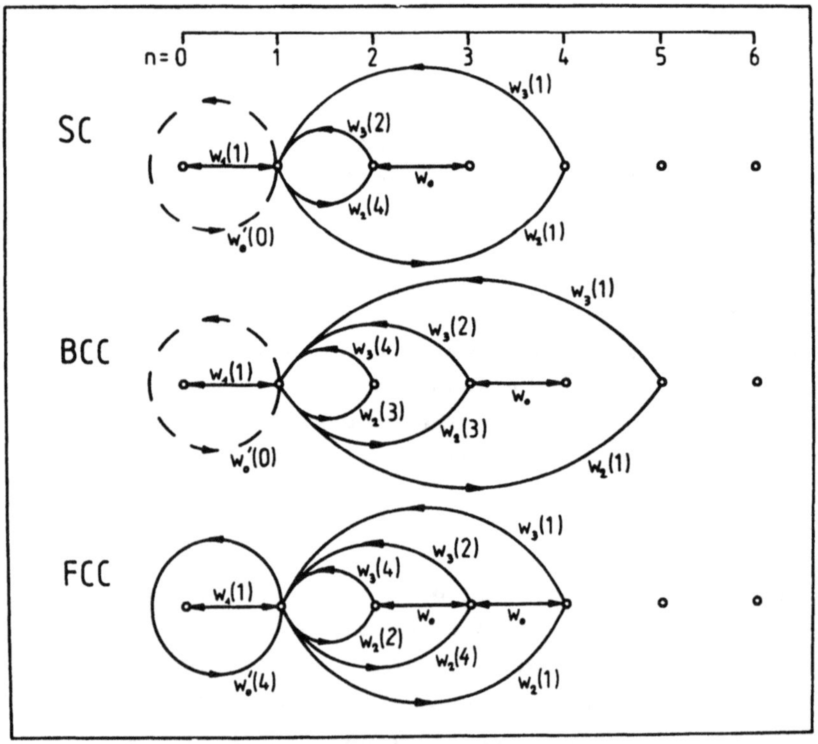

Fig. 1. Diagrams showing the main features of the five-frequency model for monovacancies. Numbers in brackets indicate the numbers of different routes between shells.

cubic lattices is a divacancy mechanism in the FCC lattice. Strongly bound means that the vacancy complex lifetime is much longer than the typical time interval of the encounter. Such a case has been analyzed in the high temperature limit for symmetrical divacancies (encounter approximation) [4]. The total self-correlation function takes on the following form

$$G_s(\bar{k},t) = \exp\{-(\omega_D' t) Z^{-1} \{1+b(Z/Z_d) - \frac{1}{3}\sum_{n=0}^{\infty} [W_n + b(Z/Z_d) W_n^d] \alpha_n(\bar{k})\}\} , \quad (3)$$

where

$$b = [4c_v/(1-12c_v)] \exp[-(Q_d+Q_b-Q_o)/k_B T] \quad (4)$$

provided the system remains at least locally at equilibrium. Here $c_v \ll 1$ stands for the total vacancy concentration, the latter being either monovacancies or members of divacancies, Q_d is a jump barrier for the monovacancy being a member of divacancy - excluding exchange with the impurity and inversion of divacancy, Q_o stands for a similar barrier of the isolated monovacancy and Q_b is the binding energy of divacancy.

Numerical Results

A Monte-Carlo method has been used to calculate the radial parts of the self-correlation functions [3,4]. The results for Z and W_o are reported here. It has been found, that:

$$Z = 1 + \left(\frac{w_e}{w_c}\right) [d_o + c_o w_r + b_o w_r^2 + a_o w_r^3], \quad (5)$$

where a_o, b_o, c_o, d_o are listed in Table I.

On the other hand:

$$Z_d = 1 + 1.419(M_m/M_i)^{1/2} . \quad (6)$$

Table I.
Numerical values for parameters a_o, b_o, c_o, d_o.

		a_o	b_o	c_o	d_o
FCC	$w_r \leqslant 3$	0	-0.00205	0.15926	0.1429
	$w_r > 3$	0	0	0.14725	0.16053
BCC	w_c=1	0.00024	-0.00448	0.18993	0.14465
SC	w_c=1	0.000175	-0.00325	0.22821	0.2002

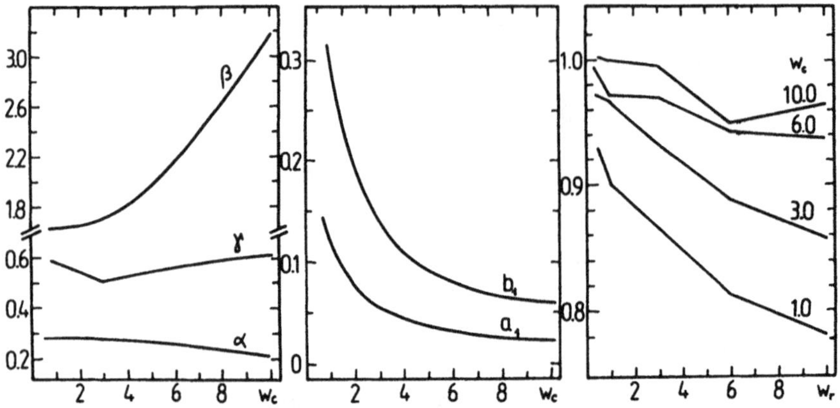

Fig. 2. Expansion coefficients plotted versus relevant model parameters. See text for details.

The scalar part W_o can be parameterized as follows:

$$W_o = A\left[1 - \exp\left(-Bw_e^{\;q}\right)\right], \text{ where} \tag{7}$$

$$A = 0.396\left[1-\exp(-1.454w_r^{-0.455}\right) \;, \quad B = 0.1215w_r+0.425 \;,$$

$$q = -0.0225w_r + 0.904 \quad \text{for SC lattice and} \tag{8}$$

$$A = 0.3725\left[1-\exp(-1.613w_r^{-0.563})\right], \quad B = -0.0018w_r^2+0.1552w_r+0.3524,$$

$$q = 0.0016w_r^2 - 0.031w_r+0.925 \quad \text{for BCC lattice.} \tag{9}$$

For divacancies:

$$W_o^d = 0.1848\left[1-\exp\{-1.264(M_m/M_i)^{0.396}\}\right]. \tag{10}$$

The expression (7) can be also used for a FCC lattice provided $w_c \gg 1.0$. However, in this case A, B and q are two parameter functions and further approximations are required. It has been found, that:

$$A(w_c,w_r) = \alpha\left[1-\exp\left(-\beta w_r^{\;-\gamma}\right)\right], \quad B(w_c,w_r) = a_1 w_r + b_1 \;, \tag{11}$$

while q remains slowly varying function of w_c and w_r. The parameters α, β, γ, a_1 and b_1 have been plotted versus w_c, while q has been plotted versus w_r for several values of w_c (see Fig. 2).

References

[1] M.C. Dibar-Ure and P.A. Flinn, Phys. Rev. B, 15 (1977) 1261

[2] D. Wolf, Appl. Phys. Lett., 30 (1977) 617

[3] K. Ruebenbauer, Hyp. Int., 14 (1983) 139

[4] K. Ruebenbauer and B. Sepioł - to be published in Hyperfine Interac.

[5] A.D. Le Claire, J. Nucl. Mat., 69 and 70 (1970) 70

[6] S. Mantl, W. Petry, K. Schroeder and G. Vogl, Phys. Rev. B, 27 (1983) 5313.

Quasielastic Coherent Scattering for Atomic Diffusion via Vacancies

K. Schroeder

Institut für Festkörperforschung, Kernforschungsanlage Jülich, Postfach 1913, D-5170 Jülich, F. R. G.

I. Introduction

Quasielastic incoherent neutron scattering has long been recognized as a tool for obtaining microscopic information on diffusion /1/. This method has been used very successfully for the study of hydrogen diffusion in metals /2/. Unfortunately, there are very few isotopes with a sufficiently large incoherent scattering cross-section to allow diffusion experiments. Thus it seems worthwhile to look into the possibilities of quasielastic coherent neutron scattering (QECS). In this paper we give a preliminary account of theoretical considerations to understand which informations about diffusion in binary substitutional alloys can be expected from QECS. For the interpretation of such experiments explicit model calculations are needed. We suggest that for such calculations an extension of our method used to describe incoherent scattering /3/ due to diffusion via vacancies in pure metals /4/ and dilute alloys /5/ can be used. This method consists of approximately solving a hierarchy of differential equations for many-particle correlation functions /4-6/. The model parameters of such calculations are the jump frequencies of vacancies with the constituents of the alloy.

First, in Section II, we shall briefly recall the procedure and results for incoherent scattering which is governed by the selfcorrelation function /1,4/.

In Section III, we treat quasielastic coherent scattering which is governed by pair correlation functions. For the simple case of diffusion in a pure substance we recover the result of Ross and Wilson /7/ which states that for non-interacting particles the quasielastic coherent scattering can be expressed by the selfcorrelation function for the diffusing vacancies. Recently, this result has been generalized by Gillan and Wolf /8/ to include lattice distortions around the mobile vacancies.

For diffusion in binary substitutional AB-alloys we use the simple model of Manning /9/ which assigns different jump frequencies w_A and w_B for the nearest neighbor vacancy exchange jump with A and B atoms, respectively, irrespective of the occupation of the other neighboring sites by A or B atoms. The different pair correlation functions for AA, BB and AB-pairs then contain the jump rates in different combinations. Thus separate determination of these jump rates becomes possible if the pair correlations can be determined separately by the method of isotope substitution which is used e.g. for the determination of the static structure of alloy glasses /10/.

II. Quasielastic incoherent scattering

The neutron scattering cross-section in an alloy can be written as a configuration average over spin and atomic arrangement /11/:

$$\frac{d\sigma}{d\omega\, d\Omega} = \frac{1}{2\pi}\, \mathrm{Re} \int_0^\infty dt\, e^{-i\omega t} \int d^3r\, e^{i\underline{Q}\underline{r}}\, \frac{1}{N} \sum_{mn} \left\langle a^m\, G^{mn}(\underline{r},t)\, a^n \right\rangle_{\substack{\text{conf}\\ \text{atomic+spin}}} \tag{1}$$

Here the double sum runs over all atoms with total number N.

$$G^{mn}(\underline{r},t) = \left\langle \int d^3r'\, \delta(\underline{r}^m(t) - \underline{r}' - \underline{r})\, \delta(\underline{r}' - \underline{r}^n(0)) \right\rangle \tag{2}$$

is the dynamic pair correlation function for the atoms m and n. The slow diffusion jumps ($w_{A,B} \simeq 10^8$-10^{11} sec^{-1}) manifest themselves in the quasielastic (small ω) region of the scattering spectrum well separated from the much faster vibrational motions ($\omega_{ph} \simeq 10^{13}$ sec^{-1}). We shall neglect the reduction of the quasielastic intensity due to the vibrations (Debye Waller factor) and treat the dynamics of the atomic positions due to diffusion in classical approximation.

For a pure substance the configurational average is over spin and isotopic configuration only. We then obtain the usual decomposition into coherent scattering

$$\left.\frac{d\sigma}{d\omega\, d\Omega}\right|_{coh.} \sim \frac{\langle a\rangle^2}{2\pi} \,\mathrm{Re} \int_0^\infty dt \int d^3r\, e^{-i(\omega t - \underline{Q}\underline{r})} \frac{1}{N} \sum_{mn} G^{mn}(\underline{r},t) \tag{3}$$

and incoherent scattering

$$\left.\frac{d\sigma}{d\omega\, d\Omega}\right|_{incoh} \sim \frac{(\langle a^2\rangle - \langle a\rangle^2)}{2\pi} \,\mathrm{Re} \int_0^\infty dt \int d^3r\, e^{-i(\omega t - \underline{Q}\underline{r})} \frac{1}{N} \sum_{n} G^{nn}(\underline{r},t) \tag{4}$$

The self-correlation function $G^{nn}(\underline{r},t)$ describes the dynamics of the particular atom No.n. Since all atoms are equivalent $G^{nn}(\underline{r},t)=G_s(\underline{r},t)$ is independent of the atomic number. We shall consider classical hopping motion via nearest neighbor (NN) vacancies only. Then $G_s(\underline{r},t)$ is different from zero only on discrete lattice sites $\underline{R}^{\alpha}$ (denoted by $\underline{\alpha}$) and obeys the following diffusion equation

$$\frac{\partial}{\partial t} G_s(\underline{\alpha},t) = \sum_{\underline{\beta}} s_{\underline{\alpha}\underline{\beta}}\, w_A \left[G_{s/v}(\underline{\beta}|\underline{\alpha},t) - G_{s/v}(\underline{\alpha}|\underline{\beta},t)\right] \tag{5}$$

which has the following meaning: The probability $G_s(\underline{\alpha},t)$ to find the atom on site $\underline{\alpha}$ increases by jumps from any NN site $\underline{\beta}$ into a vacancy at site $\underline{\alpha}$ (first term of rhs) and decreases by jumps away from site $\underline{\alpha}$ (second term). The sum is over all NN sites $\underline{\beta}$ (indicated by $S_{\underline{\alpha\beta}}$). The jump frequency with a vacancy is denoted by w_A, and $G_{s/v}(\underline{\alpha}|\underline{\beta},t)$ is the pair probability to find the atom on site α and a vacancy on site $\underline{\beta}$. It is clear that this pair probability is generally determined by the vacancy concentration c_v which is small in metals ($c_v \lesssim 10^{-4}$).

As a next step one has to consider the time dependence of the pair correlation function due to diffusion of the atom and the vacancy. The differential equation will then contain three particle correlation functions of the atom with two vacancies /4/. One can take advantage of the smallness of c_v by cutting the hierarchy of equations for many-particle correlation functions. The procedure is explained in detail in Ref. 4. The result is the description of the atomic motion in the so-called encounter model /3/. The long-range diffusion of an atom is achieved by many successively approaching vacancies, each of them inducing a few jumps with the atom during their "encounter". Because $c_v \ll 1$ the atom's neighborhood is free of vacancies most of the time and the duration of an encounter is small compared to the time between encounters. With these approximations one obtains an effective single particle equation for the self-correlation function /4/:

$$\frac{\partial}{\partial t} G_s(\underline{\alpha},t) = -\sum_{\underline{\beta}} \Lambda_{eff}^{\underline{\alpha}\underline{\beta}}\, G_s(\underline{\beta},t) \tag{6}$$

The effective jump frequency matrix

$$\Lambda_{eff}^{\underline{\alpha}\underline{\beta}} = c_v\, w_A \cdot \lambda^{\underline{\alpha}\underline{\beta}} \tag{7}$$

contains all the information about the jumps of the atom during an encounter. It is proportional to the vacancy concentration c_v, the atomic jump frequency w_A and a geometrical factor $\lambda^{\underline{\alpha\beta}}$ which can be calculated either by the iterative procedure of Ref. 4 or by Monte Carlo calculations /3,5/.

Eq. (6) is only true in the "sudden encounter" limit. It can be generalized to incorporate the dynamics during the finite duration of an encounter which can be important in dilute alloys if there is a large binding of a vacancy to an impurity atom /12/.

The quasielastic incoherent scattering law for Bravais lattices in the sudden encounter approximation is a single Lorentzian:

$$\left.\frac{d\sigma}{d\omega\, d\Omega}\right|_{\text{inch.}} \sim \frac{1}{\pi} \frac{\tilde{\Lambda}(\underline{k})}{\omega^2 + [\tilde{\Lambda}(\underline{k})]^2} \quad ; \quad \tilde{\Lambda}(\underline{k}) = \frac{1}{N} \sum_{\underline{\alpha\beta}} e^{i\underline{k}(\underline{R}^{\alpha} - \underline{R}^{\beta})} \Lambda^{\underline{\alpha\beta}}_{\text{eff}} \tag{8}$$

whose width $\tilde{\Lambda}(\underline{k})$ is the spatial Fourier transformation of the effective jump frequency matrix. The width is proportional to vacancy concentration c_V and the atomic jump rate. The $\underline{k}$-dependence is determined by the geometry of the elementary jumps and correlation effects. The same function determines the diffusion-induced line broadening of the Mössbauer line /4/.

III. Quasielastic coherent neutron scattering (QECS)

A. Pure Substance

For a pure substance the coherent scattering, Eq. (3), is determined by the sum over pair correlation functions for all atom pairs (n,m). With the definition of the microscopic density

$$\rho(\underline{r},t) = \sum_n \delta(\underline{r} - \underline{R}^n(t)) \tag{9}$$

one can write this sum as the density-density correlation function. It is clear that in our model the density, i.e. the number of atoms in a given volume, can only change by arrival or departure of a vacancy (We neglect here any distortion field around a vacancy). With the introduction of occupation probabilities for atoms and vacancies ($c^n_{\underline{\alpha}}$ = probability to find atom n on site $\underline{\alpha}$, $\hat{c}^p_{\underline{\alpha}}$ = probability to find the vacancy p on site $\underline{\alpha}$) one can recast the QECS-cross-section in terms of vacancy correlation functions. The $c_{\underline{\alpha}}$, $\hat{c}_{\underline{\alpha}}$ are time-dependent stochastic variables on the lattice site $\underline{\alpha}$ and we obtain for the sum over pair correlation functions:

$$\frac{1}{N} \sum_{nm} G^{mn}(\underline{r},t) = \sum_{\underline{\alpha\beta}} \left\langle \sum_{mn} c^m_{\underline{\alpha}}(t) c^n_{\underline{\beta}}(0) \right\rangle_{\text{thermal}} \delta(\underline{R}^{\alpha} - \underline{R}^{\beta} - \underline{r}) \tag{10}$$

Using the particle conservation condition

$$\sum_{\substack{m \\ (\text{atoms})}} c^m_{\underline{\alpha}} = 1 - \sum_{(P\,\text{vacancies})} \hat{c}^p_{\underline{\alpha}} \tag{11}$$

we obtain

$$\left\langle \sum_{mn} c^m_{\underline{\alpha}}(t) c^n_{\underline{\beta}}(0) \right\rangle = 1 - \left\langle \sum_p \hat{c}^p_{\underline{\alpha}}(t) \right\rangle - \left\langle \sum_q \hat{c}^q_{\underline{\beta}} \right\rangle + \left\langle \sum_{pq} \hat{c}^p_{\underline{\alpha}}(t)\, \hat{c}^q_{\underline{\beta}}(0) \right\rangle \tag{12}$$

The averages over single occupation numbers are time independent and equal to c_V, the pair occupation average can be calculated with the assumption that for small vacancy concentrations different vacancies move independently. Then

$$\left\langle \sum_{pq} \hat{c}^p_{\underline{\alpha}}(t)\, \hat{c}^q_{\underline{\beta}}(0) \right\rangle = c_V^2 + \sum_p \left\langle \hat{c}^p_{\underline{\alpha}}(t)\, \hat{c}^p_{\underline{\beta}}(0) \right\rangle \tag{13}$$

where the last term on the rhs is the vacancy self-correlation function $G_V(\underline{\alpha},t\|\underline{\beta}0)$. This function obeys a simple diffusion equation for a single particle (since $c_V \ll 1$)

$$\frac{\partial}{\partial t} G_V(\underline{\alpha},t\|\underline{\beta},0) = \sum_{\underline{\alpha}'} s_{\underline{\alpha\alpha}'} w_A \left[G_V(\underline{\alpha}',t\|\underline{\beta},0) - G_V(\underline{\alpha},t\|\underline{\beta},0) \right] \tag{14}$$

For the QECS cross-section we obtain

$$\left.\frac{d\sigma}{d\omega\, d\Omega}\right|_{\text{coh}} \sim \frac{(1-c_V)^2 \langle a \rangle^2}{2\pi\, N} \,\mathrm{Re} \sum_{\underline{\alpha\beta}} e^{i\underline{Q}(\underline{R}^{\alpha} - \underline{R}^{\beta})} \delta(\omega) + \frac{\langle a \rangle^2 c_V}{2\pi} \,\mathrm{Re} \int_0^{\infty} dt\, e^{-i\omega t} \sum_{\underline{\alpha\beta}} e^{i\underline{Q}(\underline{R}^{\alpha} - \underline{R}^{\beta})} G_V(\underline{\alpha},t\|\underline{\beta},0) \tag{15}$$

The first term is the elastic coherent scattering (Bragg peaks) reduced by a factor $(1-c_v)^2$, the second term is the quasielastic coherent term which has the following features: The intensity is proportional to the vacancy concentration c_v, the line is of Lorentzian form whose width is directly proportional to the vacancy jump frequency w_A and does not contain the concentration. The anisotropy of the width contains the geometry of the elementary jumps without correlation. The expression (15) for the QECS and its relation to incoherent scattering has previously been derived by Ross and Wilson /7/. It has recently been generalized by Gillan and Wolf to incorporate the distortion field around vacancies (or other mobile defects responsible for the atomic motion) /8/.

The proportionality of the intensity to the vacancy concentration c_v makes the detection of the coherent quasielastic scattering difficult if not impossible in normal metals. However, it seems possible for systems with higher defect concentrations, such as ionic conductors or (partially) ordered alloys like PdSi, FeAl, CoGa which for nonstoichiometry show a large number of structural vacancies. To interpret the results properly one has to derive the QECS cross-section for alloy systems.

B. Binary alloys

We shall discuss the case of disordered substitutional binary AB alloys whose components A and B have jump frequency with vacancies w_A and w_B, respectively. For simplicity, we shall neglect the spin and isotope dependence of the scattering length. Then the average in the cross-section, Eq. (1), is over the atomic configuration only. With the occupation probabilities introduced above, this can be written as

$$\sum_{mn} \langle a^m G^{mn}(\underline{r},t)\, a^n \rangle_{atomic} = \sum_{\underline{\alpha}\underline{\beta}} \sum_{mn} \langle a^m c^m_{\underline{\alpha}}(t)\, c^n_{\underline{\beta}}(0)\, a^n \rangle_{\substack{atomic\\thermal}} \delta(\underline{R}^{\alpha} - \underline{R}^{\beta} - \underline{r}) \quad (16)$$

The sum over atoms (m,n) contains terms referring to A atoms and terms referring to B atoms. With the concentrations c_A and c_B (which in the case of vacancy diffusion do not add up to unity) we obtain:

$$\begin{aligned}\sum_{mn} \langle a^m c^m_{\underline{\alpha}}(t)\, c^n_{\underline{\beta}}(0)\, a^n \rangle &= (c_A a_A)^2\, G_d^{AA}(\underline{\alpha},t\,\|\,\underline{\beta},0) + (c_B a_B)^2\, G_d^{BB}(\underline{\alpha},t\,\|\,\underline{\beta},0)\\ &+ c_A c_B a_A a_B \left[G^{AB}(\underline{\alpha},t\,\|\,\underline{\beta},0) + G^{BA}(\underline{\alpha},t\,\|\,\underline{\beta},0)\right]\\ &+ a_A^2 c_A\, G_s^A(\underline{\alpha},t\,\|\,\underline{\beta},0) + a_B^2 c_B\, G_s^B(\underline{\alpha},t\,\|\,\underline{\beta},0)\end{aligned} \quad (17)$$

The functions G_d^{AA}, G_d^{BB} and G_d^{AB} are partial pair correlation functions as introduced by Faber and Ziman /13/ and G_s^A, G_s^B are the self-correlation functions for A,B atoms, respectively. In our classical treatment, $G_d^{AA}(\underline{\alpha},t\,\|\,\underline{\beta},0)$ is the probability to find an A atom at time t on site α given that a different A atom was on site β at t=0. Here the pair correlation functions contain the diffusion dynamics of the atoms, e.g. G^{AA} obeys the following equation of motion:

$$\begin{aligned}\frac{\partial}{\partial t} G_d^{AA}(\underline{\alpha},t\,\|\,\underline{\beta},0) &= \delta(t)\,(1-\delta_{\underline{\alpha}\underline{\beta}})\\ &+ \sum_{\underline{\alpha}'\underline{\beta}'} \sum_{i\neq j} \sum_{\underline{\lambda}} w_A \Big\{ s_{\underline{\alpha}\underline{\lambda}} \left[G_d^{A_iA_j/V}(\underline{\lambda}\underline{\alpha}'|\underline{\alpha},t\,\|\,\underline{\beta}'\underline{\beta},0) - G_d^{A_iA_j/V}(\underline{\alpha}\underline{\alpha}'|\underline{\lambda},t\,\|\cdots)\right]\\ &+ s_{\underline{\alpha}'\underline{\lambda}} \left[G_d^{A_iA_j/V}(\underline{\alpha}\underline{\lambda}|\underline{\alpha}',t\,\|\cdots) - G_d^{A_iA_j/V}(\underline{\alpha}\underline{\alpha}'|\underline{\lambda},t\,\|\cdots)\right]\Big\}\end{aligned} \quad (18)$$

with $G_d^{A_iA_j/V}(\alpha\alpha'|\lambda,t\,\|\,\beta\beta',0)$ the triplet probability to find at time t the A atom i on site $\underline{\alpha}$, the atom j on $\underline{\alpha}'$, and a vacancy on site $\underline{\lambda}$ given that at t=0 the atoms were sitting on site $\underline{\beta}'$ and β, respectively. Similarly, the equation of motion for the other correlation functions can be written down. The next step would be to consider the time dependence of the triplet functions, etc. An approximate solution can be obtained by cutting the higher correlations and thus obtaining a closed set of equations /4/. The simplest approximation is to neglect any correlation between vacancies and atoms and assume

$$G_d^{A_iA_j/V}(\underline{\alpha}\underline{\alpha}'|\underline{\lambda},t\|\underline{\beta}'\underline{\beta},o) \cong c_V \cdot G_d^{A_iA_j}(\underline{\alpha}\underline{\alpha}',t\|\underline{\beta}'\underline{\beta},o) \tag{19}$$

Then the last two terms in Eq. (18) cancel and we obtain a single-particle-like equation of motion for $G_d^{AA}(\underline{\alpha},t\|\underline{\beta},o)$

$$\frac{\partial}{\partial t} G_d^{AA}(\underline{\alpha},t\|\underline{\beta},o) = \delta(t)\,(1-\delta_{\underline{\alpha}\underline{\beta}}) + c_V w_A \sum_{\underline{\lambda}} s_{\underline{\lambda}\underline{\alpha}} \left[G_d^{AA}(\underline{\lambda},t\|\underline{\beta},o) - G_d^{AA}(\underline{\alpha},t\|\underline{\beta},o) \right] \tag{20}$$

The effective jump frequency of an A atom which determines the width of the QECS contribution for A-A-pairs is given by $c_V w_A$ just as in the case of incoherent scattering. This simple approximation yield a QECS cross-section which consists of a sum of Lorentzians with different width due to different jump frequencies. This result will be modified by correlation effects of different origin. They can be (a) of static nature, i.e. be caused by energetic preference of like or unlike neighbors (segregation or short-range order) or (b) of dynamic nature, i.e. caused by preference of vacancy jumps for one type of atoms which also leads to percolation. Details of the concentration and jump frequency dependence have still to be worked out along the lines described in Ref. /4/ for the self-correlation function.

Different correlation functions contain different combinations of jump frequencies. Experimentally, the partial correlation functions can be separated by the isotope substitution technique known from structure determination for alloy glasses /10/.

Another possibility to decompose the QECS cross-section is due to Bathia and Thornton /14/. They define the following three correlation functions:

$\langle\rho(\underline{r},t)\rho(\underline{0},0)\rangle$ = density-density-correlation

$\langle c_A(\underline{r},t)c_A(\underline{0},0)\rangle$ = concentration-concentration-correlation

$\langle\rho(\underline{r},t)c_A(\underline{0},0)\rangle$ = mixed density-concentration-correlation

The time dependence for these functions has still to be worked out. There are some results from computer calculations /15/. However, the following general remarks can be made: The mixed correlation is different from zero only if the occupation couples to the density, e.g. if the different atomic species have different atomic volumes. If we neglect this effect, the density correlation is determined by the vacancy diffusion just as in the case of a pure substance. However, in an alloy the occupation disorder introduces fluctuations which can lead to a more complicated wave number dependence of the respective part of the cross-section. The long wavelength limit is determined by the effective vacancy diffusion constant, which for random alloys can be calculated with known techniques, e.g. coherent potential approximation (CPA). The time dependence of the concentration correlation function is determined by the interdiffusion, i.e. the occupation change of sites.

IV. Conclusions

For the simple model of a binary substitutional AB alloy with diffusion via vacancies, it should be possible to calculate the QECS cross-section. A promising procedure is the generalization previously introduced for self-correlation functions by approximately solving hierarchy equations for many-particle correlation functions. Details have still to be worked out. Experimental application should be possible in systems with high mobility (ionic conductors) and/or high defect concentrations (nonstoichiometric (partially) ordered alloys like PdSi, FeAl, CoGa).

References

1 T. Springer, Springer Tracts in Modern Physics, Vol. 64, (Springer Verlag, N.Y. 1972)

2 K. Sköld in Hydrogen in metals I (Alefeld, Völkl, Eds.) Springer Verlag, Berlin 1978

3 D. Wolf, Sol. State Comm. 23, 853 (1977)

4 O. Bender, K. Schroeder, Phys. Rev. B19, 3399 (1979)

5 K. Schroeder, D. Wolf, P.H. Dederichs in "Point Defects and Defect Interaction in Metals (J.I. Takamura, M. Doyama, M. Kiritani, Eds.) p. 570, University of Tokyo Press 1982

6 R.A. Tahir-Kheli, R.J. Elliott, J. Phys. C15, 5751 (1982) and L445 (1982); K. Kaski, R.A. Tahir-Kheli, R.J. Elliott, J. Phys. C15, 209 (1982)

7 D.K. Ross, D.L.T. Wilson, in "Neutron Scattering 1977", IAEA Vienna 1978

8 M.J. Gillan, D. Wolf, Phys. Rev. Letters 55, 1299 (1985)

9 J.R. Manning, Diffusion Kinetics for Atoms in Crystals (van Nostrand, 1968) p. 127

10 S. Steeb and P. Lamparter, in "The Structure of Non-crystalline Materials" (P.H. Gaskill, J.M. Parker, E.A. Davis, Eds), p. 441 (Taylor and Francis, London 1983); Lamparter et al. Z. f. Naturforschung 37a, 1223 (1982)

11 E. Nowak, P.H. Dederichs, Phys. Rev. B25, 875 (1982)

12 S. Mantl, W. Petry, K. Schroeder, G. Vogl, Phys. Rev. B27 5313 (1983)

13 T.E. Faber, J.M. Ziman, Phil. Mag. 11, 153 (1965)

14 A. Bathia, D.E. Thornton, Phys. Reve. B2, 3004 (1970)

15 Diffusion in Crystalline Solids, G.E. Murch and A.S. Nowick, Eds. Academic Press, N.Y. 1984

Lattice Dynamics and Interstitial Motion

H.R. Schober[1] *and A.M. Stoneham*[2]

[1]Institut für Festkörperforschung der Kernforschungsanlage Jülich, Postfach 1913, D-5170 Jülich, F. R. G.
[2]Theoretical Physics Division, AERE Harwell, Didcot, Oxfordshire, OX11 ORA, England

If an interstitial atom is introduced into a host crystal lattice, the surrounding host atoms will relax and thus lower the total energy. This relaxation energy, often called self-trapping energy, ranges typically from several 10eV for self interstitial atoms in metals to 0.1eV for H in Nb. This variation is roughly proportional to the square of the relative defect relaxation volume ($\Delta V/V_c$), "size" of the defect. On the other hand, typical lattice vibration energies, e.g. the Debye energy $k_B\Theta_D$, are of the order of 0.01eV. Comparing these energies, we conclude that the interstitial atom and the host lattice will dynamically form a strongly coupled entity, even for the smaller interstitial atoms. This strong defect-host coupling in turn causes low-frequency resonant vibrational modes of the defect and its neighbours. These resonances are especially pronounced for the "large" self interstitial atoms [1] but have also been found for the "small" H in Nb [2]. The existence of resonant vibrations indicates a local softening of the lattice. One of the consequences of this softening are the low activation energies ($E_a<0.1$) found for interstitial atoms (in metals). In a simple-minded picture the activation energy can be approximated by

$$E_a \simeq (1/8\pi^2)G_d^{-1} \tag{1}$$

where d is the jump distance and G_d the static defect Green's function in the jump direction. In the case of interstitial defects G can be expressed by a few or even by one resonance mode $G_d^{-1}=\sum_r \alpha_r m_r \omega_r^2$ where m_r and ω_r are the resonance mass and frequency [3], respectively, and α_r is its projection onto the jump vector. The classical activation energy is thus expressed in terms of a few vibrational modes. In a previous paper [4] we have shown that such a description in terms of a few dominant (resonant) vibrational modes can be used also to estimate the tunnel splitting due to orientational degeneracy of trapped heavy interstitials.

In the present paper we will try to use these ideas to derive a quantum mechanical model of the elementary step of diffusion, the thermally activated jump of the interstitial atom. As a test system we have chosen H in Nb at low concentration. This system has been studied extensively both theoretically and experimentally, see e.g. the recent review by Fukai and Sugimoto [5] on hydrogen diffusion. At low temperatures, a tunnel splitting of $\Delta E\simeq 0.2$ meV has been observed for H trapped at O-impurities in Nb [6]. This value is thought to be fairly representative also for the pure H-Nb-system, since the localized H-vibrations change only marginally upon trapping of the H at the O-impurity [7]. At higher temperatures, in the pure H-Nb-system, two distinct diffusion regimes are observed. In both regimes diffusion follows an Arrhenius law, with activation energies of 68 meV ($T<250K$) and 108 meV ($T>250K$) [8], respectively.

The standard approach in the calculation of absolute diffusion rates is that of Flynn and Stoneham [9] which is based on the small polaron model [10]. This approach involves several distinct ideas. The first is that of "sensibly localized" states between which transitions occur. The second, related idea, is that of self-trapping: the interstitial causes a distortion of the host lattice which in turn immobilizes the interstitial. The transition probability can then be found by the usual "golden rule expression"

$$W^{if} = \langle\langle \sum_n w^{if} \rangle\rangle_m \tag{2}$$

$$w^{if} = 2\pi/\hbar \ |\langle\Psi_i(m)|H-E_{im}|\Psi_f(n)\rangle|^2 \ \delta(E_{im}-E_{fn}) \tag{3}$$

where Ψ_i and Ψ_f are the initial and final state wavefunctions, m and n denote the corresponding excitational state, $\langle\langle\ \rangle\rangle_m$ denotes a thermal average over the initial states m and $E_{im}=\langle\Psi_i(m)H\Psi_i(m)\rangle$. To be able to evaluate (2) a number of further approximations are usually made. The hydrogen atom is assumed to follow the lattice vibrations adiabatically and the lattice phonons themselves are coupled linearly to the hydrogen. The matrix element in (2) is thus split into a "naked" tunneling element, J, of the hydrogen atom in the potential built by the host atoms and a dressing lattice overlap term, σ, accounting for the change of the displacement field. To simplify the evaluation even further, the dependence of the naked H-tunneling element on the state of the host lattice is neglected (Condon approximation) and one finally arrives at a simple expression

$$W^{if} = 2\pi/\hbar \ |J|^2 \ |\sigma^{if}|^2 \tag{4}$$

which yields a T^7 law at low temperatures and an Arrhenius behaviour at high temperatures

$$W^{if} = (\pi/4\hbar^2 E_c kT)^{1/2} \ |J|^2 \exp(-E_c/kT) \tag{5}$$

The activation energy, E_c, in (5) is the so-called coincidence energy, i.e. the minimal energy necessary to deform the lattice such that the initial and final configurations of H plus lattice are equivalent. This coincidence energy is different from the "classical" activation energy, which is the energy needed to transpose the H atom quasistatically from its initial to its final position.

There are numerous extensions to this simple model, mainly aimed at improving upon the concept of a constant J. Even keeping the strict splitting into a hydrogen and a lattice wavefunction J will be temperature dependent. At temperatures $T>250K$ the localized vibrations of the H-atom become noticeably excited. Since the tunneling element between excited states is independent of the exact potential shape, an order of magnitude larger than the ground state one $|J|^2$ in (4) will increase exponentially for T>250K. This leads to a sharp change in activation energy, as seen in experiment. Still within the adiabatic approximation but improving on the Condon approximation is the occurrence probability approach [11,12] where one evaluates J for all coincidence configurations and averages with respect to lattice energy and excitation of the localized H-vibrations. These approximations explain the observed H-diffusion at least qualitatively.

Whereas the approximations discussed so far relied on a weak hydrogen-phonon interaction, we want to drop this restriction, which is inconsistent with the occurrence of resonant vibrations and comparatively large self-trapping energies. In order to accommodate the resonant vibrations, which are tantamount to a strong hydrogen-phonon coupling, we treat the hydrogen and lattice degrees of freedom simultaneously, starting from (2) and (3) but avoiding the adiabatic and Condon approximations. The evaluation of the rates (3) follows closely the path outlined in [4] which can be broken down into the following stages: i) choice of interaction model, ii) calculation of structures, energies and eigenmodes, iii) evaluation of the tunnel splitting and of the transition probabilities w and finally calculation of the total rate W.

For the interaction we used the model by Sugimoto and Fukai [13] where the H-Nb interaction is via a truncated double exponential potential and the Nb-Nb interaction is harmonic, via Born-von Karman parameters [14]. This model reproduces the experimental frequencies of the localized H-vibrations and the dipole-force tensor but neglects the induced change in the Nb-Nb interaction, necessary to explain the observed change of the elastic moduli [15] and suggested by electron calculations [16].

In agreement with experiment, the stable configuration of the hydrogen interstitial is tetrahedral. We find a self-trapping energy of E_{ST}=470meV where 300meV is potential energy and the change of the zero-point vibrations contributes 170meV. The minimal coincidence configuration energy is E_c=41meV (31meV potential energy and 10meV zero-point motion) and the classical saddle-point energy is E_a=61meV (potential energy only). As eigenstates in the initial and final configurations we take the harmonic ones obtained by diagonalizing the dynamic matrix for clusters of 6 to 21 atoms. The localized frequencies are found at 42 and 27 THz and the resonant ones at 5.1 and 3 THz.

Using these eigenmodes and a correction term for the rest of the lattice [4] we find at T=0K a tunnel splitting of ΔE_t=0.01meV. Here we assume that the tunneling occurs only between a pair of adjacent sites. In general there is an additional geometric factor. The value observed experimentally in the Nb-O-H system [6] is larger (0.2meV). This may be due to a slightly different geometry in this system. Our value will be an underestimate, due to the too small overlap of the harmonic wavefunctions. Assuming a sinusoidal potential shape, one gains a correction factor of 1.3. In the spirit of the small polaron model, the tunnel splitting is usually split into a naked term J, as in (4), and a lattice dressing factor

$$\Delta E_T = J \exp(-S) \tag{6}$$

Changing the isotopic masses, we find the dressing factor as 0.15 in accordance with the usual estimates. The total tunnel splitting is, however, larger by a factor 5, as compared to a separate calculation of the two terms. This indicates a rearrangement of the contributions of the localized modes, included in J, and of the band modes, determining S. The temperature dependence of (5), gained by thermal averaging over the occupation numbers with $m_i=n_i$, is totally different from the one expected in the weak coupling model.

The total transition rate sums over all possible initial and final mode occupancies with energy conservation as the only restriction. With increasing temperature the number of terms diverges rapidly. In order to evaluate (3), we therefore divide the modes into strongly and weakly coupling ones. The latter ones we treat by the usual small polaron formalism, whereas we sum over the strongly coupled ones explicitly. As indication of the coupling strength, we take the projection of the change of displacement on the modes. This includes first of all the two localized and the two resonant modes of appropriate symmetry. Other modes were included because they couple strongly to the resonant modes. Lastly, we include the third localized mode, since its excitation reduces the transition rate due to its high energy in the saddle point configuration. In total, we include 12 modes in the summation of (2). Eq. (3) is transformed into

$$w^{if} = 2\pi/\hbar \; |\langle\psi_i(m)|H-1/2(E_{im}+E_{fn})|\psi_f(n)\rangle|^2 \; G(E_{im}-E_{fn}) \tag{7}$$

Here the δ-function is replaced by a shape function, which consists of two distinct factors. One comes from the line broadening, i.e. it accounts for the broadening of the sharp modes of the atomic cluster by the embedding lattice. We take a form

$$G_V = 1/\sqrt{\pi}\, W_V \exp[-(\omega-\Omega)^2/W_V{}^2] \tag{8}$$

with $W_V= \omega^\sigma(m^\sigma+n^\sigma)$ and $\Omega= \omega^\sigma(m^\sigma-n^\sigma)$. As widths, w, of the individual modes σ we took 0THz and 0.5THz for the localized and band modes, respectively.

The second factor in the shape function stems from the change in mean displacement of the weakly coupled modes as the hydrogen atom jumps. It corresponds to the dressing terms in (4). Its magnitude is, however, strongly reduced, since we have included the bulk of the dressing explicitly in (7). To ease the evaluation we take the high temperature limit of this term

$$G_R(\omega) = (16\pi E_R kT)^{-1/2} \exp[-(\hbar\omega+4E_R)^2/16E_R kT] \tag{9}$$

where we estimate E_R=6meV.

Figure 1 shows our results for the diffusion coefficient. Below T=100K our results agree well with experiment . The lower of the two activation energies is too low in our calculation, leading to increasing discrepancy of the diffusion constant with

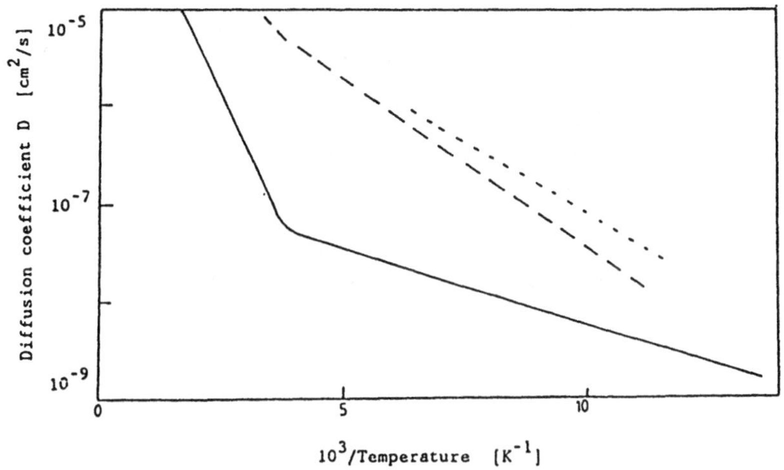

Fig. 1 Diffusion constant of H in Nb versus inverse temperature. -------- experimental value [8], ——— calculation, ... estimate with enhanced contribution of higher excited resonance states.

increasing temperature. The high-temperature activation energy agrees again with experiment. We conclude that our picture of the high-temperature activation energy being caused by the excitation of localized vibrations was correct, whereas the low temperature diffusion is activated by the band modes. A similar too low activation energy at lower temperatures was also found by Teichler and Klamt [12] in their, different, calculation. Our underestimate of this activation energy may be caused by deficiencies of the underlying interaction model, but in part probably stems also from the neglect of anharmonicity in the wavefunctions. (We have of course included the full potential shape in the Hamiltonian (2)). As we have already seen in the tunneling case, the use of harmonic wavefunctions leads to an underestimate of the transition element. This underestimate worsens progressively with higher excited states. Introducing a reasonable correction factor, one could improve the activation energy markedly (dotted line in fig. 1). We think, therefore, that our model can be improved such that it describes the hydrogen diffusion not only qualitatively but also gives reasonable quantitative values.

1. P.H. Dederichs, C. Lehmann and A. Scholz, Phys. Rev. Lett. 31, 1130 (1973)
2. V. Lottner, H.R. Schober and W.J. Fitzgerald, Phys. Rev. Lett. 42, 1162 (1979)
3. P.H. Dederichs and R. Zeller, Point Defects in Metals II, Springer Tracts in Modern Physics 87, Springer, Berlin (1980)
4. H.R. Schober and A.M. Stoneham, Phys. Rev. B 26, 1819, (1982)
5. Y. Fukai and H. Sugimoto, Adv. Phys. 34, 263 (1985)
6. H. Wipf, A. Magerl, S.M. Shapiro, S.K. Satija and W. Thomlinson, Phys. Rev. Lett. 46, 947 (1981); H. Wipf and K. Neumaier, Phys. Rev. Lett. 52, 1308 (1984); H. Wipf, A. Magerl and K.Neumaier this workshop
7. A. Magerl, J.J. Rush, J.M. Rowe, D. Richter and H. Wipf, Phys. Rev. B 27, 927 (1983)

8. Zh Qi, J. Völkl, R. Lässer and H.Wenzl, J. Phys. F 13, 2053 (1983)
9. C.P. Flynn and A.M. Stoneham, Phys. Rev. B1, 3966 (1970)
10. T. Holstein, Ann. Phys. (NY) 8, 325 (1959), ibid. 8, 343 (1959); J. Yamashita and T. Kurozawa, J. Phys. Chem. Sol. 5, 34 (1958)
11. D. Emin, M.I. Baskes and W.D. Wilson, Phys. Rev. Lett. 42, 791 (1979)
12. H. Teichler and A. Klamt, Phys. Lett. 108A, 281 (1985)
13. H. Sugimoto and Y. Fukai, Phys. Rev. B22, 670 (1980)
14. Y. Nakagawa and A.D.B. Woods, Phys. Rev. Lett. 11, 271 (1963)
15. H.R. Schober and V. Lottner, Z. Phys. Chem. NF (Wiesbaden) 114, 203 (1979)
16. G. Solt this conference

Interatomic Forces Near a Hydrogen Impurity in Metals

G. Solt and H. Beck

Université de Neuchâtel, CH-2000 Neuchâtel, Switzerland

1. Introduction

The knowledge of the interactions between impurity and host atoms is clearly essential for understanding the equilibrium lattice configuration (lattice deformation) near the impurity and the change in vibrational properties (localized and 'band' modes) related to the dissolved atom.

Experimental results for dilute hydrides indicate that the atomic configuration near the dissolved H atom cannot convincingly be explained by assuming only simple central, pairwise impurity-host forces. Indeed, the observed strain field [1], [2] around tetragonal impurity sites shows, surprisingly, a nearly cubic symmetry in BCC (Nb,Ta) hosts. Moreover, the structure in band mode spectra [3] of the same hydrides indicates a significant change of the force constants between the solvent atoms themselves near the H impurity.

From the theoretical point of view, the complexity of the atomic interactions has to be related to the complex electronic rearrangement (polarization) phenomena in the metal near the impurity. The force between two (host or impurity) ions in a metal is a sum of their electrostatic repulsion and of an indirect interaction, arising from the polarization of the gas of conduction electrons. In the simplest case, one may think of an initially homogeneous electron sea, polarized subsequently by ion A and the displaced charge interacting with ion B, which gives rise to the well known pairwise force between A and B in simple metals; linear response theory is in this case a reasonable first approximation, and the simultaneous presence of three or more ions can be neglected or, if necessary, taken into account by perturbative methods [4]. This simple picture does not apply, however, for such strongly perturbing impurities like hydrogen, first because of the substantial charge readjustment at the H^+ ion which cannot be determined, even approximately, by linear response theory [5], and second, since the electronic polarizability near the dissolved H atom is far from being that of a homogeneous electron gas. The substantial spatial variation of the polarizability near the impurity leads to a strong variation of the host-host forces (force constants), changing their pairwise character.

The simplest realistic description [5] of the hydrogen - solvent ion interaction starts with a self-consistent calculation of the charge distributed around the H^+ ion in an electron jellium. Next, this spherically symmetric charge distribution, the 'immersed impurity', is subjected to the potential of each host ion, without being as yet deformed. This results (Sect. 2.1) in a central, pairwise 'first order' impurity-host interaction, the exact shape of which is determined by the 'bare' ion-electron potential of the solvent.

The 'first order' impurity-host force has been evaluated for a number of simple metal [5]-[12] and copper [13] solvents, with the objective to find the equilibrium position of the trapped hydrogen or μ^+-meson impurity, the heat of solution, lattice strain, or the hydrogen-host force constants. The calculations [10], [11] (Sect. 2.2) explain with success the observed [14] easy delocalization of the hydrogen-like μ^+ impurities in $A\ell$ along the <111> crystallographic directions, and give reasonable results for the energetics; for alkali solvents, the experiments to check the predicted strong, highly anisotropic lattice distortion [10] are as yet lacking.

To describe the interaction between host ions A and B, realized through the deformation (polarization) of the 'immersed impurity', one has to go beyond first order. The inhomogeneity of the polarized object leads to a 'host-host' interaction, which depends not only on the distance A-B but on the whole triangular configuration A-H-B. For the calculation an appropriate linear response theory (Sect. 2.1) can be formulated, involving the exact dielectric susceptibility of the immersed H atom. The first, semiquantitative results [15] [16] show (Sect. 2.3) the main features of the impurity-induced host-host forces, namely their strongly non-pairwise character, and indicate the need for more accurate, fully self-consistent results for application in lattice statics and lattice dynamics.

2.1 The Immersed Impurity: Charge, Polarizability and Interatomic Forces

The electronic rearrangement due to the introduction of the impurity, and the resulting interatomic forces, can be analyzed in 3 steps, corresponding to subsequent orders of approximation :

i) the H atom gets immersed in an electron jellium, of density $\bar{n}$ equal to the average conduction electron density of the solvent metal. The electrons react: for metallic densities $\bar{n} \lesssim \bar{n}(A\ell)$ two electrons are trapped at a bound orbital s state and the immersed ('embedded') hydrogen atom resembles rather a H^- ion screened by 'scattered' conduction electrons. The piled-up density is spherically distributed around the H^+ ion at $\vec{R}_\mu$

$$n(\vec{r} - \vec{R}_\mu) = \bar{n} + \Delta n(|\vec{r} - \vec{R}_\mu|) \tag{1}$$

and Δn diminishes, for large $r = |\vec{r}-\vec{R}_\mu|$, in an oscillatory way; accurate non-perturbative calculations for $\Delta n(r)$ have been done [5]-[7] by the density functional method;

ii) the jellium is completed by the lattice of solvent ions. If each host ion has a charge Z and an ionic potential v, the impurity-host interaction in first order in v is

$$Q(|\vec{R}_i - \vec{R}_\mu|) = \frac{ZZ^*e^2}{|\vec{R}_i - \vec{R}_\mu|} + e\int d\vec{r}\; v(\vec{r} - \vec{R}_i)\Delta n(|\vec{r} - \vec{R}_\mu|) \tag{2}$$

where $\vec{R}_i$ is the position of the host ion and $Z^* = 1$ is the proton charge;

iii) the deformation of the spherical charge distribution $n(r)$ caused by the presence of the solvent lattice is now taken into account. The charge, displaced under the influence of host ion at $\vec{R}_j$ is in first order

$$\delta n_j(\vec{r};\vec{R}_j - \vec{R}_\mu) = \int d\vec{r}'\chi(\vec{r} - \vec{R}_\mu, \vec{r}' - \vec{R}_\mu)v(\vec{r}' - \vec{R}_j) \tag{3}$$

where χ is the response function of the jellium with the immersed impurity. The complete interaction between ions i and j is then

$$I_c(\vec{R}_i, \vec{R}_j; \vec{R}_\mu) = \frac{Z^2e^2}{|\vec{R}_i - \vec{R}_j|} + e\int d\vec{r}'\, v(\vec{r}' - \vec{R}_i)\delta n_j(\vec{r}';\vec{R}_j - \vec{R}_\mu) \tag{4}$$

which is exact in second order of v.

In the absence of the impurity I_c becomes the usual pair interaction $I_0(\vec{R}_i-\vec{R}_j)$ in the pure metal. Thus, by writing

$$I_c = I_0 + I \tag{5}$$

we have the impurity-induced host-host interaction,

$$I(\vec{R}_i, \vec{R}_j; \vec{R}_\mu) = \int d\vec{r}\ v(\vec{r} - \vec{R}_i)\ \widetilde{\delta n}_j\ (\vec{r}; \vec{R}_j - \vec{R}_\mu) \tag{6}$$

where $\widetilde{\delta n}$ is δn less the displaced charges in a homogeneous jellium.

According to (3), δn is not spherical around $\vec{R}_j$ but cylindrical along the axis $\vec{R}_j - \vec{R}_\mu$, that is why $I(\vec{R}_i,\vec{R}_j;\vec{R}_\mu)$ is really a triplet-ion interaction. The key quantity is obviously the susceptibility χ of the immersed impurity, obtainable by similar, though slightly more complicated self-consistent methods as Δn [17]-[19], or by approximate model calculations [15], [16].

It is essential that χ depends only on $\bar{n}$ but is not related to the solvent lattice : it is a characteristic of the impurity-in-jellium complex. One sees that the theory is a systematic perturbative approach as regards the solvent lattice and at the same time exact, in principle, as regards the readjustment of the electronic structure due to the impurity.

Some results for the impurity-host and induced host-host interactions are described below.

2.2. Impurity-Host Forces and Lattice Deformation in Aℓ(H) and Li(H)-Cs(H)

Out of the interstitial sites of the solvent lattice (Fig. 1), those corresponding to the energy minimum of the atomic interactions will be occupied at low temperatures

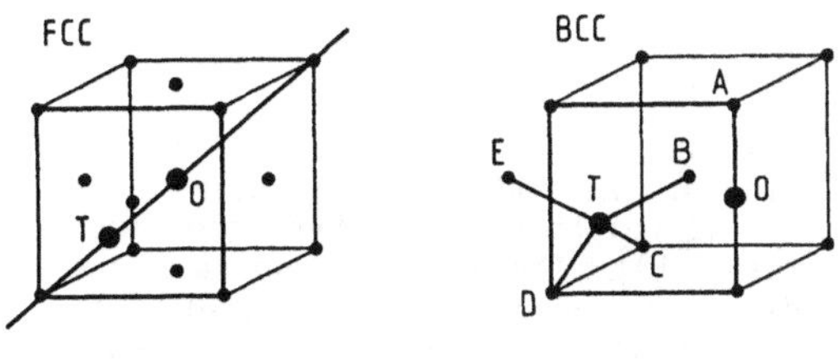

Fig. 1. Tetrahedral (T) and octahedral (O) interstitials in FCC and BCC metals. In the BCC lattice the site symmetry both for O and T is tetragonal (D_{2d} and D_{4h}).

The experimental evidence [14] for Aℓ(H) is that neither the O or the T sites are good traps for the hydrogen : μ^+ particles do not seem to be localized in either of them. The result in Fig. 2, based on a calculation according to (2) explains this fact :

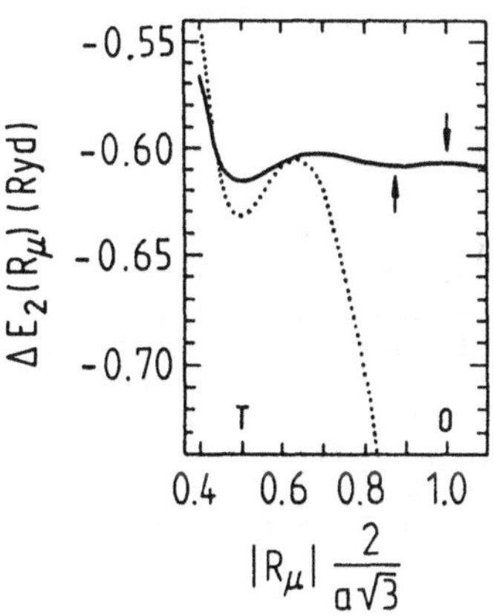

Fig. 2. The energy of the hydrogen-like impurity in aluminium as a function of its position along the cube diagonal of the undistorted FCC lattice [9]. Dotted curve: direct ion-ion repulsion, full curve : equation (2); a is the lattice constant.

The T site is clearly too exiguous from the point of view of ion-ion repulsion, but after rearrangement of the electrons it becomes slightly better than anything else - the energy surface is, however, very flat all along the cube diagonal, and makes delocalization easy.

For alkali (BCC) solvent it was suggested [8] that hydration fills up progressively the O sites, lattice relaxation tends to increase the distance OA (two neighbours) and shorten OB (four neighbours), facilitating thereby the establishing of the NaCℓ structure in stoichiometric hydrides. The first calculation [8] supported this view, though the calculated displacements were much too small. The more recent results [10] in Fig. 3 show the force acting upon the alkali neighbours, as calculated by (2), and also the atomic displacements : the ratio λ = OA/OB changes from $1/\sqrt{2}$ = 0.707 to 0.92 in Na solvent. One reason for the increase of λ is seen in the figure : for ion A the H^+ ion is too near to be well screened and repulsion prevails, but ion B 'sees' the impurity overscreened, according to the oscillatory behaviour of Δn, thus B is pulled inward. Another reason is the high elastic anisotropy of these solvents, leading to the fact that elongation in the [001] direction is necessarily accompanied by a contraction in the [110] directions.

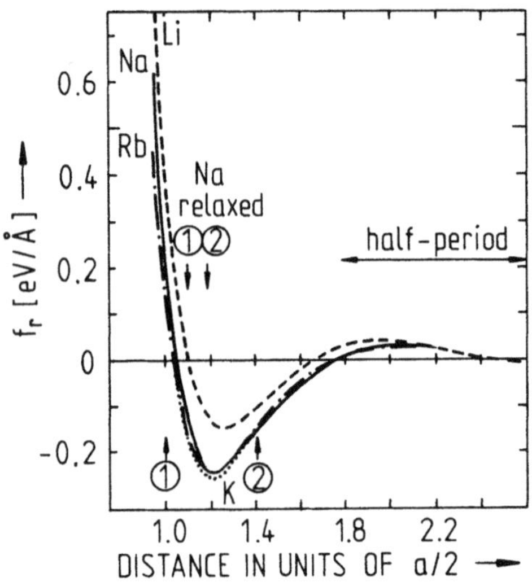

Fig. 3. The radial force acting upon alkali neighbours of a hydrogen impurity [10]. The numbers indicate the distances of 1st and 2nd neighbours in the ideal (undistorted) and also in the relaxed lattice.

Since anisotropies of the dipole force tensor and of the lattice response cooperate in this way, the 2 nearest and 4 next neighbours are strongly displaced in radially opposite directions to approach a regular (λ = 1) octahedron : the predicted λ's range from 0.88 (Li(H)) to 0.97 (Cs(H)) [10].

2.3 The Impurity-Induced Host-Host Forces

The induced interaction $I(\vec{R}_A,\vec{R}_B,\vec{R}_\mu)$ between host ions A and B near the hydrogen impurity was calculated according to (3)-(6); the geometry is shown in Fig. 4. The response function χ, showing both atomic-like and continuum properties of the immersed-atom system has to be determined self-consistently, but the essential features can be reproduced by a simpler calculation [15]-[16] based on the familiar 'square-well' model of the impurity. The result for an aluminium-like solvent ($\bar{n}\approx\bar{n}$(Aℓ), Z = 3) are plotted in Fig. 4. for two different values of the size (core radius R_c) of the solvent ions. The figure shows the central and tangential components of the induced host-host force, together with the (central) pairwise force, which is present also in the pure metal.

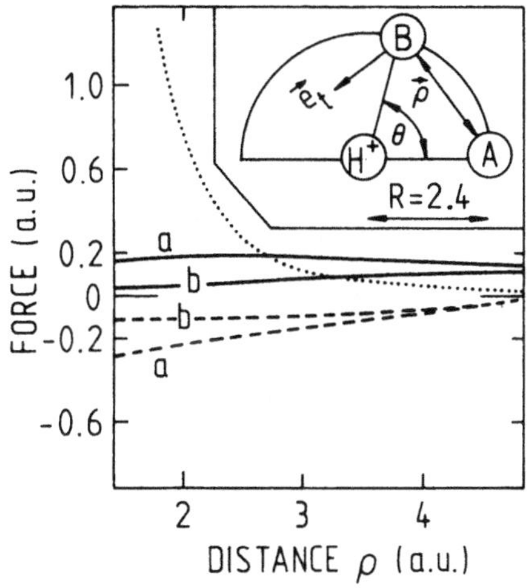

Fig. 4. The central (full line) and tangential (broken line) components f_ρ and f_t of the hydrogen-induced host-host force acting on atom B in a 3-valent host, together with the central A-B force (dotted line) present in the pure solvent; a) small core (R_c=0.4), b) large core (R_c=0.8) [16].

One sees that the induced force $\vec{f}$ is highly non-central : the tangential component is typically as large as the central one, and vanishes only at a symmetry point. Moreover, the induced force is by no means negligible beside the 'unperturbed' force acting between A and B far away from the impurity. In polar coordinates, with the H-atom as centre, one sees [16] that the radial component (along the H-B axis) of $\vec{f}$ is prevailing for $R_c \gtrsim 0.4$ au, but the angular component f_θ 'opening the angle θ is also considerable for a solvent with small R_c; it proves to be positive and decreases with increasing θ.

3. Conclusions

The first order theory, using only central impurity-host forces, was seen to reproduce the observed delocalization of the hydrogen-like μ^+ particle in aluminium; an experimental check of the predicted anisotropic deformation in alkali hydrides would be of great interest.

Besides the pairwise impurity-host force, the impurity-induced host-host interaction near the dissolved H atom is equally important. This fact, indicated by experiment, was confirmed by the calculation: the induced host-host interaction for near neighbours of the H atom is typically as large as their interaction in the pure solvent. The hydrogen-induced interaction proves to be strongly non-central, with comparably large tangential and central components.

The results show that the polar component of the induced host-host force, at the geometry of Fig. 4, tends to open the angle AHB, and even more the smaller this angle is. This may be relevant for the case of hydrogen at a T site in a BCC host : the force opening the angle DTE on pushing neighbours B and E of Fig. 1. upwards (and D,C downwards) tends, at the same time, to form a dipole force tensor approaching cubic symmetry.

Our calculation of the hydrogen-induced host-host force in the aluminium-like solvent is essentially qualitative in nature, since the response function is calculated in a simplified model and not in a self-consistent way. To be more quantitative, one has to proceed according to Sect. 2.1 but with the self-consistently determined χ for the given electron density $\bar{n}$ of the host; such results for χ begin to become available [18].

Even if our simple local potential v, adequate for simple metal hosts, cannot represent the ionic potential of a transition metal solvent, the qualitative features of the induced interaction will possibly be similar in transition metal hosts, since the mechanism, described in Sect. 2.1. remains the same. We believe, therefore, that the observed nearly cubic dipole-force tensor in BCC hydrides is a manifestation of the 'angle-opening' component of the hydrogen-induced host-host interaction.

References

1. G. Bauer, E. Seitz, H. Horner and W. Schmatz, Sol. St. Comm. 17, 161 (1975)
2. H. Metzger and H. Peisl, J. Phys. F: Met. Phys. 8, 391 (1978)
3. V. Lottner, H.R. Schober and W.J. Fitzgerald, Phys. Rev. Lett. 42, 1162 (1979)
4. E.G. Brovman and Yu. Kagan in Dynamical Properties of Solids, vol. 1, ed. G.K. Horton and A.A. Maradudin (North-Holland, Amsterdam 1974)
5. Z.D. Popovic and M.J. Stott, Phys. Rev. Lett. 33, 1164 (1974)
6. D.S. Larsen and J. K. Norskov, J. Phys. F: Met. Phys. 9, 1975 (1979)
7. M. Manninen and R.M. Nieminen, J. Phys. F: Met. Phys. 9, 1333 (1979)
8. F. Perrot and M. Rasolt, Phys. Rev. B23, 6534 (1981) and B25, 7327 (1982)
9. G. Solt, M. Manninen and H. Beck, J. Phys. F: Met. Phys. 13, 1379 (1983)
10. G. Solt, M. Manninen and H. Beck, Sol. St. Comm. 46, 295 (1983)
11. S. Estreicher and P. Meier, Phys. Rev. B27, 642 (1983)
12. M. Kaneko, K. Tsuchiya, K. Ohashi, Y.H. Ohashi and M. Fukuchi, J. Phys. F: Met. Phys. 14, 1095 (1984)
13. H. Teichler, Phys. Lett. 67A, 313 (1978)
14. K.W. Kehr, D. Richter, J.M. Welter, O. Hartmann, E. Karlsson, L.O. Norlin, T.O. Niinikoski and A. Yaouanc, Phys. Rev. B26, 567 (1982)
15. G. Solt and H. Beck, Sol. St. Comm. 51, 883 (1984)
16. G. Solt and H. Beck, J. Phys. F 1985 (in press)
17. E. Zaremba and M.J. Stott, Phys. Rev. A21, 12 (1980)
18. A. Zangwill and P. Soven, Phys. Rev. A21, 1561 (1980)
19. E. Zaremba in Density Functional Theory, ed. J. Keller and J.L. Gazquez (Springer, Berlin 1983)

H Tunneling and Local Diffusion in NbO_xH_y

H. Wipf[1], *A. Magerl*[2], *and K. Neumaier*[3]

[1]Institut für Festkörperphysik, Technische Hochschule Darmstadt, D-6100 Darmstadt, F. R. G.
[2]Institut Laue-Langevin, F-38042 Grenoble, France
[3]Walther-Meissner-Institut für Tieftemperaturforschung, D-8046 Garching, F. R. G.

1. Introduction

The presently best understood example of low-temperature H tunneling in metals is represented by H interstitials in Nb that are trapped (below ~150 K /1,2/) by impurity atoms such as O or N. The tunneling behavior has been investigated in a great number of specific heat, ultrasonic (anelastic) and neutron spectroscopic measurements (see e.g. the studies of /3-11/ and references therein). The relevant tunneling process observed in the experiments is most likely to occur between two nearest-neighbor tetrahedral trap sites located in the neighborhood of an O or N trap center (for a detailed discussion, see /2,5/). In the temperature range below ~5 K, the tunneling matrix element was found to be ~0.2 meV for the isotope H /3,5-7,9/ and ~0.02 meV for D /9/.

In this paper, we address two problems investigated in recent neutron spectroscopic studies /12,13/. First, the temperature dependence of the tunneling states and, second, the question of to what extent conduction electrons directly influence the tunneling behavior.

2. Temperature Dependence of the H Tunneling States

The temperature behavior of the tunneling process was subject of the neutron scattering study of Ref. 12. The experiments were performed on a $NbO_{0.002}H_{0.002}$ sample at the spectrometer IN6 at the ILL in Grenoble. Fig. 1 presents four spectra taken at temperatures between 0.05 and 9.8 K (Q-range from 1.1 to 1.8 A^{-1}; elastic-energy resolution 0.055 meV (FWHM)).

The spectra in Fig. 1 show for the two lowest temperatures (0.05 and 5.0 K) inelastic intensity peaks at about 0.2 meV. The temperature behavior is essentially controlled by detailed balance. Such a behavior is in agreement with the existence of two eigenstates for the H (with an energy splitting of ~0.2 meV) as expected for two-site tunneling. With rising temperature, the inelastic line seems to broaden and to shift towards lower energies as a result of an increasing destruction of the tunneling states. At 9.8 K, the spectrum is nearly quasi-elastic, which indicates the existence of a local jump diffusion process for the H.

For a more quantitative description of the temperature behavior, fits were made in Ref. 12 to the data with the help of a scattering law that represents a generalization of that proposed in an earlier paper /3/. Since a detailed description of the fit procedure is given in Ref. 12, only the most important characteristics of the underlying tunneling model will be summarized as follows:

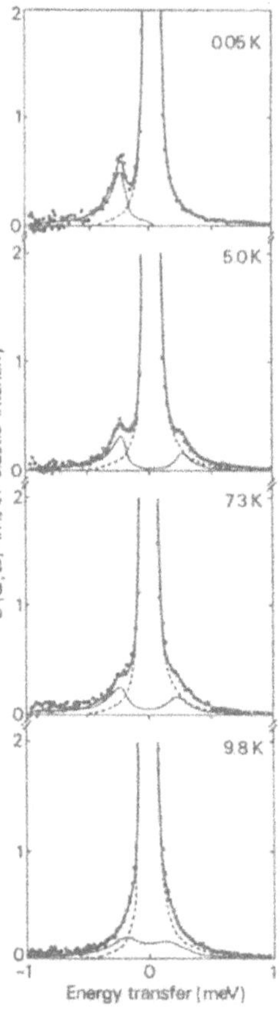

Fig. 1 Temperature-dependent neutron spectra for a $NbO_{0.002}H_{0.002}$ sample. See text for details.

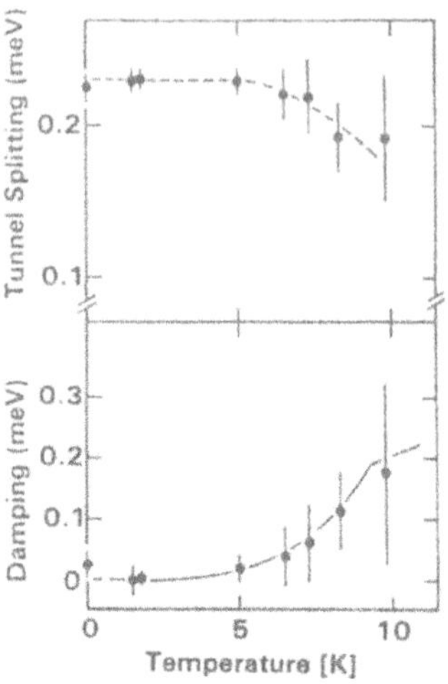

Fig. 2 Fit results for the temperature behavior of the tunneling matrix element J_0 (tunnel splitting) and of the damping constant Γ /12/.

(i) The trapped H is assumed to tunnel between two sites which are slightly shifted against each other by an energy value ε, as a result of elastic interaction effects between different O-H complexes. The energy difference between the two tunneling eigenstates is then given by

$$E = \sqrt{J^2 + \varepsilon^2} \quad (1)$$

where J is the tunneling matrix element.

(ii) In view of the statistical spatial arrangement of the O-H complexes, a distribution function needs to be taken into account for the energy shifts ε. This distribution function was assumed to be Lorentzian. Similarly, a distribution function was presupposed for the value of the matrix element (average value J_o, width ΔJ).

(iii) A dynamical coupling of the tunneling states to phonons and/or electrons which limits the lifetime of the states, and causes their destruction above ~5 K, was not considered in the scattering law of Ref. 3. In order to take account of this effect, the δ-functions in Ref. 3 were replaced by oscillator functions

$$O(\hbar\omega) = \frac{1}{\pi} \frac{\Gamma E^2}{[(\hbar\omega)^2 - E^2]^2 + (\Gamma\hbar\omega)^2} \tag{2}$$

with a damping constant Γ.

The fit curves obtained for the spectra in Fig. 1 are shown by bold solid lines, thin solid lines and broken lines which represent the intensities of total scattering, inelastic scattering and elastic scattering, respectively. At low temperatures (T = 1.5 K), the fit result for the matrix element was $J_0 = 0.23 \pm 0.005$ meV which is in reasonable agreement with previous studies /3,7,9/. The distribution width of the matrix elements was $\Delta J = 0.015 \pm 0.008$ (i.e. $\Delta J/J_o = 0.065$). The fit results for the temperature dependence of the two most important quantities that describe the tunneling dynamics, J_o and Γ, are reproduced in Fig. 2. The figure indicates a decrease of the matrix element (tunnel splitting) J_o with rising temperature, and at the same time an increase of the damping constant Γ. Although the error bars become large for the higher temperatures, which in part is due to a strong correlation between the two fit parameters, the data clearly reflect the drastic changes that occur in the tunneling behavior.

3. Direct Interaction with Conduction Electrons

The experiments reported in Section 2 cannot definitely answer the question as to what extent the thermal destruction of the tunneling states is the result of a direct interaction with conduction electrons. For an experimental investigation of this problem, it is of importance that Nb is a superconductor which (below the superconducting transition temperature of 9.2 K) can be held in both normal- and superconducting state, depending on whether a sufficiently large magnetic field is switched on or off. Experiments in the presence and absence of a magnetic field allow, therefore, an unambiguous clarification of the rôle of conduction electrons for the tunneling behavior.

The first evidence for the electronic influence on the H tunneling states is reported from ultrasonic relaxation measurements /10/. These measurements demonstrated a drastic increase of the relaxation rate in the normal-conducting state. Very recently, a detailed neutron spectroscopic study was performed on a $NbO_{0.0002}H_{0.0002}$ sample in both the presence and absence of a magnetic field of 0.7 T (Instrument IN 6 at the ILL, temperature range below ~ 10 K) /13/. The preliminary results of this study can be summarized as follows:

(i) At the lowest temperatures investigated (below 0.2 K), the tunneling matrix element measured in the normal-conducting state is reduced (renormalized) by about 15% as compared to its value in the superconducting state.

(ii) The inelastic linewidth is, also below ~ 0.2 K, considerably broader in the normal-conducting than in the superconducting state.

(iii) Because of the electronic interaction, the tunneling states become destroyed (and a transition to a quasi-elastic scattering behavior is observed) at much lower temperatures in the normal-conducting state.

The results above are clear evidence for a drastic electronic influence, both on the value of the matrix element and on the lifetime of the tunneling states. In particular, the dominant influence on the lifetime indicates that transport processes of the H interstitials are controlled at low temperatures by direct interaction with conduction electrons. This demonstrates, finally, that the commonly assumed adiabatic Born-Oppenheimer approximation is inapplicable for a theoretical description of such transport processes.

References

/1/ G. Pfeiffer, H. Wipf, J. Phys. F.: Metal Phys. 6, 167 (1976)

/2/ A. Magerl, J. J. Rush, J. M. Rowe, D. Richter, H. Wipf, Phys. Rev. B 27, 927 (1983)

/3/ H. Wipf, A. Magerl, S.M. Shapiro, S. K. Satija, W. Thomlinson, Phys. Rev. Lett. 46, 947 (1981)

/4/ G. Cannelli, R. Cantelli, G. Vertechi, J. Less-Common Met. 88, 335 (1982)

/5/ H. Wipf, K. Neumaier, in Electronic Structure and Properties of Hydrogen in Metals, eds. P. Jena, C. B. Satterthwaite, Plenum Publishing Corp., New York and London 1983, p. 485

/6/ G. Bellessa, J. Physique (Lett.) 44, L387 (1983)

/7/ H. Wipf, K. Neumaier, A. Magerl, A. Heidemann, W. Stirling, J. Less-Common Met. 101, 317 (1984)

/8/ D. B. Poker, G. G. Setser, A. V. Granato, H. K. Birnbaum, Phys. Rev. B 29, 622 (1984)

/9/ H. Wipf, K. Neumaier, Phys. Rev. Lett. 52, 1308 (1984)

/10/ J. L. Wang, G. Weiss, H. Wipf, A. Magerl, in Phonon Scattering in Condensed Matter, eds. W. Eisenmenger, K. Lassmann, S. Dottinger, Springer-Verlag, Berlin-Heidelberg-New York 1984, p. 401

/11/ F. K. Huang, A. V. Granato, H. K. Birnbaum, Phys. Rev. B 32, 2178 (1985)

/12/ A. Magerl, A. J. Dianoux, H. Wipf, K. Neumaier, I. S. Anderson, Phys. Rev. Lett., in print

/13/ H. Wipf, D. Steinbinder, K. Neumaier, P. Gutsmiedl, A. Magerl, A. J. Dianoux, to be published

Jahn-Teller Relaxation Effects in the Inelastic Neutron Scattering Spectra of Hydrogen Metal Systems

A. Klamt

Max-Planck-Institut für Metallforschung, Institut für Physik,
Heisenbergstr. 1, Postfach 800665, D-7000 Stuttgart 80, F.R.G.

Abstract: Jahn-Teller relaxations due to degenerate excited states of hydrogen in metals are shown to be of considerable importance for a consistent interpretation of the optical mode part of the inelastic neutron scattering spectra of hydrogen - metal systems.

1. Introduction

The local-mode part of the inelastic neutron scattering (INS) spectra of hydrogen-metal systems usually is analysed in terms of a model that treats the hydrogen nuclei as isolated oscillators. The potential well seen by a hydrogen nucleus is formed by the interaction with the metal atom cores surrounding the occupied interstitial site. The motion of the lattice is described by harmonic vibrations around the equilibrium configuration, and any dependence of the lattice motion on the vibrational state of the hydrogen nucleus is neglected. The positions of the local-mode peaks in the INS spectra correspond with the vibrational quanta of the hydrogen nucleus in its potential well. From them informations on the shape of this well may be deduced [1].

Model calculations on hydrogen in bcc metals [2,3,4] have shown that in these systems the situation is more complicated. The equilibrium configuration of the nearest neighbour atoms appears to depend considerably on the vibrational state of the hydrogen nucleus. Especially the equilibrium configurations of the two first harmonics, which are degenerate due to the tetragonality of the tetrahedral interstices in bcc lattices, differ strongly from the groundstate equilibrium configuration. Hence the excitation of these modes will be connected with a relaxation of the lattice.

In section 2 it is shown that this relaxation can be understood as a Jahn-Teller (JT) relaxation [5] caused by the degeneracy of the excited hydrogen states. Due to this insight, possible JT effects also are considered for other hydrogen-metal systems. In section 3 the consequences of the JT relaxation to the INS spectra are discussed and compared with experimental data. This article can only report some main ideas. A brief discussion of the subject has been given in two recent papers [6,7]. For a detailed introduction to the theory of the JT effect, the reader is referred to the books of Englman [8] and Bersuker [9] .

2. The Jahn-Teller Effect of Excited States of Hydrogen in Metals

The theory of the JT effect has been developed to describe the behaviour of molecules and crystals in which some localized electrons occupy degenerate states. Due to this degeneracy the electronic states are mixed by the vibrations of the atom cores surrounding the electron, and hence the electronic and vibrational motions are coupled. Although analytical solutions of JT problems can only be given in rare cases, asymptotic and numerical solutions have been obtained for the JT problems of most importance [8,9]. Jahn and Teller [5] proved the general theorem that the occupation of a degenerate electronic state in a molecule or crystal always causes a relaxation which removes the degeneracy, with the only exception of Kramers degeneracies. The lattice modes which contribute to the relaxation are said to be JT active, the energy gain due to the relaxation is called JT energy E_{JT}.

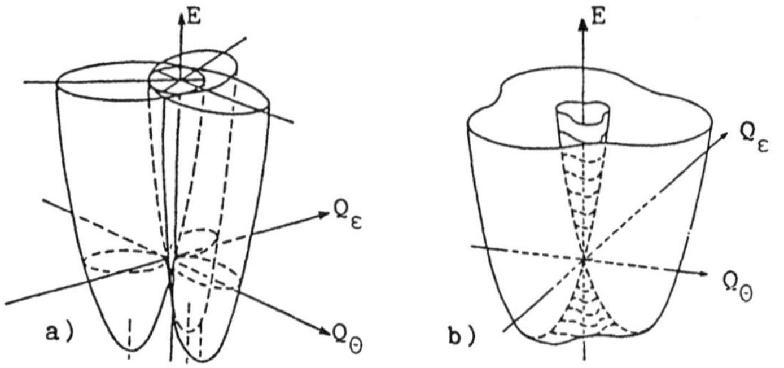

Fig. 1: Adiabatic potentials for a) the T-e problem (adiabatic) and b) the E-e problem (non-adiabatic)(from I.B. Bersuker [9])

In the theory of the JT effects the JT problems are classified by group theory. The classes are characterized by the group theoretical symbol of the degenerate states and by the JT active modes coupling to them. For example, T-(e+t) denotes an electronic triplet T, coupling to a set e of two-fold degenerate modes and a set t of three-fold degenerate modes.

In large molecules and in crystals the number of JT active modes usually is rather large. Fortunately the main features of such multimode problems equal those of the so-called ideal problems, which only take into account a few atoms near the JT center. Thus the great number of JT problems may be approximated by a few ideal problems, which have been investigated intensively. In some of the ideal problems the vibrational-electronic (vibronic) Hamiltonian may be diagonalized exactly by an appropriate choice of the basis functions of the degenerate multiplet. These problems are called adiabatic. The corresponding adiabatic potential sheets $\varepsilon^{\nu}(\underline{Q})$, which are given by the electronic energy levels at a fixed configuration $\underline{Q}$, are just f identical, but displaced, paraboloids, where f is the original electronic degeneracy. For the adiabatic T-e problem the adiabatic potentials are illustrated in Fig. 1a. The total energy levels of adiabatic problems are the vibrational states in the paraboloids. Since equivalent vibrational states in the f paraboloids are degenerate, the JT effect does not induce any additional splitting for adiabatic problems. For non-adiabatic problems the adiabatic potentials are more complicated (Fig. 1b). The vibronic equations can no longer be solved exactly. In the cases of weak and strong coupling, where E_{JT} is much less and much greater than a vibrational quantum of the coupling modes, respectively, analytical approximations are available. In the weak coupling limit the vibronic interactions are treated as perturbation, while in the strong coupling limit vibrations around the JT minima become quasi-stationary solutions, and transitions between these minima may be taken into account by perturbation theory. Additional splitting of the total energy levels occurs in the full range of JT coupling strength.

The entire theory of the JT effect is based on group theory and on the fact that the motion of the electrons is fast compared to that of the atom cores. No further properties of the electrons are used. Hence it may as well be applied to other light particles in matrices of heavy atom cores, e.g. hydrogen nuclei in metals. Surprisingly, JT effects due to degenerate states of hydrogen nuclei in metals have not been discussed so far. This may be due to the fact that the ground state of hydrogen nuclei in the interstices of metals always is non-degenerate and that hence JT effects are of minor importance in thermal equilibrium. Yet JT effects due to degenerate excited states may influence spectroscopic experiments considerably.

Now let us first turn to the above-mentioned systems of hydrogen nuclei occupying tetrahedral sites in bcc metals. Due to the tetragonality of the T-site in the bcc structure two of the three lowest excited states are degenerate. The corresponding ideal JT problem is $E-(b_1+b_2)$, where b_1 and b_2 are two non-degenerate modes coupling to different linear combinations of the E-states. Since the coupling to one of the b-modes dominates, the problem reduced to E-b, which is an adiabatic JT problem. Thus the

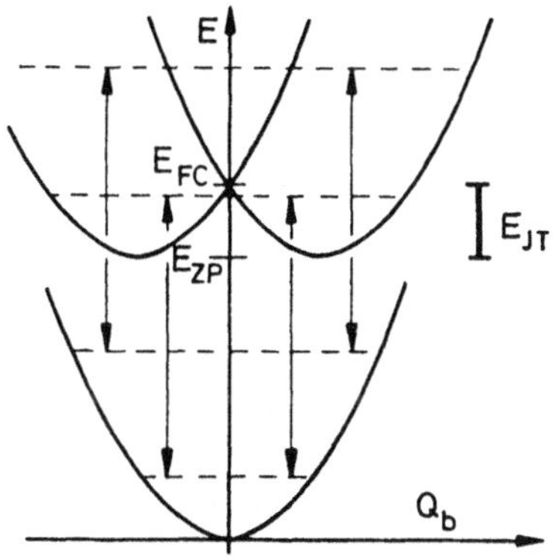

Fig.2: A→E transitions in the case of weak E-b coupling

adiabatic potentials with respect to this b-mode are two identical but displaced parabolas, intersecting at $Q_b = 0$, which corresponds to the tetragonal symmetry (Fig. 2). But this is exactly the situation found by the model calculations on hydrogen in bcc metals. From these we know that the relaxation energy is of the order of 30 meV for H in Nb and Ta, which is 20% of the corresponding excitation energy. Therefore we have to realize that JT relaxations due to degenerate excited states of hydrogen in metals may be of non-negligible magnitude. Hence it is useful to look for possible JT relaxations even in other systems.

The octahedral site in bcc structures has tetragonal point symmetry, too. The ideal problem corresponding to the lowest excited doublet again is E-(b_1+b_2) and may be reduced to E-b. Thus the situation is very similar to that of tetrahedral site occupation in bcc metals.

In fcc metals octahedral and tetrahedral interstices, the likely sites of hydrogen nuclei, possess cubic symmetry. Hence the lowest excitation is threefold degenerate. The corresponding ideal problem is T-(e+t). The coupling to the e-modes can be assumed to be much weaker than the t-mode coupling in both cases [7]. Thus the problem is T-t. This is non-adiabatic and hence more complicated than the E-b problem. The total energy levels of the T-t problem have been calculated numerically by Caner et al. [10].

In some ordered hydride phases bcc and fcc structures are slightly distorted. As long as the corresponding splitting of the degenerate states is much smaller than the JT energy, the JT effect may even appear in these systems.

The interstitial sites of hcp lattices are of low symmetry in general, but their first coordination shells are regular tetrahedrons and octahedrons, respectively, if the c/a-ratio is ideal. Therefore in hcp metals the T-t JT problem may be realized approximately.

3. Consequences to the INS Spectra

Inelastic scattering of neutrons by hydrogen-metal systems causes transitions between the groundstates of the hydrogen nuclei and excited states. As long as any JT coupling is neglected the interpretation of the corresponding INS spectra is very simple. Since the phonon system is not effected by the transitions of the hydrogen nuclei, the energy losses of the neutrons just correspond to the bare excitation energies of the hydrogen nuclei. Hence a degenerate excited state gives rise to a single peak in the INS spectrum, positioned at the excitation energy in the groundstate equilibrium configuration. This energy will be called Franck-Condon energy E_{FC}, since it is just the peak position proposed by the semi-classical Franck-Condon principle even in the case of nonvanishing JT relaxation.

From the theory of electronic transitions in colour centers it is well-known that for weak and moderate JT coupling ($E_{JT} \lessgtr \hbar\omega$) the dominant feature in the corresponding optical spectrum is the zero-phonon line, caused by transitions between equivalent phonon states of the groundstate and the excited state adiabatic potentials [11]. This result even holds for the INS spectra of hydrogen metal systems [6]. The zero-phonon line is centred at $E_{ZP} = E_{FC} - E_{JT}$. If the final vibrational states are split by the JT effect the zero-phonon line gets internal structure.

Due to the adiabaticity of the E-b JT problem the phonon states are not split by the JT effect for hydrogen in bcc metals. In agreement with the experiments a single peak due to transitions from the groundstate to the E-states is therefore expected in the INS spectra of these systems, but its position now has to be interpreted as zero-phonon energy. The corresponding Franck-Condon energy, which is required for a calculation of the shape of the potential well seen by the hydrogen nuclei in the groundstate equilibrium configuration, thus differs from the INS peak position by the JT energy (Fig.2), which is of the order of 30 meV for H in Nb and Ta [3,4]. Thus we have to realize that hitherto all calculations of the potential wells of these systems are based on a wrong interpretation of the INS peaks, since JT relaxations are neglected.

The interpretation of the INS peaks as zero-phonon lines is consistent with the fact that so far Stokes' shifts between energy-gain and energy-loss spectra have not been observed [2]. The JT effect may even explain the frequently observed broadening of the degenerate local-mode peak in the INS spectra of bcc metals if quadratic JT coupling is taken into account. The broadening of the zero-phonon line due to quadratic coupling in multimode systems has been pointed out by Silsbee [12].

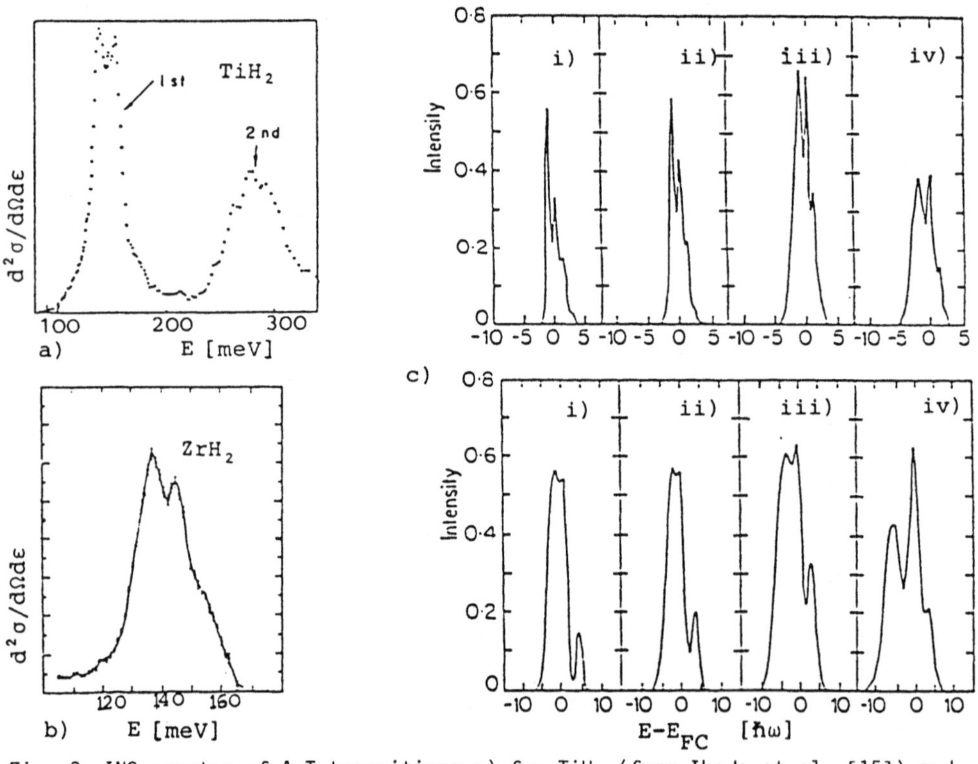

Fig. 3: INS spectra of A→T transitions a) for TiH_2 (from Ikeda et al. [15]) and b) for ZrH_2 (from Couch et al. [16]), both at T = 300 K. In c) computed spectra (from Englman et al. [13]) are shown for E_{JT} = 2/3 $\hbar\omega$ (upper row) and E_{JT} = 3,5 $\hbar\omega$ (lower row) and for four temperatures: i) T=0, ii) T=0.5, iii) T=1 and iv) T=4 (in units of $\hbar\omega/k_B$)

Englman et al. [13] computed the shape of the zero-phonon line due to an electronic A→T transition for different strengths of T-t coupling. Since they used the Condon approximation, their results may be applied to the INS spectra of hydrogen in fcc metals. A comparison with experimental spectra shows that the peak structures observed in the first optical mode peaks of TiH_2 and ZrH_2 may be explained by the T-t JT effect (Fig. 3). Thus the splitting is a natural consequence of the degeneracy and has not to be accounted for by additional assumptions as in the central-force calculations of Slaggie [14].

4. Summary

From the above considerations it should have become clear that the JT effect may be of considerable importance for the interpretation of the INS spectra of hydrogen metal systems. The local-mode peaks of degenerate states in bcc metals have to be interpreted as zero-phonon lines. Hence for the calculation of the shape of the groundstate potential well additional information on the JT energy is needed. Even for hydrogen in other metals a more careful interpretation of the spectra appears to be required.

References

1 J. Eckert, J.A. Goldstone, D. Tonks, D. Richter: Phys. Rev. B 27, 1980 (1983)
2 H. Sugimoto, Y. Fukai: J. Phys. Soc. Jap. 50, 3709 (1981)
3 A. Klamt: Diploma thesis, Universität Göttingen (1984)
4 A. Klamt, H. Teichler: submitted to phys. stat. sol. (b) (1985)
5 H.A. Jahn, E. Teller: Proc. R. Soc. London A 161, 220 (1937)
6 A. Klamt: submitted to J. Phys. F (1985)
7 A. Klamt: submitted to J. Phys. F (1985)
8 R. Englman: The Jahn–Teller Effect in Molecules and Crystals (Wiley, New York 1972)
9 I.B. Bersuker: The Jahn–Teller Effect and Vibronic Interactions in Modern Chemistry (Plenum Press, New York 1984)
10 M. Caner, R. Englman: J. Chem. Phys. 44, 4054 (1966)
11 M. Lax: J. Chem. Phys. 20, 1752 (1952)
12 R.H. Silsbee: Phys. Rev. 128, 1726 (1962)
13 R. Englman, M. Caner, S. Toaff: J. Phys. Soc. Jap. 29, 306 (1970)
14 E.L. Slaggie: J. Phys. Chem. Sol. 29, 923 (1968)
15 S. Ikeda, N. Watanabe, K. Kai: Physica 120 B, 131 (1983)
16 J.G. Couch, O.K. Harling, L.C. Clune: Phys. Rev. B 4, 2675 (1971)

Localized Vibrations of Hydrogen in Metals

D. Richter[1], *R.Hempelmann*[2], *J.J. Rush*[3], *J.M. Rowe*[3], *J. Eckert*[4], *J.A. Goldstone*[4], *and D. Tonks*[4]

[1]Institut Laue-Langevin, F-38042 Grenoble, France
[2]Institut für Festkörperforschung, Kernforschungsanlage Jülich, Postfach 1913, D-5170 Jülich, F.R.G.
[3]National Bureau of Standards, Washington D.C., USA
[4]Los Alamos National Laboratory, Los Alamos, NM, USA

We report inelastic neutron scattering experiments on H and D in β-$TaH_{0.5}$ and $Nb_{99}Ti_1H(D)_1$. On $TaH_{0.5}$ higher harmonics of the hydrogen vibrations were found up to 400 meV. Qualitatively the observed anharmonicity agrees with recent ab initio frozen phonon calculations on NbH. Spectroscopic results on the impurity system NbTiH(D) allow for the first time a determination of the H-potential for a dilute metal-H- system. The observed anharmonicity is twice as large as in a corresponding hydride phase.

I. Introduction

Hydrogen dissolved in a metal takes part in two classes of vibrational motions: (i) at low frequencies the H moves according to the displacements of the host atoms and more or less mirrors its density of states (band modes), (ii) at high frequencies the H vibrates against its metal neighbours. Then, at low H-concentration and/or small direct H-H interactions the H can be considered as a 3-d Einstein oscillator moving in a cage formed by the slowly vibrating metal atoms. This contribution is entirely concerned with these local H vibrations. They reveal important information on metal-H systems: (i) Their investigation, including higher harmonics and isotope effects, yields the effective H-single particle potential in great detail. With the availability of first ab initio calculations - so far performed for NbH(1) - a comparison with theoretical concepts becomes meaningful, and an in-depth understanding of the H-Me forces may come in sight. (ii) The point symmetry of the H-site determines the degeneracy of the local H-modes: e.g. two different fundamentals with a universal sequence of degeneracy exist for H-sites on tetrahedral (T) or octahedral (O) sites in bcc structures. The relation between point symmetry and local vibrations has been used successfully in order to perform site assignments by neutron spectroscopy (2,3). (iii) Since the metal-H interaction is short range, the H is very susceptible to distortions of its neighbourhood and thus constitutes a sensitive local probe in order to study changes in the proton environment (3,4).

In this paper we report new data on H excitations in the hydride $TaH_{0.5}$ and in the impurity systems $Nb_{99}Ti_1H(D)_1$. In the hydride it was possible to observe higher harmonics up to 400 meV leading to substantial improvements in the determination of the H-potential. The measurements on the trapped protons in $Nb_{99}Ti_1H(D)_1$ revealed the first observations of higher harmonics in dilute bcc-metal-H-systems.

II. Theoretical Considerations

The double differential neutron cross-section $\partial^2\sigma/\partial\omega\partial\Omega$ for an harmonic 3-d oscillator can be given analytically and has the form (5):

$$\frac{\partial^2\sigma}{\partial\omega\partial\Omega} = \frac{\sigma^{tot}}{4\pi}\frac{k_f}{k_i}\exp[-2W(\underline{Q})]\exp[\tfrac{1}{2}\hbar\omega\beta] \; * \qquad (1)$$

$$\sum_{\substack{n= \\ m=-\infty \\ \ell=}}^{+\infty} I_n(y_x)I_m(y_y)I_\ell(y_z)\delta(h\omega-\ell h\omega_x-mh\omega_y-nh\omega_z)$$

with $$y_i = \frac{hQ^2}{6M\omega_i} \operatorname{csch}(\tfrac{1}{2}h\omega_i\beta)$$

where σ_{tot} is the total cross-section, M the H(D) mass $\beta = 1/kT$ and $I_n(y)$ is the modified Bessel function of the first ind. For intensity calculations we remark that (i) expansions of the Bessel function are often sufficient; (ii) in general the anisotropic Debye-Waller factor W(Q) has to be taken into account; and (iii) for polycristalline averages it is important to perform joint averages of the DBW-factor and the Bessel function.

Now we consider small anharmonic disturbances of the H-potential on a T-site in a bcc-metal (point group $D_{2d}(x=y\neq z)$). The possible cubic and quartic terms under this symmetry can be found by group theory and have the form (6)

$$H_1 = ez(x^2-y^2) + C_{4z}z^4 + fx^2y^2 + gz^2(x^2+y^2) + C_{4x}(x^4+y^4) \tag{2}$$

The vibrational corrections due to the third and fourth order terms are calculated in second and first order perturbation theory respectively and yield excitation energies $e_{\ell mn} = E_{\ell mn} - E_{000}$, where $E_{\ell mn}$ is the perturbed energy of the vibrational level corresponding to the (ℓ,m,n) harmonic state. The prevailing anharmonicity parameters are (i) the anharmonic disturbances within a given well β_x and β_z and the anharmonic coupling between the vibrations in different directions ζ_{xz} and ζ_{xy}. Using the notation of ref. (6) they can be written as

$$\beta_z = \frac{3C_{4z}\hbar^2}{M^2\omega_z^2}\,; \quad \zeta_{xz} = \frac{g\,h^2}{M^2\omega_x\omega_z} - \lambda' - \lambda''$$

$$\beta_x = \frac{3C_{4x}\hbar^2}{M^2\omega_x^2} - \frac{1}{2}\lambda'' + \frac{1}{2}\lambda' - 2\lambda \tag{3}$$

$$\zeta_{xy} = \frac{f\hbar^2}{M^2\omega_x^2} + 2\lambda$$

The λ-parameters are the second order perturbations resulting from the third order term in the Hamiltonian. Using the above parameters, the anharmonic correction for an excitation $e(\ell, m, n)$ is given as:

$$\Delta e(\ell, m, n) = \frac{1}{2}\beta_z\, n(n+1) + \frac{1}{2}\beta_x\,\{m(m+1) + \ell(\ell+1)\} \tag{4}$$

$$+ \frac{1}{2}\zeta_{xy}[2\ell m + \ell + m] + \frac{1}{2}\zeta_{xz}[(2n+1)(\ell+m) + 2n]$$

Since the anharmonic corrections scale with 1/M while ω_i scales with $1/\sqrt{M}$, the anharmonicity parameters also cause deviations from the harmonic isotope effect. For the z-vibration e.g. we get:

$$\frac{e^H_{001}}{e^D_{001}} = \sqrt{2}\,\frac{e^H_{001}}{e^H_{001} - (1-\sqrt{2})(\beta_z+\zeta_{xz})} \tag{5}$$

Deviations from the cubic crystal symmetry due to hydride formation cause slight deviations from the tetragonal site symmetry. Therefore, Eqn. (2) - (5) are valid only approximately. However, since the splitting of the upper fundamentals due to deviations from tetragonal symmetry is small (7), its influence on the anharmonicity corrections is considered to be of higher order and neglected.

III. Experiments on $TaH_{0.46}$

The inelastic neutron scattering experiments were performed on the crystal-analyser spectrometer (6) at the pulsed spallation neutron source (WNR) at the Los Alamos National Laboratory. Using a graphite analyser in combination with a cooled Be-filter in time focussing geometry the instrument offers an energy resolution between 4% and 7% (FWHM) and a structureless background.

Before H-loading out of the gas phase the Ta sample was annealed in UHV at 2300 C for 24 h in order to remove interstitial impurities. The H-concentration c=46% was determined by weighting.

The TOF data were corrected for background and the intensities were normalized to the incident neutron spectrum. The experiment was carried out at 10 K. Fig. 1 represents the experimental results above 200 meV after transformation to energy scale. In this high energy regime well above the fundamentals occurring at 121.3 and 163.4 meV we observe at least 5 peaks ranging between 227 and 356 meV. At energies higher than that the intensity trails off and no further excitations were detected. The data were fitted with a superposition of 5 Gaussians in order to determine the peak positions. The resulting excitation energies are given in Fig. 1.

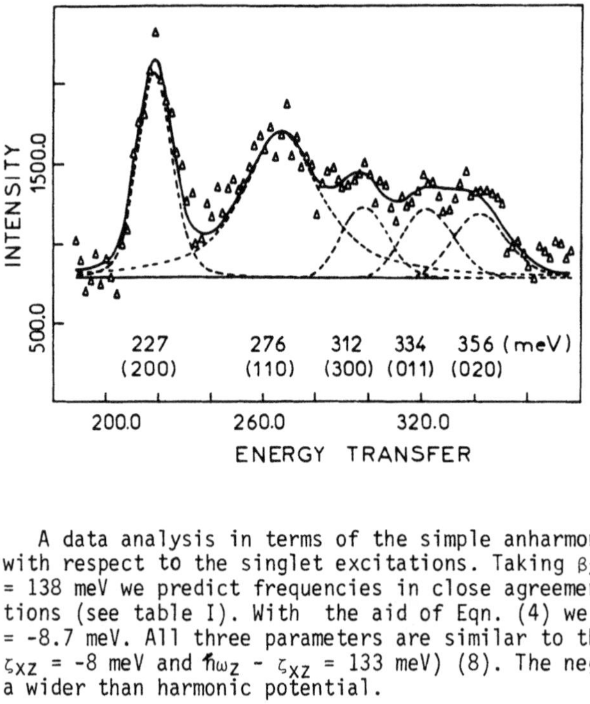

Fig. 1: Hydrogen excitations in $TaH_{0.46}$ above 200 meV.

A data analysis in terms of the simple anharmonic model presented above succeeds with respect to the singlet excitations. Taking $\beta_z = -16.5$ meV and $\hbar\omega_z - \zeta_{xz} = 138$ meV we predict frequencies in close agreement to the experimental observations (see table I). With the aid of Eqn. (4) we find $e_{110} - e_{010} - e_{100} = \zeta_{xz} = -8.7$ meV. All three parameters are similar to those found for Nb($\beta_z = -12$ meV; $\zeta_{xz} = -8$ meV and $\hbar\omega_z - \zeta_{xz} = 133$ meV) (8). The negative value of β_z indicates a wider than harmonic potential.

Table I: Vibrational frequencies in β-$TaH_{0.5}$ in [meV]

excitation	e^H_{001}	e^D_{001}	e^H_{002}	e^D_{002}	e^H_{003}	e^H_{001}
experiment	121.3	88	227	168	312	276
anharmonic model	121.5	89	226.5	170	314	276

In the case of the doublet the fundamentals are at 163.4 meV for the proton and at 119 meV for the deuteron (7). In contrast to the singlet case, both observed higher harmonics (334 meV, 356 meV) are above twice the fundamental frequency, indicating a steeper than harmonic potential. This observation agrees with recent frozen phonon calculations of the H-vibrations in NbH which revealed a downward shift of the second harmonic for the singlet and an upward shift for doublet (1). Other than for the singlet, however, an analysis with the anharmonic model appears to fail. It is not possible to understand the H and D vibrations in terms of a common set of parameters. The relatively poor quality of data in the high-frequency regime, however, advises caution in drawing further reaching conclusions, such as on isotope dependent potentials or on a proton Jahn-Teller effect (9). New measurements preferentially at small Q-values are called for.

IV. Experiments on $Nb_{99}Ti_1H(D)_1$

The experiments were carried out at the BT4 triple axis spectrometer at the NBS research reactor using Cu220 as a monochromator and a cold Be-filter as analyser. Applying 40' collimation before and after the monochromator the resolution varied between 6.5 meV at $\hbar\omega$ = 100 meV and 9.5 meV at 170 meV. The sample was produced by HF induction melting of MARZ-grade Nb of 99.99% nominal purity which was decarbonated and subsequently outgased in UHV and 99.999% pure Ti. The dopant concentration was controlled by atomic absorption spectroscopy and determined to 0.998 ± 0.021% Ti. H and D was loaded out of the gas phase. The gas content was monitored by weight increase. The achieved concentrations were C_H = 0.92% and C_D = 0.94% slightly below the Ti content.

Figs. 2(a) and (b) present experimental spectra from NbTi H(D) at 10 K. Fig. 3(a) and (b) display experimental results in the frequency region above the fundamental vibrations.

Table II shows the obtained frequencies and linewidths.

Table II: Vibrational frequencies in $Nb_{99}Ti_1H(D)$ at 10 K.

Sample	e_{001}[meV]	FWHM[meV]	e_{100}[meV]	FWHM[meV]	e_{002}[meV]
NbTiH	107 ± 0.5	8 ± 1	153 ± 1	20 ± 2	188 ± 2
NbTiD	82.5 ± 0.5	6 ± 1	116 ± 1	16 ± 2	151 ± 2
cosine pot.	108; 81		----		194; 151

The observed frequencies are distinctly different from the hydride phase vibrations which occur at 122 (87) and 166 meV (121) in the proton (deuteron) case (8). Thus Ti provides a strong trap for H in Nb and prevents hydride phase separation. The vibrational pattern resembles closely that of H in pure Nb and immediately shows that the protons occupy tetrahedral sites. The unusually broad second fundamental (for comparison Γ = 4.2 meV for $NbD_{0.85}$ at 10 K (10)) indicates an unresolved splitting of the doublet. Most likely the H occupies a T-site nearest neighbour to the Ti which then disturbes the tetragonal symmetry. This situation is quite different to that for interstitial traps in Nb where the protons appear to occupy 4th nearest neighbour sites (4). While there trapping most likely is strain related, for H-Ti clusters in Nb strain plays only a minor role ($\Delta a/a|_{\%} = 6.10^{-3}$ for N in Nb but $\Delta a/a|_{\%} = -0.32\cdot10^{-3}$ for Ti) and trapping is most likely an electronic effect. Recently Manninen et al (11) calculated binding energies for H at a Ti impurity in Nb in terms of an effective medium theory. Their value at 110 meV is in the right order to explain the suppression of the hydride phase, the enthalpy of formation of which is 120 meV (12).

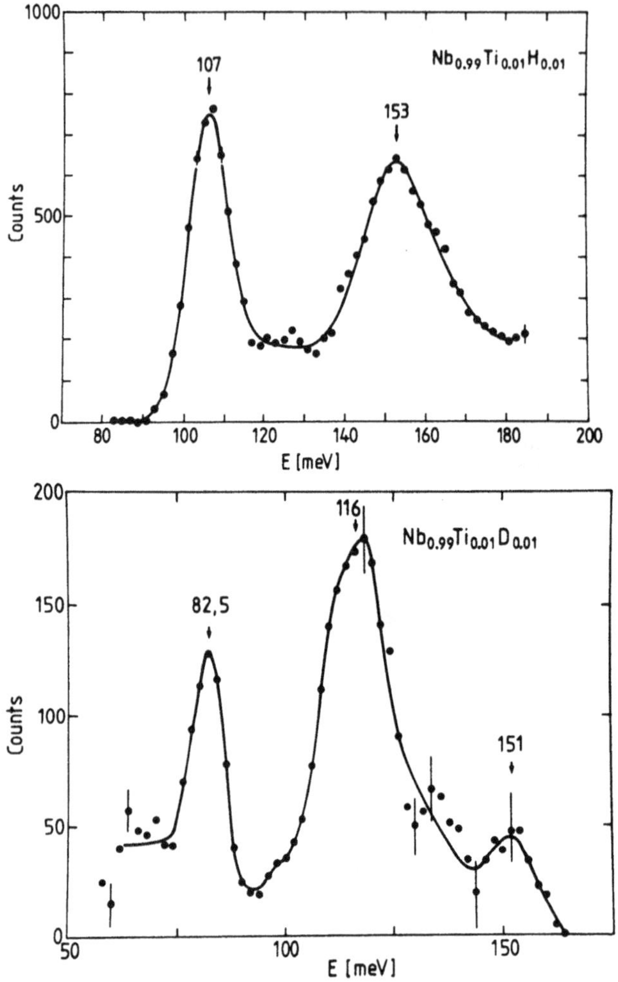

Fig. 2a + b: Local mode spectrum of (a) H and (b) D dissolved in $Nb_{0.99}Ti_{0.01}$

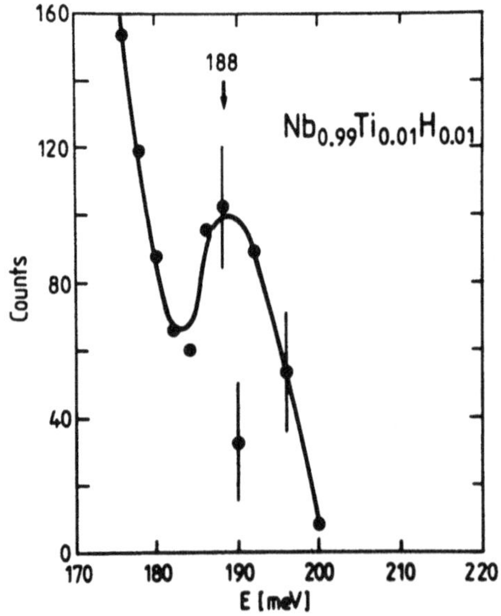

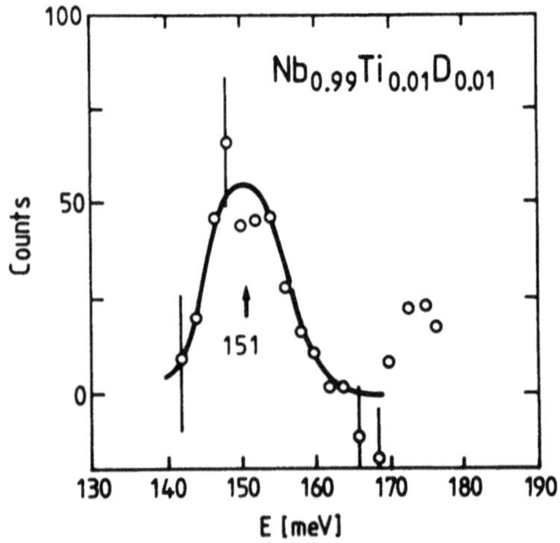

Fig. 3a + b: Second harmonic of the lower fundamental of (a) H and (b) D in $Nb_{.99}Ti_{0.01}$

The anharmonicity of the singlet excitation can be evaluated in terms of the anharmonic model of section II. The shifts of the second harmonic from their harmonic positions reveal $\beta_z^H = -26 \pm 3$ for the proton and $\beta_z^D = -14 \pm 3$ for the deuteron. Within the experimental error they fulfil the theoretical relation $\beta_z^H = 2\beta_z^D$. Using the isotope shift Eqn. (5) leads to $\zeta_{xz} = -7 \pm 5$. While the anharmonic coupling ζ_{xz} is practically identical to the value in the different Niobium hydrides (ζ_{xz} = -6...-8 meV(8)), the anharmonicity in z-direction is twice as big (β_z^H = -9 .....-13 meV) in the dilute system. Its large value renders the application of the low order perturbational approach questionable. Therefore, alternatively, we have studied a simple cosine potential $V = V_0(1-\cos\frac{2\pi}{d}x)$(d is the distance between T-sites) for which the oscillation energies are known to high order perturbation theory (13). For V_0 = 260 meV table II compares the predicted frequencies with the experimental observations. With the exception of e^H_{002} which is very close to the top of the potential (E^H_{002} = 253 meV) and presumably is shifted further downwards than revealed by the treatment of Das, we find very good agreement with the 1-parameter model.

V. Conclusion and Outlook

The experimental results presented here demonstrate the capabilities as well as the limitations of high-energy neutron spectroscopy applying filter methods where the momentum transfer rises ~ $\sqrt{\hbar\omega}$. In the case of TaH_x the increasingly larger Debye Waller factor in connection with multiphonon contributions limits the range of observation to below 400 meV. The data on $Nb_{99}Ti_1D_1$ taken on an optimally designed filter set-up present the limit with respect to signal to noise. Future progress in both directions is anticipated for small Q-TOF measurements at spallation sources, where background and multiphonon contributions can be kept significantly smaller. Higher resolution at the same time will allow a better separation of high-frequency excitations.

With the present experiments, new information on the H-potential could be obtained in both cases. (i) While the H-potential for the singlet is wider than harmonic, the doublet vibrations experience a steeper than harmonic potential. Qualitatively, the observed anharmonicities agree with ab initio calculations on NbH. (ii) In the case of $Nb_{99}Ti_1(H)D_1$ a first determination of the singlet H potential in a dilute system was possible. Its anharmonicity is distinctly larger than that found in hydride phases.

References

(1) K.M. Ho, H.J. Tao, X.Y. Zhu, Phys. Rev. Lett. 53, 1586 (1983)
(2) D. Richter, J.J. Rush, J.M. Rowe, Phys. Rev. B27, 6227 (1983)
(3) R. Hempelmann, D. Richter, G. Eckold, J.J. Rush, J.M. Rowe, M. Montoya, J. Less Common Metals 104, 1 (1985)
(4) A. Magerl, J.J. Rush, J.M. Rowe, D. Richter, H. Wipf, Phys. Rev. B27, 927 (1983)
(5) W. Marshall, S.W. Lovesey, Theory of Thermal Neutron Scattering (Clarendon, Oxford 1971)
(6) J. Eckert, J.A. Goldstone, D. Tonks, D. Richter, Phys. Rev. B27, 1980 (1983)
(7) R. Hempelmann, D. Richter, A. Kollmar, Z. Phys. B44, 159 (1981)
(8) D. Richter in Proceedings of the 1984 workshop on High-Energy Excitations in Condensed Matter, Los Alamos, New Mexico, Feb. 13-15 (1984), Conf. LA 10227
(9) A. Klamt, this conference
(10) D. Richter, S.M. Shapiro, Phys. Rev. B22, 599 (1980)
(11) M. Manninen, M.J. Puska, R.M. Nieminen, P. Jena, Phys. Rev. B30, 1065 (1984)
(12) G. Pfeiffer, H. Wipf, J. Phys. F6, 167, (1976)
(13) T.P. Das, J. Chem. Phys. 25, 896 (1956)

One-, Two- and Three-Dimensional Disorder of D-Atoms in Non-Stoichiometric TaD_x

*J. Hauck and K. Parlinski**

Institut für Festkörperforschung, Kernforschungsanlage Jülich,
Postfach 1913, D-5170 Jülich, F. R. G.
*Permanent address: Institute of Nuclear Physics, ul. Radzikowskiego 152,
PL-31-342 Krakow, Poland

Introduction

The bcc tantalum metal can absorb H_2 or D_2 gas to a ratio H(D)/Ta $\sim$ 1. Tetrahedral sites are occupied randomly at temperatures $\sim$ 300 K [1,2]. Below this temperature the H or D atoms order in β-$TaD_{0.5}$ and δ-TaD crystal structures with the intermediate non-stoichiometric solid solutions β-$TaD_{0.5+x}$ and δ-TaD_{1-x}. The phase boundary between the two structures is either a second order phase transition at the approximate composition $TaD_{0.75}$ [1] or a miscibility gap with a first order phase transition [2]. Upon further cooling to $\sim$208 K a new ordered ζ phase is formed at composition $TaD_{0.75-0.80}$ [1,2]. In this investigation the phase transitions from α to ζ phase at approximate composition $TaD_{0.75}$ will be considered. Based on the close relationship between the three ordered structures and on the large disorder of D atoms in β and δ phase, one can deduct an atomistic diffusion mechanism of D atoms between (0,4) fourth and (0,6) sixth nearest neighbor positions with translations T_4(a/2, a/2, 0) and T_6(a/2, a/2, a/2) of bcc lattice with lattice constant a. This mechanism should also be valid for α-phase at high D content, if thermodynamic data are considered. The calculated configurational entropy agrees with experimental values, if a blocking of first, second and third nearest neighbor positions is considered [1,3]. A group theoretical analysis of the crystal structures shows that second order phase transitions are possible for $\beta \rightarrow \delta$, $\beta \rightarrow \zeta$ and $\delta \rightarrow \zeta$ phase transitions. It will be argued that the mechanisms of all phase transitions relay on the diffusion process of deuterium.

The suggested diffusion mechanism is different from the diffusion in α-phase at low H content [4,5]. Inelastic neutron scattering and NMR data on β-$NbH_{0.9}$ [6], which is isomorphous to δ-TaD_x indicate an increased activation energy U = 0.23 eV, a reduced jump frequency of $\sim 10^8$ s^{-1} and an increased jump distance. The NMR data on $NbD_{0.78}$ sensitive to the electric field gradient at a site suggest that the D atoms jump only between one type of lattice sites [5,6].

Crystal structure of β-$TaD_{0.5}$, δ-TaD and ζ-$TaD_{0.75}$

The number of tetrahedral interstices in bcc Ta exceeds the number of Ta atoms by a factor of 6, so that only 12.5 % of the tetrahedral sites are occupied by D atoms at composition $TaD_{0.75}$. The positions of the D atoms for β-, δ - and ζ-TaD_x are shown in fig. 1 and the crystal structure data as obtained by neutron scattering [1,7] are listed in table 1. The three ordered structures exhibit D atoms only at fourth and sixth nearest neighbor positions with translations T_4(a/2, a/2, 0) and T_6(a/2, a/2, a/2). The crystal structures of non-stoichiometric β- and δ-phases at composition $TaD_{0.75}$ are not well known. Based on the structural relationship of the three phases, one can predict the kind of disorder by a simple structural model. This structural model, which was obtained from a large number of interstitial alloys [8], predicts an equal distribution of interstitial atoms within the metal matrix and a repulsive interaction between the interstitial atoms. In β-$TaD_{0.5}$ each Ta atom exhibits two D atoms as nearest neighbors, in ζ-$TaD_{0.75}$ three and in δ-TaD four neighbours. The repulsive interactions between D atoms are caused by the charge transfer $Ta^{\delta+}$ and $D^{\delta-}$ between Ta and D and the repulsive Coulomb interaction between $D^{\delta-}$. The Madelung part of the lattice energy of $Ta^{\delta+}D^{\delta-}$ with D atoms in different interstices of first, second, third,

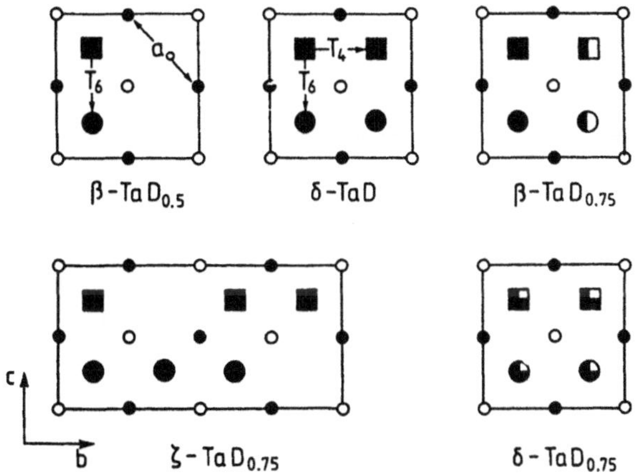

Fig. 1. The ordering of D atoms in β-TaD$_{0.5}$, δ-TaD, ζ-TaD$_{0.75}$, β-TaD$_{0.75}$ and δ-TaD$_{0.75}$ is indicated by the projection height of the atoms: ■, ◧ filled or half-filled D position at x = 0.75; ●, ◐ D at x = 0.25; • Ta at x = 0; o Ta at x = 0.5

Table 1: Space group, lattice parameters of unit cell compared to a_0 of bcc metal lattice and positional parameters of α, β, δ and ζ-TaD$_x$

structure	cell	space group	atom positions
α-TaD$_x$	$a \approx a_0$	$Im\bar{3}m$ (229)	Ta at (2a) 000 D at (12d) $\frac{1}{4}\ 0\ \frac{1}{2}$
β-TaD$_{0.5}$	$a \approx a_0$, $b \approx c \approx a_0\sqrt{2}$	C222 (21)	Ta at (4K) $\frac{1}{4}\ \frac{1}{4}\ z$, $z \approx \frac{1}{4}$ D at (2a) 000
δ-TaD	$a \approx a_0$, $b \approx c \approx a_0\sqrt{2}$	Cccm (66)	Ta at (4e) $\frac{1}{4}\ \frac{1}{4}\ 0$ D at (4a) $0\ 0\ \frac{1}{4}$
ζ-TaD$_{0.75}$	$a \approx a_0$, $b \approx 2\,a_0\sqrt{2}$, $c \approx a_0\sqrt{2}$	I222 (23)	Ta at (8 K) $\sim\frac{1}{4}\ \frac{1}{8}\ \frac{3}{4}$ D at (4 g) 0y0, $y \approx \frac{1}{4}$ D at (2a) 0 0 0

fourth and sixth nearest neighbor positions increases to a maximum value, if the D atoms are as far apart as possible (fig. 2). An occupation of positions further apart is impossible for the density of D atoms at composition TaD. The Madelung energy E_M is proportional to the Madelung factor MF, which is independent of the charge δ which is not determined experimentally. The lattice energy of δ-TaD with fourth and sixth nearest neighbors increases to a maximum if the lattice is slightly distorted with b/c ∿1.01 (fig. 1), so that the D atoms at T_4 translations are further apart. Calculations [8] at composition TaD$_{0.75}$ exhibit a stronger distortion for δ- than for ζ- or β-TaD$_{0.75}$ because of the larger fraction of D atoms at T_4 translations in δ-TaD$_{0.75}$. The maximum Madelung energies of β- and δ-TaD$_{0.75}$ are smaller by 6.5 and 9.3 % than the value of ζ-TaD$_{0.75}$, so that the structures should be stabilized by the additional entropy of disorder [8].

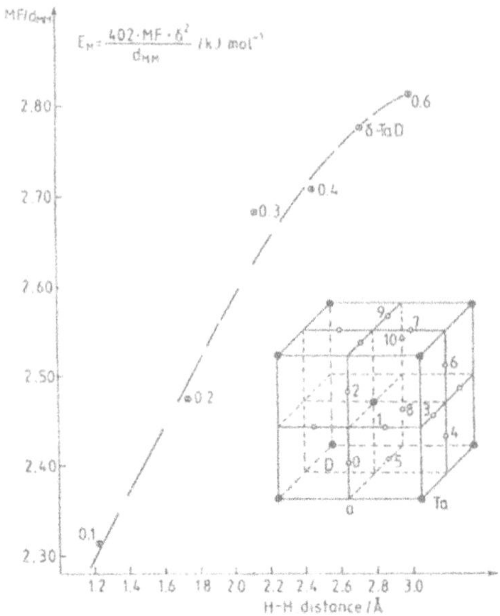

Fig. 2. Madelung factors MF for occupation of different interstitial sites 0,1; 0,2; 0,3; 0,4 or 0,6.

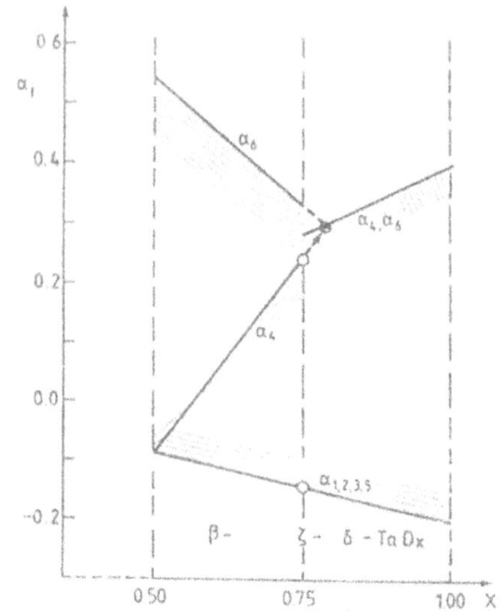

Fig. 3. Cowley short-range order parameters α_i of i = 1-6 nearest neighbor D positions in β-Ta$D_{0.5+x}$, δ-TaD_{1-x} and ζ-Ta$D_{0.75}$ (o). The hatched area indicates the values for further disorder at elevated temperatures.

Disorder and atomistic diffusion process in non-stoichiometric β- and δ-Ta$D_{0.75}$

In δ-TaD_x at composition x = 0.75 25 % of D positions are vacant. The site symmetry of the crystal structure suggests an occupation of the (4a) position (table 1) by 75 % (fig. 1). At the high fraction of vacancies the D atoms of a single domain can jump on T_4 translations in b-direction and on T_6 translations in the (a,c)-plane, which is perpendicular to this direction, so that all D-sites are occupied with equal probability.

At lower D contents the D atoms are concentrated in every second plane with T_6 translations in β-Ta$D_{0.5}$ structure (fig. 1), while T_4 translations in b-direction are omitted. The excess D atoms (x < 0.25) of β-Ta$D_{0.5+x}$ are distributed on (2d) positions [1] of the intermediate planes with a 50 % occupancy at composition Ta$D_{0.75}$ (fig. 1). These D atoms are two-dimensional disordered and can jump by T_6 translations in (a,c)-plane or they can exchange with atoms of the filled planes by T_4-translations.

Below 208 K the disordered D atoms of (2d) position of β-Ta$D_{0.75}$ are ordered on alternating occupied and vacant positions in ζ-Ta$D_{0.75}$ (Ta_4D_3) structure (fig. 1). In this ordered structure the diffusion should be strongly reduced because of the smaller amount of disorder. The jump process of T_6 translations should be smaller because of the larger distance between D atom positions in the half-filled (a,c) planes.

The atomistic jump process can be related to the number of D atoms at T_4 and T_6 translations within a single domain and the occupation probability of these positions. The number of fourth and sixth nearest neighbors $[T_4]$ and $[T_6]$ can be correlated to the Cowley short-range-order parameter:

$$\alpha_i = 1 - \frac{p_i^V}{x_V} \qquad \text{with} \qquad p_i^V = \frac{n_i - [T_i]}{n_i}$$

which depends on the probability p_i^v to find a vacancy at translation T_i for composition TaD_x, where $x_v = \frac{6-x}{6}$ tetrahedral positions are vacant. The p_i^v-values of ordered or partially ordered structures can be obtained from the number of D atoms $[T_i]$ compared to the total number of all available sites n_i, where n_i = 4, 2, 8, 4, 4, 8 for i = 1-6, respectively.

The α_i values for i = 1, 2, 3, 5 change continuously within the range $0.5 \leq x \leq 1$ of TaD_x while α_4 and α_6 exhibit a discontinuity at x = 0.75 (fig. 3). The values would be continuous for a triple point of β, δ, ζ-phase transition at x = 0.79 as outlined by the dotted lines. A ζ-phase with x = 0.79 would exhibit an excess of 4 % D atoms in the half-filled (a,c) plane at (2d) position. This composition of ζ-phase is close to the value x = 0.80 determined by SCHOBER and WENZL [2]. Another possibility for continuous variation of α_4 and α_6 would be at additonal disorder which decreases the α_6 values of β-TaD_x and the $\alpha_{4,6}$ values of δ-TaD_x within the hatched area of fig. 3. For a continuous phase transition the occupation of the various D positions in β- and ζ-$TaD_{0.75}$ should decrease or increase, respectively, to equal values like the 75 % in δ-$TaD_{0.75}$.

The disordered D atoms at T_1, T_2, T_3 and T_5 positions reduces the size of the single domains of β- and δ-TaD_x, which are oriented in different directions depending on the T_i translations. At the β → α or δ → α phase transitions the α_i values change discontinuously in a step towards $\alpha_i = 0$ for a statistical distribution of D atoms.

The measurement of the short-range-order in the similar α-VD_x system [9] and the determination of the configurational entropy [1,3] show a blocking of T_1, T_2 and T_3 positions, so that the structure of α-$TaD_{0.75}$ might still be described by very small units of β and δ-TaD_x structure, which are disordered in different directions. The atomistic diffusion process in T_4 and T_6 translations should be still of some importance in α-$TaD_{0.75}$ because of the preferential occupation of fourth nearest neighbor atoms and other positions further apart. The T_4 diffusion in α-$TaD_{0.75}$ is two-dimensional while T_6 translations can be threedimensional.

Group theoretical analysis of α, β, δ, ζ-TaD_x phase transitions

We have performed the group theoretical analysis of the phase transitions (PT) which allowed to establish the active irreducible representation (IR) of the order parameter [10] and accompanying secondary order parameters (SOP) [11]. In the next step of analysis we have written down the invariants of Landau free energy expansion and judged from the absence of the third order terms that a given PT can be of second order. It is well known that even in the absence of third order term the PT may become of the first order when coupling of the primary order parameter to the SOP is strong enough. This coupling leads to negative fourth order terms in the free energy expansion. Although there is no strict rule for that, the probability of having second order PT increases when no SOP exists. In each analysed case we have checked that the hydrogens occupying tetrahedral sites can form the active IR of the order parameter. Below, the IR are given in Kovalev notation [12]. To be consistent with a,b,c of Table 1 the a and c lattice constants have to be permuted.

The reduction of symmetry in the PT α → β is caused by simultaneous action of two IR τ^5 and τ^6, and is given by the diagram $Im\bar{3}m$ (Z = 1) → ($k_9 = (\pi/a, \pi/a, 0)$, $c_2 = c_3 = c_4 = c_5 = c_6 = 0$; $\tau^5 + \tau^6$) → C222 (Z = 2). The IR's τ^5 and τ^6 separately reduce the $Im\bar{3}m$ space group into Cccm (Z = 2) and Cmma (Z = 2), respectively, and their intersection Cccm ∩ Cmma = C222 form the resulting space group. Such PT is called triggered and must be of the first order. The PT α → δ is also experimentally first order [1,2] with symmetry reduction $Im\bar{3}m$ (Z = 1) → ($k_9 = (\pi/a, \pi/a, 0)$, $c_2 = c_3 = c_4 = c_5 = c_6 = 0$; τ^5) → Cccm (Z = 2). It is accompanied by two SOP's, namely $Im\bar{3}m$ → ($k_{11} = 0$, E_g, $c_1 = c_2$) → I4/mmm and $Im\bar{3}m$ → ($k_{11} = 0$; T_{2g}, $c_1 = c_2 = 0$) → Fmmm. Both representations E_g and T_{2g} characterize the shear components on the crystal $(2V_{zz} - V_{xx} - V_{yy})$, $(V_{xx} - V_{yy})$ and V_{xy}, respectively. The symmetry of δ-

TaD_x phase is further reduced at the PT $\delta \rightarrow \beta$ with symmetry lowering Cccm (Z = 2) $\rightarrow (k_{14} = 0, A_u) \rightarrow$ C222 (Z = 2). No SOP's occur and the invariants of the free energy expansion allows for second order PT. The A_u representation is not related with the shear. A similar situation occurs in the PT $\beta \rightarrow \zeta$, where the symmetry of the order parameter is given by the diagram C222 (Z = 2) $\rightarrow (k_{17} = (\pi/a, 0, 2\pi/c); \tau^1) \rightarrow$ I222 (Z = 4). Again the invariants allow for the second order PT and no SOP's occur. At last, the PT $\delta \rightarrow \zeta$ is described by the diagram Cccm (Z = 2) $\rightarrow (k_{17} = (\pi/a, 0, 2\pi/c); \tau^1, c_2 = 0) \rightarrow$ I222 (Z = 4). The free energy expansion allows for the second order PT and only one SOP which is not a strain may occur, namely Cccm $\rightarrow (k_{14} = 0; A_u) \rightarrow$ C222.

Conclusion

The close structural relation between β-, δ- and ζ-TaD_x let suggest a one-dimensional jump process of D atoms with $T_4(a/2, a/2, 0)$ translations to fourth nearest neighbors and a two-dimensional jump process with $T_6(a/2, a/2, a/2)$ translations in a plane perpendicular to T_4-direction. The fraction of diffusing atoms can be correlated to the short-range order parameters α_i and the fraction of vacancies in the disordered lattice. The atomistic jumps in T_6-translations exhibit a minimum at x = 0.75 (or 0.79), while one-dimensional T_4 jumps should increase continuously with D-content if the phase transition $\beta \rightarrow \delta$ is second order. The distortion of the lattice b/c > 1 increases continuously [1] and can be correlated to the increasing repulsive interactions of fourth nearest $D^{\delta-}$ atoms at increasing α_4. The diffusion process on T_4 translations can be compared to the jump process of nearest neighbor H atoms in $TiH_{1.55}$ [5]. $TiH_{1.55}$ exhibits the fluorite structure with 23 % vacancies, which can be related to δ-$TaD_{0.75}$ (fig. 1), if the $\underline{a}$ lattice constant is increased by ~40 % and all H positions at x = 0.25 and x = 0.75 are occupied. The T_6 translations in bcc lattice would be second-nearest neighbor positions in fcc fluorite lattice. The atomistic jump process to positions further apart like the translations T_8(a 0 0) or T_{10}(a, a/2, 0) are less probable because of the large jump distance. The fraction of vacancies in the disordered lattice of β- and δ-TaD_x exhibit a maximum at x = 0.75. The diffusion rate should be strongly reduced in ζ-$TaD_{0.75}$ because of the smaller fraction of vacancies in the ordered structure.

References

1 H. Asano, Y. Ishino, R. Yamada, M. Hirabayashi, J. Solid State Chem. 15, 45 (1975)
2 T. Schober, H. Wenzl in: Hydrogen in Metals (G. Alefeld and J. Völkl, eds.), Springer, Berlin (1978)
3 W.A. Oates, J.A. Lambert and P.T. Gallagher, Trans. Met. Soc. AIME 245, 47 (1969)
4 K. Sköld in same ref. as 2
5 R.C. Bowman, Jr., in Metal Hydrides (G. Bambakidis, ed.) Plenum, New York (1981)
6 B. Alefeld, H.G. Bohn and N. Stump: Proc. Int. Meeting H in Metals, Jül Conf. 6 Vol. 1 (KFA Jülich 1972), p. 286
7 J. Hauck, Acta Cryst. A34, 389 (1978)
8 J. Hauck, J. Less-Common Met. 77, 99 (1981)
9 A. Magerl, H. Zabel and H. Wipf, J. Less-Common Met. 103, 91 (1984)
10 O.V. Gurin, V.N. Syromiatnikov, W. Sikora, J. Less-Common Met. 101, 529 (1984)
11 K. Parlinski, Z. Phys. B -Condensed Matter 58, 245 (1985)
12 O.V. Kovalev, Irreducible Representations of the Space Groups, Gordon and Breach Science Publishers, New York, London, Paris, 1965

Neutron Radiography Studies of Spinodal Decomposition in NbH_x

G. Steyrer[1] *and J. Peisl*[2]

[1]Institut Laue-Langevin, 156 X, F-38042 Grenoble Cedex, France
[2]Sektion Physik der Universität München, D-8000 München 22, F.R.G.

1 Introduction

The phase diagram of hydrogen dissolved in niobium shows great similarity to that of a one-component fluid. However, there are also striking differences, and unusual features due to the long-ranged elastic interaction which is responsible for the lattice gas (α-phase) - lattice liquid (α'-phase) phase transition [1]. This elastic interaction originates from the strain field a proton produces whilst acting as a dilatation center in the host lattice. This strain is then felt by the other interstitials. Concentration fluctuations are accompanied by coherency stresses and therefore critical fluctuations of short wavelength are strongly suppressed compared to fluctuations on a macroscopic scale. The long-ranged strain field leads to spinodal decomposition with smoothly varying density profiles. By means of neutron radiography we have mapped the projections of the hydrogen distribution in a niobium sample, and hence reconstructed the distribution in arbitrary planes of the crystal by computerized tomography. These were then compared to the predictions of the well-established theory of Wagner and Horner [2].

2 Spinodal Decomposition in the NbH_x System

If a crystal loaded with hydrogen in the one-phase lattice gas region (i.e. with a homogeneous hydrogen distribution) is quenched below the phase boundary of the α-α' transition, the coherency stresses in the host lattice that arise from the build-up of differences in the density of the dissolved interstitials increase the free energy of the system. This allows only fluctuations with wavelengths in the dimensions of the sample size to develop, instead of creating sharp interfaces between the two phases. The NbH_x system undergoes spinodal decomposition well known in theory [3] as counterpart to nucleation processes but rarely observed in experiments. Wagner and Horner [2] gave special emphasis to the long-ranged nature of the elastic interaction. Applying Landau's mean field theory, they found that the resulting hydrogen distribution can be expressed in terms of macroscopic density modes as long as coherency is maintained in the crystal lattice. These modes can be regarded as the normal modes of a macroscopic strain field caused by internal stress. They form a discrete spectrum below the phase boundary, and only a few of them are excited near the critical point. The hydrogen density distribution below the phase transition is then given by

$$\rho(\mathbf{r}) = \rho_o + \sum_L \rho_L(T,\rho_o)\cdot\psi_L(\mathbf{r}) \qquad (1)$$

Here ρ_o is the mean concentration, and the long wavelength fluctuations are given by the macroscopic modes ψ_L and corresponding amplitudes ρ_L.

It turns out that the macroscopic modes are sensitive to the shape of the crystal via elastic boundary conditions. For spherical crystals the modes can be expressed in terms of spherical harmonics, and not until recently have there been found analytic solutions for a rotational ellipsoid [4] . This shape can be regarded as a good approximation to the cylinders used in the experiments. The amplitudes are determined by minimizing the free energy of the system at a given mean concentration and temperature. Fig. 1 shows schematically the spinodals and the incoherent phase boundary, the latter being the final stable state of demixing, after dislocations have more or less released the coherency stresses in the crystal.

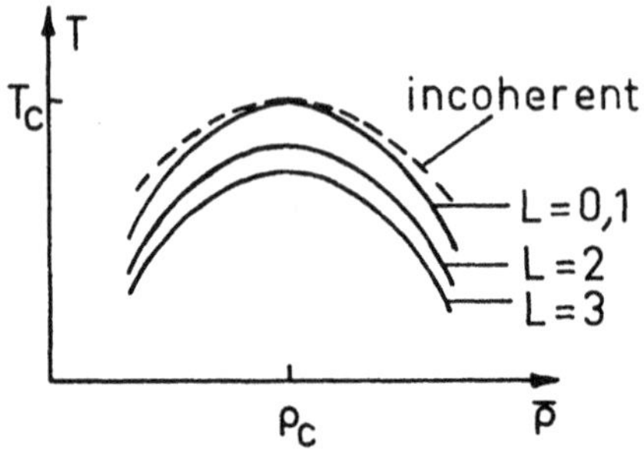

Fig. 1
Spinodals and coexistence curve in NbH_x [2]

3 Neutron Radiography

The locally varying hydrogen density below the critical point (T_{cr} = 171° C, c_{cr} = 0.31 H/Nb) was determined by neutron radiography in a most direct manner. A beam of thermal neutrons passing through a sample of niobium loaded with hydrogen is attenuated according to an exponential law.

$$I = I_o \cdot \exp[-N \cdot d \cdot (\sigma_{Nb} + c \cdot \sigma_H)] \tag{2}$$

σ = total cross-section of niobium/hydrogen
N = number of Nb atoms per volume
d = thickness of the sample
c = hydrogen concentration [H/Nb]

The high incoherent cross-section of the hydrogen gives rise to a strong attenuation of the incoming beam compared to the absorption of the niobium host lattice and makes neutron radiography very sensitive to changes of the local hydrogen density in the bulk of the sample.

4 Sample and Experimental Set-Up

A cylindrical single crystal of niobium (h=2.5cm, Ø=1.2cm) was loaded in situ with a hydrogen concentration of 0.29 H/Nb at 600° C and cooled down slowly (ΔT=0.2°) below the phase boundary. At each temperature a series of 4 radiographs were taken rotating the sample in steps of 45° around its z-axis. The neutron beam with a cross-section of 3x5cm² coming out of a neutron guide tube was monochromized with a perfect germanium crystal in order not to worsen the excellent collimation of the beam. The use of monoenergetic neutrons made it possible to get concentration profiles on an absolute scale. After passing through the sample, the image was recorded on X-ray photographic film using a gadolinium converter foil as back screen in the camera. The exposure time for each radiograph was 40 min. The processed films were analyzed with a microdensitometer (Université Paris-Sud, Orsay) and corrected for background and inhomogeneous intensity distribution of the incoming beam. In order to get the real 3-dimensional hydrogen distribution in the sample, computer programs for tomography were developed [5]. In this way, cuts through the crystal in arbitrary planes could be reconstructed from the projections and displayed.

5 Results and Discussion

The initially homogeneous density distribution above the phase boundary is shown in the radiograph in Fig. 2a, where the cylindrical shape of the sample is reflected. At 169.2° C (2 degrees below the incoherent phase boundary) a density fluctuation develops with half a wavelength along the z-axis of the sample (Fig. 2b). The amplitude grows with decreasing temperature, as shown in Fig. 3. There the hydrogen density along the z-axis at 3 temperatures below the phase boundary is compared to the distribution at the critical temperature.

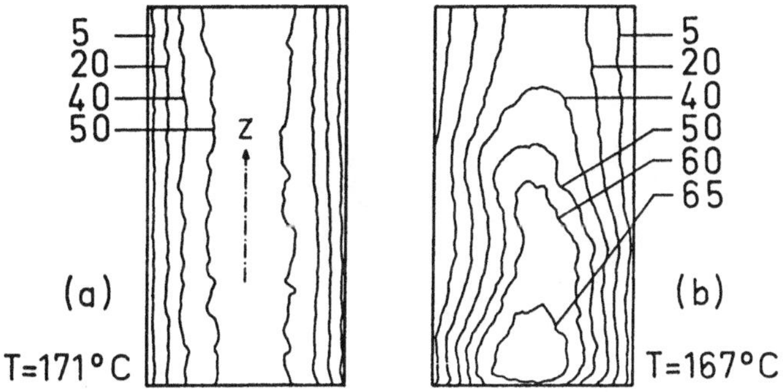

Fig. 2: Absorption levels (arbitrary units) before and after spinodal decomposition

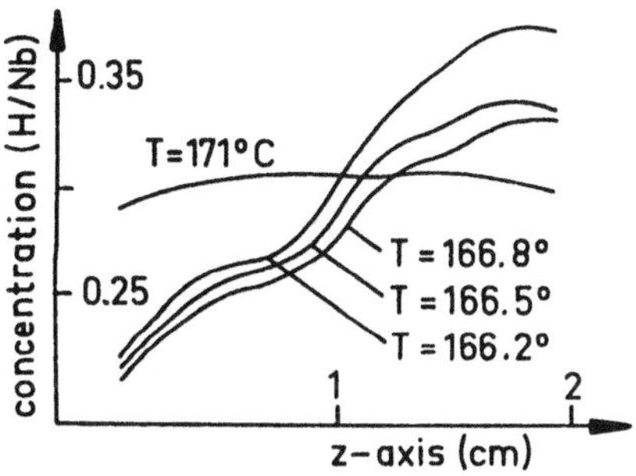

Fig. 3
Hydrogen density fluctuations along the z-axis of the crystal

At 163°C the coherency of the lattice was lost and no more changes in the hydrogen distribution occurred. At the same temperature cracks in the crystal were visible on the radiographs. The profile of the density distribution is mainly determined by a linear gradient in axial direction, which corresponds to the linear macroscopic mode (L=1). Other modes with low L are also present. A comparison with theoretical predictions [6], where 3 axially symmetric inhomogeneous modes (L = 1,2,3) were taken into account, is made in Fig. 4. The amplitudes of the modes were calculated for a rotational ellipsoid with an a/b ratio of 0.5. Clearly the measured hydrogen distribution is due to the excitation of macroscopic modes. Even on an absolute scale the accordance with theory is good, moreover as the whole crystal is concerned and complete coherency is certainly not maintained.
The projections at different angles show a nearly symmetric, smoothly varying density profile in radial direction, therefore tomography with

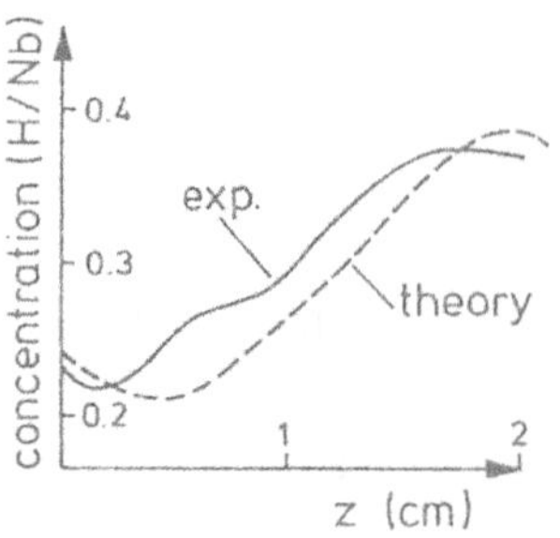

Fig. 4
Comparison theory [6]
and experiment:
fluctuations along z-axis
T=166.2° C c=0.29 H/Nb

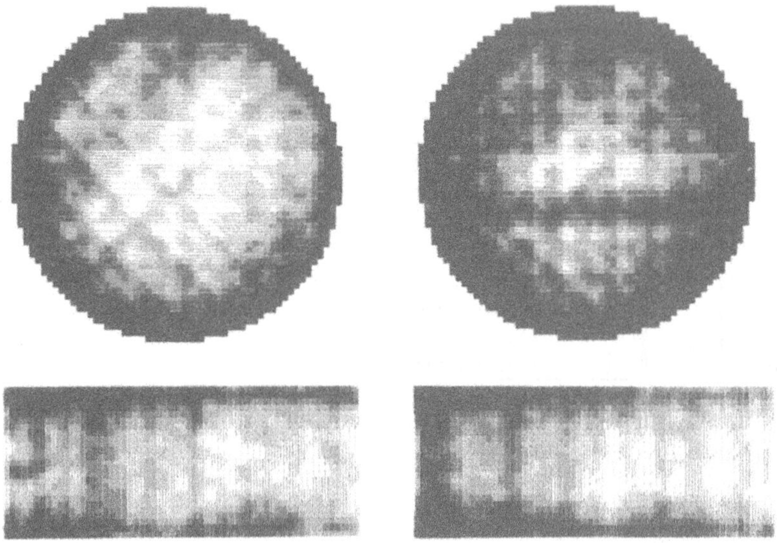

Fig.5 Top: transversal tomography T=166°C c=0.29 H/Nb
Bottom: axial tomography

a limited number of projections was possible. The Fourier reconstruction method as proposed in [5] was used. An example is given in Fig. 5 (T = 166° C and c = 0.29 H/Nb).

Tomographic reconstructions of the hydrogen distribution are displayed in 2 axial cuts perpendicular to each other, and 2 transversal cuts perpendicular to the z-axis at the top and the bottom of the crystal. There the hydrogen density is represented by different gray values; high density as light shaded and lower density as dark shaded pixels. These tomographs demonstrate clearly the variation of the concentration in axial direction, and give a view of the overall 3-dimensional hydrogen distribution in the sample.

6 Conclusion

Neutron radiography combined with computerized tomography is a very direct method to determine the macroscopic local hydrogen distribution in a niobium sample. Spinodal decomposition at the critical point of the α-α' phase transition has been observed, and the experimental results fit well into the frame of theory. The predominant role of the elastic interaction for this transition and the concept of the macroscopic modes have been confirmed.

7 Acknowledgements

We thank H. Maier-Bötzel for valuable discussions and the BMFT for financial support.

8 References

[1] Alefeld G., Phys. Stat. Sol. 32 67 (1969)
[2] Wagner H., Horner H., Adv. Phys. 23, 587 (1974)
[3] Cahn J.W., Acta Met. 10, 907 (1962)
Cahn J.W., Hilliard J.E., J. Chem. Phys. 28 258 (1958)
[4] Maier-Bötzel H., Platzer W., Phys. Rev. B27, 5173 (1983)
[5] Shepp L.A., Logan B.F., IEEE Trans. Nucl. Sci. 21, 3 (1974)
[6] Maier-Bötzel H., private communication

Part V

Point Defects, Radiation Damage, Voids, and Bubbles

Irradiation Damage in Metals: A Story of But Also's

Y. Quéré

C.E.N., B.P. n° 6, F-92265 Fontenay-aux-Roses Cedex, France

The effects of an irradiation on metals or alloys are briefly described along the two lines of instantaneous processes (fundamental aspects) and of structural evolution (practical applications).

Irradiation creates defects in metals, **but also** anneals them out ; induces hardening, **but also** softening; introduces disorder, **b.a.** order... We shall hereunder describe-and in some cases explain - some of these **b.a.**'s. We shall, in a first part, recall briefly the instantaneous processes taking place in a metal under flux. A second part will be devoted to the evolution of the metal being irradiated. Obviously, this first section will have a more or less low temperature fundamental character, the second one considering rather the high-temperature experiments, with a definite technological flavour.

1. INSTANTANEOUS PROCESSES.

1.1. Creation and nature of point defects.

Point defects are created as soon as the energy received by atoms exceeds a threshold value. These defects are essentially vacancies, interstitials in equal numbers (Frenkel pairs : F.P.) and various clusters of them. The interstitial has been the subject of a large literature. In the case of the f.c.c. structure, two configurations have been proposed : the <100> split interstitial [1], **b.a.** the <110> crowdion [2]. The former has been clearly identified in a number of f.c.c. metals by X-ray scattering [3-4] and diffuse scattering [5]. In b.c.c. metals, the observed entity is a <110> split configuration [6].

Point defects are surrounded by a recombination volume which prevents creating a new defect in the recombination volume of the antimorphic species. Instantaneous recombinations make it impossible to exceed a saturation value (typically 10^{-2} - 10^{-3} atomic fraction) in the concentrations of point defects. Complete athermal, low-temperature saturations are easily observed when the irradiation flux is high (e.g. case of uranium irradiated by fission fragments via thermal neutrons [7]).

All physical parameters are altered by the presence of point defects : transport properties like electrical conductivity [8], Hall effect [9], magnetoresistivity [10], thermoelectric power [11], elasto-resistivity [12], carrier concentrations of semi-metals [13], superconductivity [14] ... ; magnetic properties [15] ; elastic moduli [16] ; internal friction [17] ; stored energy [18] ; positron lifetime [19] ; volume, which generally increases [20], **b.a.** may decrease if the F.p. formation volume is negative (fig. 1)...

The case of alloys contains the previous points, **b.a.** some other specific features. In an alloy A-B, the creation of defects depends on A-A, A-B, B-A and B-B collisions and on the presence of two, generally distinct, threshold energies. This must be considered when one calculates the total number of displaced atoms [22]. Such double thresholds may be observed e.g. in alloys of A-15 structure [23], in ordered alloys [24], in carbides [25]...In the case of organic metallic

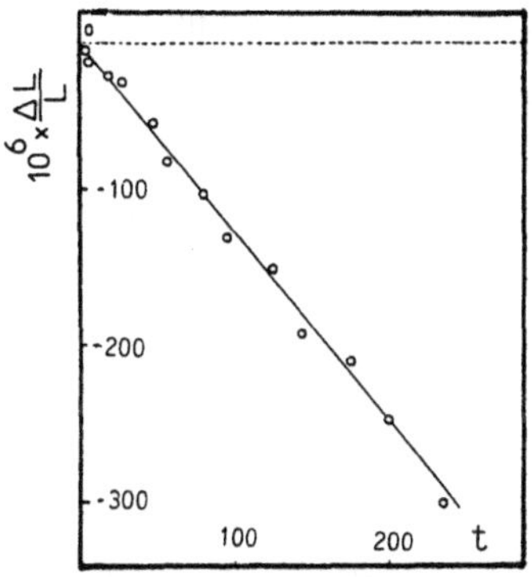

Fig. 1 - Volume of β-plutonium (here, length along an arbitrary direction) decreases under irradiation. The same occurs in c.f.c. δ-plutonium. This decrease is due to point defects, not to some phase change [21].
t : time of "self-irradiation" of plutonium, maintained at 4.2 K (hours).

"alloys", point defects which are created as well by photons as by particles [26] have a chemical nature [27]. They allow to extend the previous list of physical parameters altered by point defects to optical properties [28] and to pinning of charge density waves [29].

1.2 Subthreshold events.

During irradiation, a large number of collisions yield to atoms energies below the threshold. These subthreshold (s.t.) collisions may play a significant, or even dominant, role in the final result of an irradiation.

s.t. collisions may recombine close F.p. previously created, which decreases the real damage rate [30]. Moreover, when a F.p. is spatially related to an excited nucleus, it may be recombined by recoil-induced s.t. collisions, exhibiting a fundamentally finite lifetime (fig. 2).

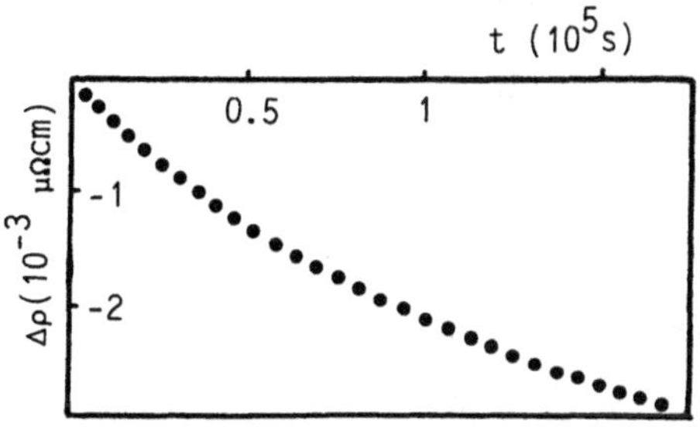

Fig. 2 - After an irradiation of iridium by thermal neutrons at low temperature (20 K), the irradiation-induced resistivity $\Delta\rho$ decreases with time t, due to i-v recombinations induced by subthreshold recoils of excited Ir nuclei [31].

s.t. events recombine previous defects, **b.a.** can produce new ones, when a s.t. collision is initiated on a lattice defect, in particular on a free surface or on a grain boundary , where the threshold energy is greatly reduced. For example, a s.t. irradiation of a thin T.E.M. sample creates fresh vacancies [32]. Moreover these vacancies are mobilized by s.t. collisions, either by direct s.t.-induced jumps of one of the nearest neighbours of the vacancy,or by indirect mechanisms where a virtual F.p. is created in the neighbourhood of a stable previous one [33]. In both cases, s.t.-induced migration, which creates mobility at the lowest temperatures, is a by-product of s.t. defect creation on previous - defect sites.

1.3 Clustering.

When point defects are created in a large energy cascade, collisions along dense atomic rows tend to separate point defects into i/ a vacancy rich core, or depleted zone [34] and ii/ groups of interstitials out of the depleted zone and along the dense directions emerging from the zone. These separated clusters are then ready to get stabilized into new defects : vacancy microvoid or dislocation loop for vacancies, interstitial loops for interstitials.

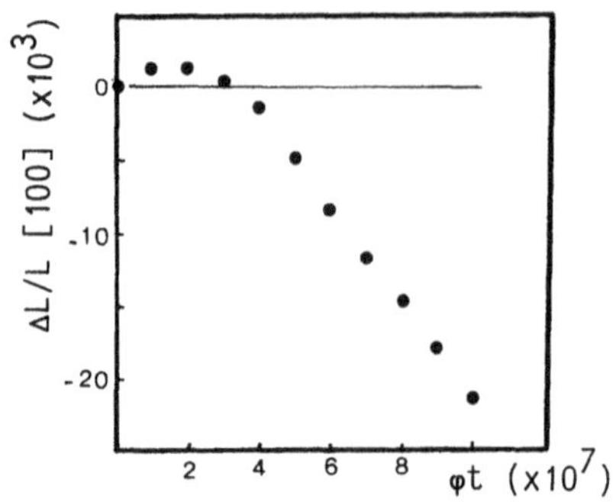

Fig. 3 - Length along direction [100] of an α-uranium single crystal decreases under irradiation [38] whereas it increases along [010].
Irradiation temperature : 4.2 K
φt : "burn-up" = fraction of fissioned nuclei.

The depleted zones have been observed in detail by field ion microscopy in a number of ion-bombarded metals [35]. In particular, for a given energy deposited in a given target, the denser the zones, the higher the mass of the projectile ion [36].

To observe unambiguously the clustering into loops, it may be useful to remove the degeneracy of various equivalent Burgers vectors to induce a polarization of the loops and create macroscopic effects. A classical case is that of anisotropic metals, like uranium, in which the strong anisotropy of elastic moduli and/or thermal expansion creates such a polarization. Schematically, vacancy loop vectors are polarized along one direction ([100]) and interstitial ones along another one ([010]), giving rise to changes of shape (radiation-induced growth). This phenomenon is magnified at low temperatures [37-38], showing that it is intrinsic and owes nothing to thermal diffusion. Fig. 3 shows the decrease of length along [100].

Knowing that volume does not diminish, this decrease gives a good signature of an athermal clustering and collapse into vacancy loops. Moreover, considering(fig.3)the initial increase of length, one is able to deduce that the superposition of a small number (typically 2) of cascades on a given point is necessary for the collapse to occur, producing the incubation period visible on fig. 3 [39] (for Fe,see [29']).

This removal of degeneracy may be created, in isotropic metals like W, by a non-hydrostatic - e.g. uniaxial - stress,which polarizes dislocation loops at their very nucleation. This is the principle of stress-induced irradiation growth [42-44] which is essentially a low-temperature creep.

Not least, but last, all these events created by high-energy collisions displace a large number of atoms from their initial site. In pure metals, or in disordered alloys, these site interchanges have little or no consequence. In an ordered alloy, they produce disorder, often by addition of individual disordered domains(visible by electron microscopy, fig. 4),**b.a.** by a more diffuse continuous mechanism [46].

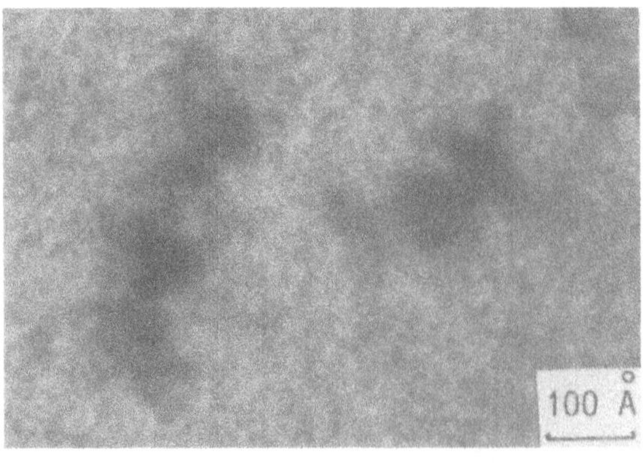

Fig. 4 - Transmission electron microscopy of an ordered Cu_3Au alloy irradiated by ≈ 3 x 10^{15} Ar^{16+} ions of high energy (1.8 GeV), in G.A.N.I.L. accelerator [45].
Large disordered domains are visible.

2. EVOLUTION UNDER IRRADIATION.

In some cases, one of the major sources of evolution is the creation of new chemical species due to nuclear reactions (by fast ions, by thermal neutrons...). For example, fission produces $\simeq$ 20 Kg of rare gases (Kr, Xe) per day in the french PWR nuclear reactors. How this gas agglomerates, provoking a swelling of fuel elements, is a question of obvious importance.

We shall concentrate here on those properties which are directly influenced by the evolution of point defects and clusters: their migration, their recombination, their agglomeration, their interaction with other defects... This evolution, along time t, may itself be described by a number of chemical reactions, and of reaction equations, the simplest set of which, due to Lomer [47], may be written as:

$$dC_i/dt = G - K\,(D_i + D_v)\,C_i\,C_v - \Sigma_\alpha\; k_{\alpha i}\,D_i\,C_i \qquad (1.a)$$

$$dC_v/dt = G - K\,(D_i + D_v)\,C_i\,C_v \quad \Sigma_\alpha\; k_{\alpha v}\,D_v\,C_v \qquad (1.b)$$

(1.a) describes the evolution of the (atomic) concentration C_i of interstitials (subscript v stands for vacancies) as the sum of a constant creation, at rate G, minus mutual (di-molecular) recombination, with constant K, minus absorption by various defect sinks of type α, with absorption constants $k_{\alpha i}$. Any terms including concentration gradients ∇C are omitted in (1) which implies uniformity, i.e. a mean distance between sinks smaller than the mean defect separation, or $C_{i,v} < \rho a^2$ in the typical case where sinks are dislocations of density ρ (a = interatomic distance).

The D's in (1) are the diffusion coefficients of both point defects. There is a well-founded postulate that interstitials are faster than vacancies ($D_i > D_v$). This is most frequently true [48], **b.a.** sometimes wrong : in alloys A-B, the trapping of interstitials on B atoms may be strong enough to reverse the tendency (c.f. the case of Ag-Zn alloys [49-50]).

Equations (1) are non-linear, non-mutually symmetric in i and v. Moreover "constants" like G, k .. are in fact generally **time-dependent** : G because instantaneous recombination (see § 1.1), which increases with the C's, tends to decrease slightly the production rate ; the k_α's because the sinks (morphology, concentration ...) change , sometimes drastically, all along the irradiation etc. So, only numerical solutions of system (1) are possible. Fig. 5 shows a typical case of evolution of C_i and C_v in a simple case, where "constants" are kept constant with plausible numerical values.

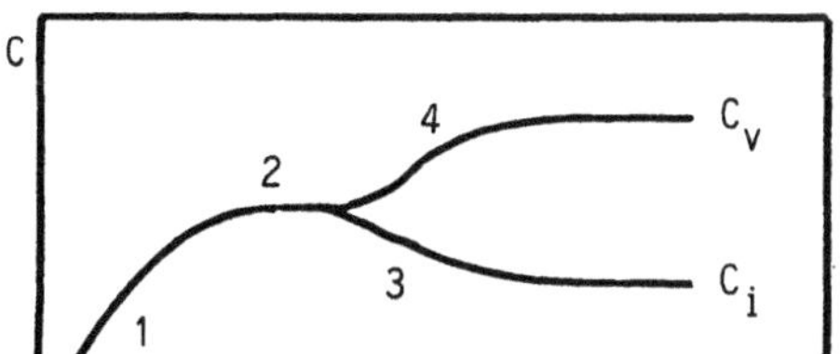

Fig. 5 - Schematical evolution of concentrations of i's and v's versus irradiation dose φt in the case of a low sink density.
1 : build-up of C's
2 : i-v recombination dominates
3 : annihilation of i's on sinks
4 : annihilation of v's on sinks.

2.1 Enhanced diffusion.

The presence of concentrations C_i and C_v generally higher (C_v) or much higher (C_i) than equilibrium concentrations, increases atomic diffusion under irradiation, at least if temperature is such that D_i and/or D_v are not zero (for a review, see [51]). This enhanced diffusion is of the order of $D_i\,C_i + D_v\,C_v$. In steady state conditions (see fig. 5), and at a given temperature, this diffusion term is constant.

Enhanced diffusion was detected first by observing enhanced short-range ordering rates in alloys such as Cu-Zn [52] or austenites [53] ; then by discovering radiation-provoked long-range order in alloys like FeNi [54] ; last by measuring longe-range radiation-induced atomic diffusion in diffusion couples like Al/Mg [55], or in self diffusion experiments [56].

2.2 Void swelling.

In some cases, vacancies agglomerate into voids, giving rise to an increase of volume V called swelling g : $g = \Delta V/V$. This phenomenon has been studied in great detail [57].

If we suppose that voids may nucleate (e.g. by stabilization of depleted zones with - or without - help of gas atoms), swelling corresponds to the growth of these cavities, slightly more vacancies than interstitials joining them. This slight disequilibrium is due to the behaviour of edge dislocations which, because of elastic interactions, attract the i's more than the v's, whereas voids capture both equally.

Let us describe this biased situation in a very simple case. Consider, among the α's in (1), the two contributions of dislocations (α = D) and voids ($\alpha = v$). We write

$$\Sigma k_{\alpha i} D_i C_i = [\rho(1 + \varepsilon) + k_v] D_i C_i \quad (2a)$$

$$\Sigma k_{\alpha v} D_v C_v = [\rho + k_v] D_v C_v . \quad (2b)$$

where k_{Di}, which is $\simeq \rho$ (density of dislocation taken as a constant), has been considered slightly larger (by small factor ε, e.g. $\varepsilon \simeq 0.01$) than k_{Dv} ; whereas one puts $k_{vi} = k_{vv} = k_v$. Considering the growth of n (per unit volume) voids of mean radius R, one writes the balance s of the number of v's minus the number of i's, both joining the cavities per unit time, as :

$$s = dg/dt = k_v (D_v C_v - D_i C_i) \qquad \text{and from (2) :}$$

$$s = D_i C_i \rho \varepsilon k_v (\rho + k_v)^{-1} . \quad (3)$$

Conventional theory of precipitation gives :

$$k_v \simeq n 4 \pi R . \quad (4)$$

Taking the situation where a stationary state is reached ($dC_i/dt = dC_v/dt = 0$ in (1)) and moreover where recombination is either negligible (K small enough ; k_v and k_D large enough) or constant, reducing G to an apparent lower value G', we write, from (1a) and (2a) :

$$G' = [\rho(1 + \varepsilon) + k_v] D_i C_i .$$

Combining this and (3), it comes :

$$s \simeq G' \rho \varepsilon k_v (\rho + k_v)^{-2} . \quad (5)$$

In terms of a void distribution (n, R), with swelling $g = 4\pi nR^3/3$, we combine (4) and (5) into :

$$(\rho + 4\pi nR)^2 R dR = \rho \varepsilon G' dt \quad (6)$$

from which function g(t) can be explicitly derived in two extreme cases:

i/ a first regime ($4\pi nR \ll \rho$) where (6) gives :

$$g \simeq 12\, n\, (\varepsilon G'/\rho)^{3/2}\, t^{3/2} ; \quad (7a)$$

ii/ a later regime ($4\pi nR \gg \rho$) where the swelling slows down as :

$$g \simeq 0.25\, (\rho \varepsilon G')^{3/4}\, n^{-1/2}\, t^{3/4} . \quad (7b)$$

Expressions (7) give the general expected trend of swelling versus irradiation time t (fig. 6) : an acceleration, then a deceleration, ending in fact towards a saturation when voids are close enough to completely erase the influence of dislocations as sinks : essentially all point defects (i's and v's) are absorbed by voids ($\rho \simeq 0$ in 7b) which, then, cease to grow. This saturation regime is some-

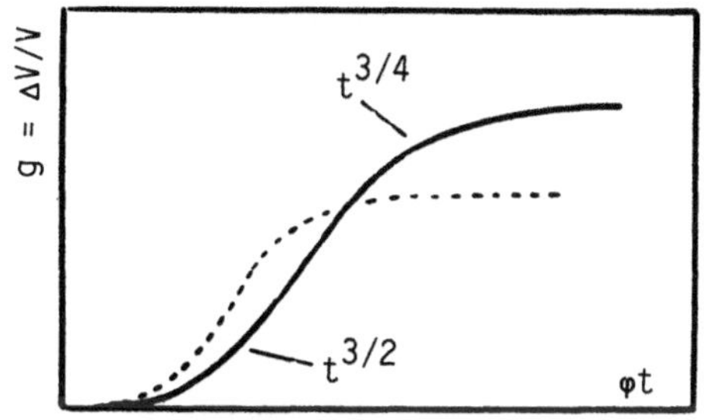

Fig. 6 - Schematical evolution of swelling g versus irradiation dose φt at given temperature (see 7a, b). Parameter n (concentration of voids) is larger for the dotted line than for the full curve (see point v/).

times accompanied by an ordering of the voids [58],of same crystal structure and orientation as the host metal. This ordering is expected to be stabilized [59] if a one-dimensional motion of interstitials (in the excited crowdion configuration (see § 1.1)) along the densest atomic rows has a life time long enough to reach the closest void before returning to the split interstitial ground state, which implies a void density large enough.

Obviously, starting from the same ingredients (eq. (2a, b), one can prepare everything between fast-food type results like (7a, b), and exquisite "Michelin 3 étoiles" meals. To transform (2) into better and possibly delicious cooking,one must :

i/ take into account all the variety of possible sinks, and in particular impurity atoms [60] which, quite generally, decrease the swelling by trapping point defects, then favouring i-v recombination at these trapping sites, and possibly contributing to decrease bias factor ε [61] (see for instance the role of P in austenic steels [62]) ;

ii/ examine the exact nature of dislocations (their Burgers vector ; their nature : glissile, sessile) hereabove averaged in the "blind" constant ρ ; this probably explains the swelling resistance of ferritic steels as compared to austenitic ones [63] ;

iii/ establish rigorous formulations for sink strengths k_α [64] ;

iv/ introduce the time dependance of these sink strengths. For example, in cascade-producing irradiations, depleted zones may convert into vacancy dislocation loops (see § 1.3) which, both, store a large number of vacancies and favour recombination, explaining why neutron irradiation, at given temperature,produces much less swelling - by possibly one order of magnitude - than "equivalent" (i.e. producing the same number of displaced atoms) electron irradiations [65]. Also for example, the general tendency to develop interstitial dislocation loops tends to increase term ρ as irradiation proceeds (fig. 7) ;

v/ consider the temperature dependance of all "constants" in (2.a, b)and particularly that of void density n (see fig. 6) [57];

vi/ take into account the existence of an incubation time before which no obser-

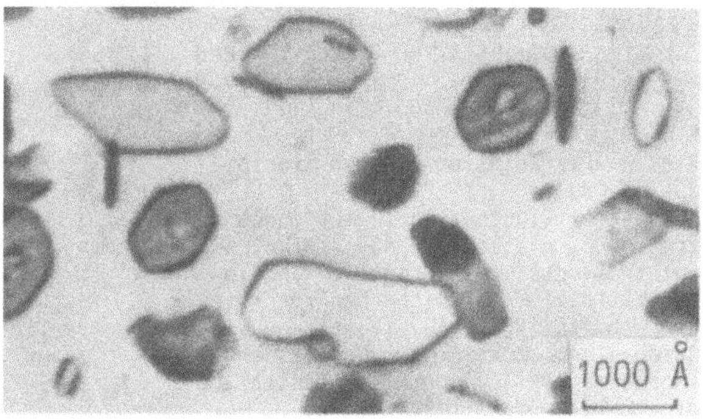

Fig. 7 - Interstitial dislocation loops of various structures (perfect, imperfect ; simple, double ...) in an austenitic steel irradiated by 0.5 MeV Ni^+ ions at 650°C [62].

vable swelling takes place. This is due to the necessity of nucleating and stabilizing the microvoids and also is implicitly contained in laws of type $t^{3/2}$ (see (7a)). This incubation time is of the order of one year in the structure-materials of french fast neutron breeder PHENIX [66].

2.3 Irradiation creep.

We have already mentioned (see § 1.2) a low-temperature irradiation-induced creep mechanism [42-44] due to a stress-induced preferred nucleation of dislocation loops (S.I.P.N.).

At higher temperatures (D_i and/or $D_v \neq 0$), creep is most generally increased by irradiation(**b.a.** sometimes decreased at intermediate temperatures). Several models have been proposed to explain this increase.

In the S.I.P.A. (A for absorption) model, considering the inhomogeneity interaction between point defects and dislocations [67], it is argued that the external stress σ favours the climb of those edge dislocations $D_{/\!/}$ – or the growth of those interstitial loops $D_{/\!/}$ – whose Burgers vector lies in sympathy (e.g. parallel to a tensile stres σ) with σ as compared to other dislocations $D_{\perp}$ [68, 69]. Starting again as in § 2.2 from system (2a, b), a bias factor is introduced, proportional to σ, which essentially favours $k_{D/\!/ i}$ compared to $k_{D\perp i}$ [69]. The same antimorphic arguments are carried out with the v's. The result is a multiplication of extra "half planes" perpendicular to σ, producing creep. This S.I.P.A. creep rate is proportional to σ (like Nabarro creep), and to G (see 2a, b).

Another mechanism [70, 71], which probably works together with the previous one, consists in successive periods of irradiation-enhanced climb, then climb-enabled glide, then climb again .. (S.I.C.G.), vacancies being lost either on voids (creep + swelling [70])or on grain boundaries [72]. Creep deformation, here, is due to the glide component of the mechanism. The climb component may be due to a long-term uniform flow of point defects to dislocations, or to discrete cascade-induced fluctuations of point defect concentrations [73].

2.4 Phase stability.

Consider an alloy A-B. According to temperature T and concentration c of B, the equilibrium state may consist either of two phases α and β, or of one α phase containing solute atoms B. If this alloy is irradiated, nearly everything may occur – depending on the parameters of the irradiation and especially on the damage rate (number of displaced atoms per unit time) – including: 1/ ballistic dissolution of the 2-phase alloy, **b.a.** 2/ radiation-accelerated precipitation in the 2-phase alloy [74], **b.a.** 3/ precipitation [75, 76] and even coarsening [77] in the one-phase alloy ; 4/ amorphization of the alloy, **b.a.** 5/ partial crystallization of the initially amorphous solid [78]. All these phenomena are of crucial importance for the actual behaviour of nuclear reactor fuel- and structure- elements, for the future behaviour of nuclear waste confinement materials, for the final state of ion-implanted alloys or devices ...

Among these various possibilities, point 3/ is probably the most unexpected and puzzling one. In normal solid solutions such as Ni-Si (2-6 at %) [75], Al-Zn (2 at %) [79] ..., precipitation (of respectively Ni_3Si, Al-Zn G.P. zones ...) may be observed if the damage rate φ is high enough. In φ-T coordinates, "phase diagrams" may be determined, defining domains of 2- or 1-phase steady state equilibrium.

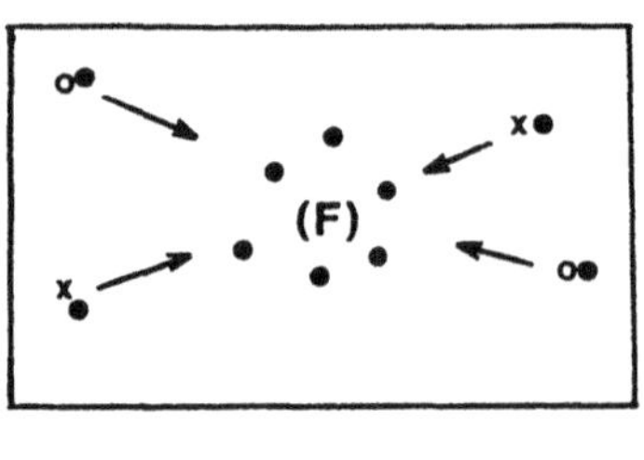

Fig. 8 - Irradiation-induced precipitation [80].
A positive fluctuation (F) of concentration c_B in alloy A-B may attract point defects : vacancies **o** and interstitials **x** , which then recombine preferentially at (F). If v's and i's sweep along atoms B ●, c_B goes on increasing at (F), until precipitation of a B-rich phase takes place.

This radiation-induced precipitation is now well understood [80]. The idea, in essence, is the following. If a fluctuation increases locally – in a region (F) – the concentration c of B in solid solution A-B, point defects created by irradiation at rate φ are generally attracted towards (F), (e.g. by its elastic strain field). The probability of i-v recombination is then increased in (F). If now the coupling between defect- and B- fluxes is positive, i.e. if the diffusion of defects to (F) drags along atoms B, concentration of B in (F) increases without possibility of irradiation-induced back diffusion because of i-v recombination (F). Concentration c thus increases in (F), up to precipitation. The instability of the system, provided by the recombination-induced energy dissipation, obviously entails a value of φ high enough to ensure a flux of B to (F) higher than the back flux of B from (F) due to thermal vacancies, (fig. 8).

All the various effects of phase destabilization under irradiation are not as well understood as the previous one. However, a law of corresponding states has been recently proposed which should help understanding - and sometimes foreseeing- a number of possible situations in irradiated alloys[81]. Based on the introduction of ballistic effects together with diffusion, and on the use of the Lyapunov functional formalism, this law consists in reading in the equilibrium phase diagram at temperature T', the expected state of the alloy irradiated at temperature T, with $T' = T(1 + \Delta)$. The "dilatation" Δ of temperature, due to irradiation, increases with φ and decreases with T.

REFERENCES.

1 H.B. Huntington, F. Seitz : Phys. Rev. 61, 315 (1942).
2 H. Paneth : Phys. Rev. 80, 708 (1950).
3 P. Ehrhart, W. Schilling : Phys. Rev. 8, 2604 (1973).
4 P.H. Dederichs : J. Phys. F 3, 471 (1973).
5 H.G. Haubold : Rev. Phys. Appl. 11, 73 (1976).
6 P. Ehrhart : J. Nucl. Mat. 69-70, 200 (1978).
J.P. Guillot, J.P. Rivière, M.F. Beaufort-Richard : J. Physique 44,651 (1983).
7 G. Burger, K. Isebeck, H. Wenzl, J.C. Jousset, Y. Quéré : Phys. Letters 16, 240 (1965).
8 T.H. Blewitt, R.R. Coltman, C.E. Klabunde, T.S. Noggle : J. Appl. Phys. 28, 639 (1957).
9 K. Böning, W. Mauer, K. Pfändner, P. Rosner : Rad. Eff. 29, 177 (1976).
10 A. Dunlop, J. Dural, D. Lesueur : J. Phys. F 10, 1913 (1980).
11 G. Sieber, G. Wehr, K. Böning : J. Phys. F 7, 2503 (1977).
12 J. von Stebut : Rapport CEA-R-4558 (1973).
13 M. Le Goff, J. Heremans : J. Phys. F. 14, 399 (1984).
14 F. Rullier-Albenque : Rapport CEA-R-5291 (1984).
15 P. Moser, D. Dautreppe, P. Brissoneau : Comptes Rendus Ac. Sc. 250,3963 (1960).
16 W. Schilling : J. Nucl. Mat. 69-70, 465 (1979).
17 P. Moser : Mém.-Scient. Rev. Métal. 63, 431 (1966).
18 A.V. Granato, T.G. Nilan : Phys. Rev. 137, 1233 (1965).
19 M. Eldrup, D.E. Morgensen, J.H. Evans : J. Phys. F 6, 4909 (1976).
C. Corbel : Rapport CEA-R-5334 (1985).
20 R. Vook, C. Wert : Phys. Rev. 109, 1529 (1958).
21 J. Jacquemin : Rapport CEA-R-4386 (1973).
22 D. Lesueur : Phil. Mag. 44, 905 (1981).
23 F. Rullier-Albenque, Y. Quéré : Phys. Letters 81A, 232 (1981).
24 A. Alamo, C.H. de Novion, G. Désarmot : Rad. Eff. 99 (1985).
25 D. Gosset : Rapport CEA, in press (1986).
26 L. Zuppiroli : Rad. Eff. 62, 53 (1982).
27 N. Mermilliod, N. Sellier : J. Physique 44, C3-1353 (1983).
28 L. Zuppiroli, C.S. Jacobsen, K. Bechgaard : J. Physique, 46, 799 (1985).
29 S. Bouffard : Rapport CEA-R-5149 (1982).
29' A. Dunlop, Rapport CEA-R-5053 (1981).

30 J. Dural, J. Ardonceau, J.C. Jousset : J.Physique 38, 1007 (1977).
31 M.H. Gély, A. Dunlop, Y. Quéré : J. Physique Lettres 44, 219 (1983).
32 D. Cherns, F.J. Minter, R.S. Nelson : Nucl. Inst. Meth. 132, 369 (1976).
33 K. Urban, A. Seeger : Phil. Mag. 30, 1395 (1974).
34 A. Seeger : 2nd Geneva Conf. on atomic energy 6, 250 (1958).
35 D. Seidman : J. Phys. 3, 393 (1973).
36 C.Y. Wei, D. Seidman : Appl. Phys. Lett. 34, 622 (1979).
37 Y. Queré, J. Doulat : Comptes Rendus Ac. Sc. 252, 1305 (1961).
38 B.A. Loomis, S.B. Gerber : Phil. Mag. 18, 539 (1968).
39 C. Weinberg : Rapport CEA-R-4315 (1972).
40 C. Weinberg, Y. Quéré : to be published.
41 J. Ponsoye : Rad. Eff. 8, 13 (1971).
42 J.L. Pouchou, L. Zuppiroli, J. Dural, J. Leteurtre : Rad. Eff. 29, 91 (1976).
43 L. Zuppiroli, J.L. Pouchou, A. François, J. Leteurtre, Y. Quéré : Phil. Mag. 35, 853 (1977).
44 J.Y. Bounaud : Rapport CEA-R-5037 (1980).
45 A. Barbu, G. Martin, M. Toulemonde, J.C. Jousset : Comptes Rendus Ac. Sc. 299, 409 (1984).
46 J. Bloch : J. Nucl. Mat. 3, 327 (1961).
J. Bloch, J. Doulat, J. Blin : J. Nucl. Mat. 3, 327 (1961).
47 W.M. Lomer : A.E.R.E. Report T/R 1540 (1954).
48 H. Ullmaier, W. Schilling : in Physics of modern materials, I.A.E.A. Vienna, 361 (1980).
49 M. Halbwachs, J. Hillairet : Phys. Rev. B 18, 4927 (1978).
50 P. Vaessen, B. Lengeler, W. Schilling : Rad. Eff. 81, 277 (1984).
51 R. Sizmann : J. Nucl. Mat. 69-70, 386 (1978).
52 A.C. Damask : A.I.E.A. Symposiom,Venice (1962).
53 C. Dimitrov, C. Corbel, P. Moser : Ann. Chim. Fr. 9, 89 (1984).
54 J. Paulevé, D. Dautreppe, J. Laugier, L. Néel : Comptes Rendus 254,965 (1962).
55 G. Moreau : Rapport CEA-R-3916 (1969).
56 W. Schüle, E. Lessmann, R. Scholz : Internat. Conf. on Fundam. aspects of rad. dam. in metals, Gatlinburg (1975).
57 S.D. Harkness, Che-Yu Li : Metall. Trans. 2, 1457 (1971).
P.T. Heald, M.V. Speight : Acta Met. 23, 1389 (1975).
L.K. Mansur : Nucl. Technol. 40, 5 (1978).
A.D. Brailsford, R. Bullough : J. Nucl. Mat. 69-70, 434 (1978).
58 J. Brimhall, G.L. Kulcinski : Rad. Eff. 20, 25 (1973).
59 C.H. Woo, W. Frank : to be published in J. Nucl. Mat. 137 (1985).
60 O. Dimitrov : Ann. Chim. Fr. 9, 59 (1984).
61 J. Weertman, D.M. Parkin : J. Nucl. Mat. 99, 66 (1981).
62 L. Boulanger : Science et Recherche (CEA) 1, 63 (1980).
63 E.A. Little, R. Bullough, M.H. Wood : A.E.R.E. R-9678 (1980).
64 A.D. Brailsford, R. Bullough : Phil. Trans. Roy. Soc. 302, 87 (1981).
65 R. Bullough, B.L. Eyre, K.Krishnan : Proc. Roy. Soc. 346, 81 (1975).
66 J.L. Seran : Ann. Chim. Fr. 9, 463 (1984).
67 J.D. Eshelby : Proc. Roy. Soc. A 241, 376 (1957).
68 R. Bullough : Ann. Chim. Fr. 9, 179 (1984).
69 P.T. Heald, M.V. Speight : Phil. Mag. 29, 1075 (1974).
70 J.H. Gittus : Phil. Mag. 25, 345 (1972).
71 L.K. Mansur : O.R.N.L./T.M. - 6443 (1978).
72 P.T. Heald, J.E. Harbottle : J. Nucl. Mat. 67, 229 (1977).
73 L.K. Mansur, W.A. Coghlan : Yamada Conf. V, Un. Tokyo Press, 907 (1982).
74 L. Boulanger : J. Nucl. Mat. 64, 179 (1977).
75 A. Barbu, G. Martin : Scripta Met. 11, 771 (1977).
76 P.R. Okamoto, L.E. Rehn : J. Nucl. Mat. 83, 2 (1979).
77 K. Urban, G. Martin : Acta Met. 30, 1209 (1982).
78 D. Lesueur : Rad. Eff. 24, 101 (1975). Rapport CEA-R-4502 (1973).
79 R. Cauvin, G. Martin : J. Nucl. Mat. 83, 67 (1979).
80 G. Martin : Phys. Rev. B 21, 2122 (1980).
81 G. Martin : Phys. Rev. B 30, 1424 (1984).

Determination of Helium Densities in Bubbles Using a Contrast Variation Method

D. Schwahn, W. Kesternich, S. Spooner, H. Schroeder, H. Ullmaier, and J. Schelten*

Kernforschungsanlage Jülich GmbH, Postfach 1913, D-5170 Jülich, F. R. G.

1. Background

Because of the vanishingly small solubility of noble gases in solids, metal-helium "alloys" cannot be made by conventional means. In nuclear devices, however, helium is produced by the interaction of fast particles with the nuclei of fuel or structural materials and can lead to drastic property changes already at low concentrations /1/. An example is given in Fig. 1 showing the decrease in creep life of a stainless steel specimen at 1023 K into which helium has been implanted in different ways. This so-called high temperature helium embrittlement is e.g. regarded as a life-time limiting effect for the first wall of future fusion reactors /2,3/. Similar effects have been observed or are expected in fast breeder reactors, spallation neutron sources and tritium storage materials. Typical helium productions rates in these devices are given in table 1 together with the introduction rates attainable by α-particle implantation which is often used for basic studies of helium effects.

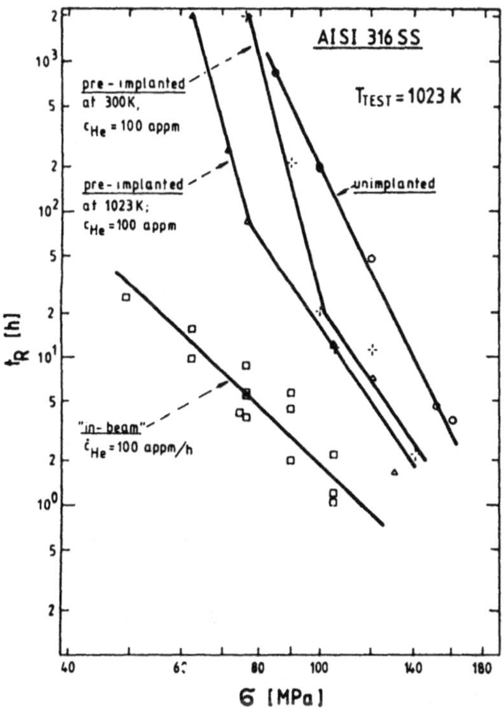

Fig. 1: Stress dependence of the creep rupture times t_R of AISI 316 stainless steel at 1023 K for differently implanted specimens and He-free controls /3/.

*)Permanent adress: Metal and Ceramics Division, Oak Ridge Nat. Lab., Oak Ridge, TN 37831, USA

Table 1: Typical Helium Production Rates in Nuclear Technology

(N, α)	Fast Breeder Reactors		5 - 15 PPM/A
	Fusion Devices		50 - 300 PPM/A
(T,He^3)	Tritium Storage	up to	100 PPM/D
(P, α)	Spallation Sources	up to	100 PPM/D
α	Implantation	up to	several % Hr

Microstructural observation has shown that almost all macroscopic helium effects (embrittlement, blistering and flaking, hardening, etc.) are consequences of the helium bubbles forming under supersaturation. A detailed understanding of the nucleation and growth of helium bubbles is therefore the key for a theoretical modelling of helium effects. One of the most important parameters which govern these processes is the density (or the pressure) of helium in the bubbles.

Until recently this quantity could only indirectly be determined by taking the size distributions observed by TEM and assuming thermal equilibrium, i.e. a balance between the internal helium pressure and the surface tension force where p is the

$$p = 2\gamma/R \tag{1}$$

helium pressure, γ is the specific surface energy and R is the bubble radius. Other attempts employed the pressure shift of the electronic energy levels of helium measured by EELS /4/ and UVAS /5/. These methods are, however, restricted to very thin specimens and high helium concentration.

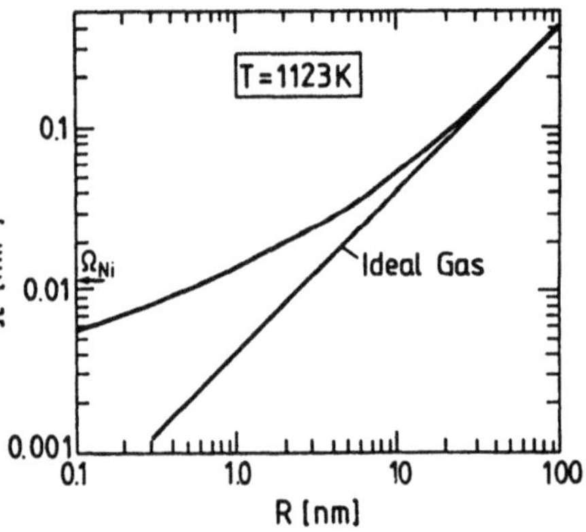

Fig. 2: Atomic volume Ω_{He} of helium in equilibrium bubbles in Ni(γ = 2N/m) as a function of their radius R. For R < 20 nm the realistic equation of state /6/ strongly deviates from the ideal gas behavior.

We have therefore applied neutron scattering to obtain information about helium densities (in the following expressed as the inverse of the atomic volume Ω_{He}) in bulk specimens and without presupposing the validity of Eq. (1). Such an approach appears promising, since the atomic volume of helium in bubbles of around nm-size is comparable with the atomic volume of the surrounding metal matrix, as shown in Fig. 2. In the following we present a contrast variation method which allows a direct determination of Ω_{He} in bubbles by SANS and describe the procedure for obtaining the size distribution n(R) of the bubbles. We then briefly give some experimental details and finally present first results on $\Omega_{He}(R)$ and n(R) of helium bubbles in nickel specimens.

Helium Density Determination by SANS Contrast Variation

The SANS cross-section $d\Sigma/d\Omega$ as a function of the scattering vector Q is given by /7/

$$\frac{d\Sigma}{d\Omega} = n\, V^2\, F(Q)\; \Delta\rho^2 \qquad (2)$$

where n is the number density, V the volume and F(Q) the structure function of the helium bubbles. Since the expected helium densities in nm-sized bubbles are of the same order of magnitude as the matallic densities, the contrast factor $\Delta\rho$

$$\Delta\rho = \frac{b_M}{\Omega_M} - \frac{b_{He}}{\Omega_{He}} \qquad (3)$$

sensitively depends on the helium density $1/\Omega_{He}$. The unknown quantities n, V and F can be eliminated by using two chemically identical specimens with different $\Delta\rho$. This is achieved by employing either different metal or different helium isotopes. Then the ratio of $d\Sigma/d\Omega$ from both isotopes I1 and I2

$$\alpha(\omega) \equiv \frac{d\Sigma/d\Omega\,(Q,I1)}{d\Sigma/d\Omega\,(Q,I2)} \qquad (4)$$

is a function of $\omega = \Omega_M/\Omega_{He}$ only. Two examples of the function $\alpha(\omega)$ are given in Fig. 3. The interesting range for ω is between zero and two. $\omega = 0$ is a void and $\omega > 2$ is unphysical because such a He-pressure would exceed the strength of the matrix. The He^4 in Ni^{58}/Ni^{60} combination exhibits $\alpha = \infty$ at $\omega \simeq 1$ (i.e. bubbles of this density have no contrast in Ni^{60}) and is therefore not suited for studying bubbles in the 0.1 to 1 nm range. The situation is improved for He^3/He^4 in natural Ni and even better in Ni^{58}. A contrast variation by different He-isotopes is preferable also for another reason, since it allows an application of the method to alloys where a variation by different matrix isotopes is in general not possible.

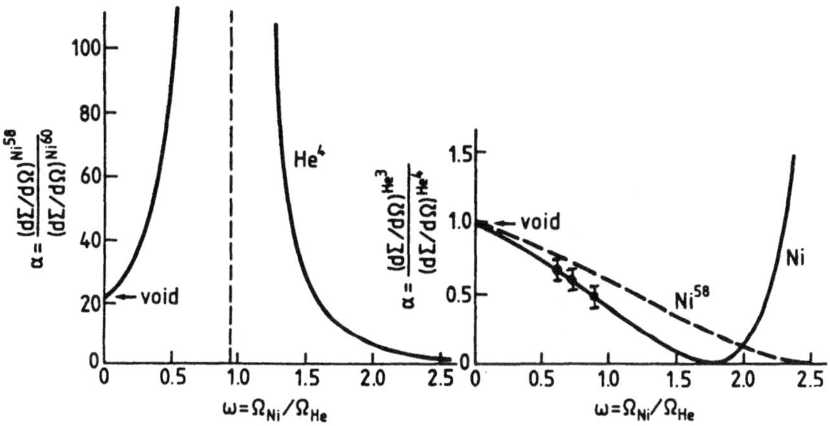

Fig. 3: Ratio of scattering cross-section α vs. ratio of atomic volumes ω for He^4 in Ni^{58}/Ni^{60} (left part, /8,9/) and for He^3/He^4 in natural nickel and Ni^{58}, respectively (right part /10/).

3. Evaluation of Bubble Size Distribution

In the case of equilibrium bubbles a complete analysis of the SANS data is possible. The scattering law for a m-model log-normal size distribution of spherical shape particles is given by /8/.

$$\frac{d\Sigma}{d\Omega}(Q) = \sum_{i=1}^{m} \int_0^\infty \frac{n_i}{R\,\beta_i\sqrt{(2\pi)}}\, V^2\, F_R(Q)\,|\Delta\rho(R)|^2 \exp[-\ln^2(R/\bar{R}_i)/2\beta_i^{\,2}]\,dR \qquad (5)$$

$\bar{R}_i$ is the geometrically averaged radius and β_i the relative width of the size

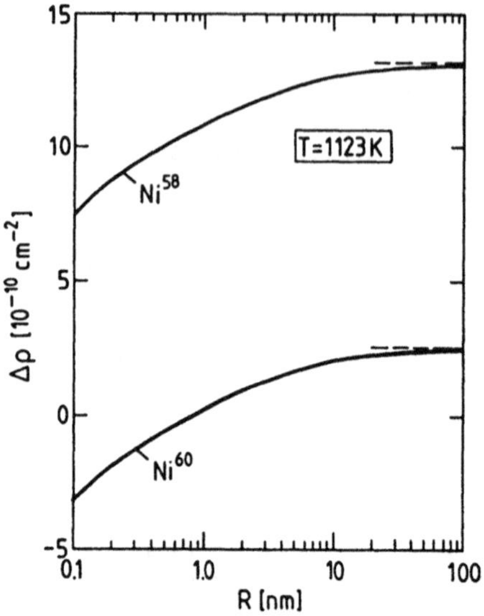

Fig. 4: Scattering contrast $\Delta\rho$ for He^4 in Ni^{58} and Ni^{60} as a function of the bubble radius R

distribution. The application of equ. 5 to the scattering data is standard and e.g. thoroughly discussed in Ref. /8/. However, it should be mentioned that - in contrast to solid precipitates - there is a size dependence of $\Delta\rho$ in our case since Ω_{He} depends on R. For the case of equilibrium bubbles this can fully be taken into account (see Equ.(1) and Fig. 2). As an example $\Delta\rho(R)$ is given in Fig. 4 for He^4 in in Ni^{58} and Ni^{60}, respectively.

4. Experimental

Foil specimens of annealed nickel of about 100 μm thickness and 8 mm diameter were implanted at room temperature with α-particles and He^3 ions, respectively. Continously varying the particle energy from 0 to 28 MeV ensured a spatially homogeneous deposition of He atoms whose final concentration was around 1000 appm. Then isochronal (2 hrs) anneals for each pair of specimens were performed in high vacuum. After each annealing step SANS measurements were carried out. After the last annealing step at 1123 K and the last SANS measurement the specimens were thinned for TEM observation of the bubble structure. The SANS investigation was performed at the SANS facility in Jülich /8,9/ and in Oak Ridge /10/. A more detailed description of the experiments is given in Ref. /8/.

5. Results

(a) He^4 in Ni^{58}/Ni^{60} /8,9/

Fig. 5a shows the neutron scattering results after final anneal of the Ni^{58} and Ni^{60} specimen at 1123 K. The two observed Q^{-4} (Porod)-regions of the Ni^{58} specimen suggest a bimodal size distribution of the bubbles.

In the Ni^{60} specimen only the scattering from the large-sized bubbles can be measured and an α-value of 27 ± 0.2 which is close to the calculated value of a void is found (Fig. 3a). As expected (see Fig. 3a), the smaller bubbles give no measurable contrast, so that only a lower limit for α -value could be determined to be larger than 100, which gives an ω -value of 0.7 ± 0.2 (Table 2). This value is compatible with the ω-values obtained from the averaged radius $\bar{R}$ (determined from SANS and TEM) and assuming equilibrium bubbles (see Fig. 2 and Table 2).

Table 2: TEM and SANS results of He^4-bubbles in Ni^{58} and Ni^{60} specimens after final anneal at 1123 K

	ω_{exp}	ω_{cal}	$\bar{R}$(nm)	σ(nm)	$N_0(m^{-3})$	He_{cal}	He_{impl}
SANS	0.7 ± 0.2	0.7. ± 0.1	1.2	0.5	9 10^{22}	910	1000
TEM		0.6 ± 0.1	1.4	0.5	7 10^{22}		

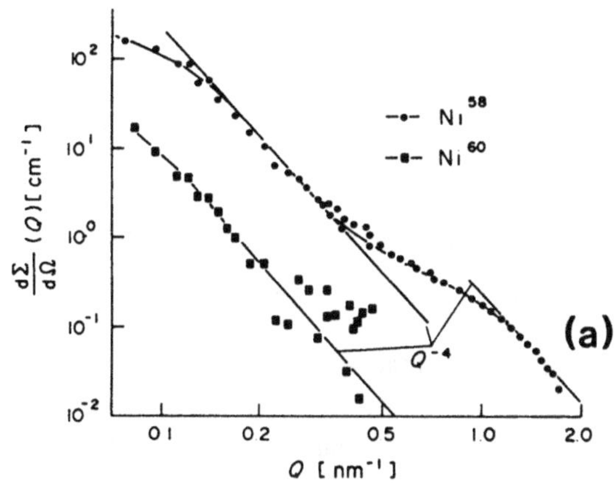

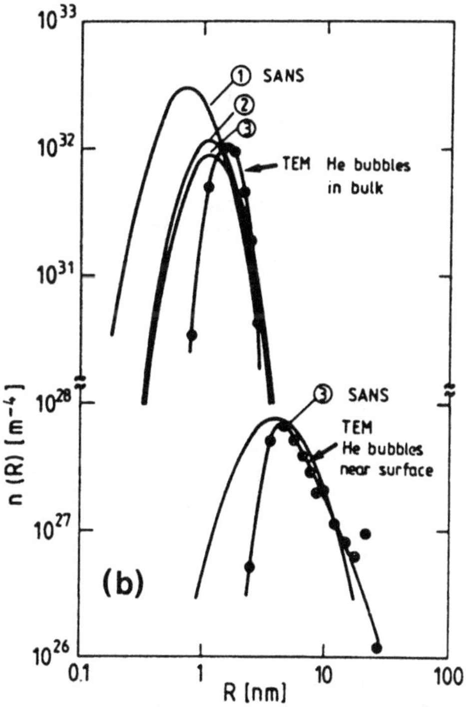

Fig. 5: (a) Scattering curves of He^4 bubbles in Ni^{58} and Ni^{60} specimens after final anneal at 1123 K /8/.

(b) Size distribution of He^4 bubbles in Ni^{58} after annealing at 1073 K for 2 h (curve 1), 6.75 h (2) and 50 h (3) determined by SANS. Also shown are the TEM results for the specimen annealed for 50 h /9/.

A detailed TEM study /9/ has shown that the bubbles are inhomogeneously distributed in the sample. Whereas the smaller bubbles are observed throughout the bulk of the specimens, the larger ones are situated only near the surface,with great probability.

(b) He^3/He^4 in Ni /10/

Here the α-values have been determined for three different annealing temperatures and yield ω -values via Fig. 3 (right part). A simultaneous evaluation of the scattering curves yieldes the corresponding size distributions and mean radii $\bar{R}$.*)

*) For the last annealing temperature of 1123 K the size distribution from SANS has been confirmed by TEM

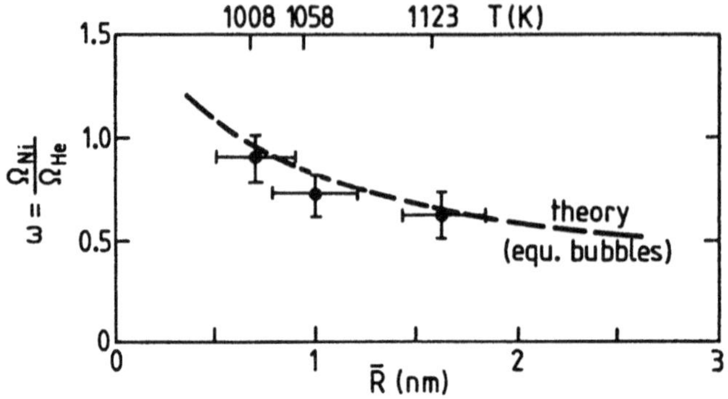

Fig. 6: Atomic volume ratio ω as function of mean radius $\bar{R}$ of He bubbles in Ni annealed at three different temperatures for 2 h. The dotted line gives the theoretical relation assuming equilibrium bubbles.

Combining this information allows to plot ω as a function of $\bar{R}$ as shown in Fig. 6. Also shown is the theoretical relationship between ω and R assuming equilibrium bubbles (Equ. 1), γ = 2N/m for Ni and Trinkaus' equation of state /6/. Considering the fact that no adjustable parameter is involved, the agreement between theory and experiment is satisfactory and justifies the assumption of equilibrium bubbles. These findings have been confirmed on He^3/He^4 in Ni^{58} specimens which will be presented in detail in Ref. /10/.

6. Conclusion

(1) First experiments on Ni specimens have demonstrated that SANS contrast variation can yield direct information on the helium density in bubbles if
   (a) 2 specimens of different matrix/helium isotopes with identical bubble structure are available
   (b) there is non-zero and continuously varying contrast in the region of interest ($0 < \omega < 2$)

(2) Combination of TEM and SANS yields detailed and reliable information on bubble size distributions.

In the future we plan a systematic investigation on He^3/He^4 bubbles in Ni^{58} with narrow annealing steps and better counting statistics in order to determine helium densities and size distributions particularly in the early states of bubble formation. This region is not accessible to TEM and would yield valuable information on the nucleation stage.

In the further future polarized neutrons will be applied. In such a study only one specimen is needed, which would eliminate the difficulties of providing two specimens with identical bubble structure. The method is, however, limited to ferromagnetic materials.

References

/1/ Proceedings of the Intern. Symp. on Fundamental Aspects of Helium in Metals, Ed. H. Ullmaier, Rad. Effects 78, 1 - 426 (1983)

/2/ H. Ullmaier: Nucl. Fusion 24, 1039 (1984)

/3/ H. Schroeder, W. Kesternich and H. Ullmaier: Nucl. Eng. and Design/Fusion 2, 65 (1985)

/4/ W. Jäger, R. Manzke, H. Trinkaus, Grecelius, R. Zeller, J. Fink, and H.L. Bay: J. Nucl. Mat. 111&112, 674 (1982) and J.C. Rife, S.E. Donnelly, A.A. Lucas, J.M. Gilles and J.J. Risko: Phys. Rev. Lett. 46, 1220 (1981)

/5/ S.E. Donnelly, J.C. Rife, J.M. Gilles and A.A. Lucas: J. Nucl. Mat. 93&94, 767 (1980)

/6/ H. Trinkaus: in Ref. /1/ p. 189

/7/ A. Guinier and G. Fournet: Small-Angle Scattering of x-rays (J. Wiley, New York (1955)

/8/ D. Schwahn, H. Ullmaier, J. Schelten and W. Kesternich: Acta Met., 31 2004 (1983)

/9/ W. Kesternich, D. Schwahn and H. Ullmaier: Scripta Met; 18, 1011 (1984)

/10/ D. Schwahn, W. Kesternich, S. Spooner, H. Schroeder and H. Ullmaier: to be published

Studies of Stress- and Fatigue-Induced Defects Using Small-Angle Neutron Scattering

J.R. Weertman

Northwestern University, Department of Materials Science and Engineering and Materials Research Center, 2145 Sheridan Road, Evanston, IL 60201, USA

The technique of small-angle neutron scattering (SANS) is proving to be of great value in the study of deformation-induced defects in various materials. Frequently SANS is the only practicable method for obtaining detailed, quantitative information on the kinetics of defect production. In crystalline material, the following defects produced by deformation are most amenable to study by SANS: bulk microcracking in brittle material; voids, bubbles and vacancy clusters; dislocations. If a broad definition is adopted for the types of defects to be considered, we can include the accelerated coarsening of particles and precipitates produced by high-temperature deformation. To date the greatest success with SANS in the study of defects associated with stress has been achieved with cavities (voids, etc.) and microcracks. The presence of dislocations can be detected by neutron scattering, but more extensive data on their nature, disposition and density usually can be obtained readily by TEM examination of thin foils. Accordingly, emphasis in this review will be on those defects for which neutron scattering has yielded detailed information not readily available by other methods.

1. Cavitation in Ceramics

Many ceramic materials maintain their strength to higher temperatures than do metals and alloys. Therefore ceramics frequently are used in applications in which elevated temperatures are encountered. A common mode of failure for ceramics under such circumstances involves the nucleation, growth and coalescence of grain boundary cavities. Not only is it of scientific interest to understand the processes by which pores are initiated and by which they grow in various materials, but a quantitative knowledge of the dependence of nucleation and growth rates on deformation parameters is essential in deriving accurate life prediction relationships. The group at Southwest Research Institute and their collaborators at Oak Ridge National Laboratory and the United States National Bureau of Standards have demonstrated the value of the SANS technique in elucidating the cavitation processes which occur in different types of ceramics [1-4]. The knowledge obtained from these experiments has permitted the critical assessment of existing cavitation models and development of new ones. For example, in a recent theory [5] of grain boundary failure in ceramic material, it is assumed that the voids nucleate at grain boundary triple points and grow outward across grain facets in a crack-like fashion, one crack per facet. In the absence of any data to the contrary this is a reasonable assumption. One of the great virtues of the SANS technique is that, in suitable systems, information can be obtained on both the number density and the total volume fraction of voids or pores.

In a series of ongoing experiments [1-4], PAGE and his colleagues are using SANS to study the kinetics of nucleation and growth of grain boundary voids in two types of ceramics. In one case the grain boundaries are "clean", containing no second phase. In the second class, because of processing requirements, the grain boundaries are covered with a glassy film. At elevated temperatures such films become viscous. It has been postulated that, in the former case, voids grow by vacancy diffusion, in the latter, by viscous hole growth. Sintered alumina doped with MgO (Lucalox) was used to study cavitation behavior in material with clean boundaries [1,3]. Samples of the material were crept in compression at 1600°C under a stress of 140 MPa for varying lengths of time. Flat specimens with parallel faces for SANS measurements were cut from the crept material, such that the creep axis was contained in the plane of a specimen. Extensive examination of the material revealed little in the way of second phase particles. Therefore it was assumed that the usual subtraction process involving an undeformed specimen was unnecessary and that the small angle scattering from as-received material may be interpreted as coming from pores initially present in the material. All SANS data from the crept and uncrept specimens were radially averaged and extrapolated at small and large values of q using the Guinier and Porod approximations [6]. Thus the differential scattering cross-section, $d\Sigma/d\Omega$, was found for all values of q. (The scattering vector $q=4\pi \sin \theta/\lambda$, where 2θ is the scattering angle and λ is the neutron wave length.) With this information it is possible to calculate values for the Porod radius R_p, the Guinier radius R_g, and the void volume fraction $\Delta V/V$ [6]. A measure of the void number density N_p/V also can be obtained from the relationship $(4/3)\pi R_p^3 N_p/V=\Delta V/V$. It was found that both Np/V and $\Delta V/V$ increase linearly with creep time. Thus the individual void growth rate dv/dt=0. In this type of ceramic the nucleation of voids is continuous but, after nucleation and perhaps a brief period of transient growth [7], further growth ceases. Therefore failure is nucleation-controlled. Distributions of void sizes (Fig. 1) were calculated using methods developed by Schmidt and co-workers [8,9]. It can be seen that the number of voids increases with increasing creep time (strain) but that the distributions do not move to higher sizes. This behavior is consistent with the conclusion that dv/dt=0. Examination by TEM of two stage replicas of fractured specimens revealed the presence of rows of round, approximately equi-sized and equispaced grain boundary voids. The void sizes agreed well with the SANS results. A model of cavity formation consistent with the SANS and TEM observations was developed which envisages voids forming at jogs on grain boundary ledges. Grain boundary sliding provides the high tensile stress normal to the jog surfaces which is necessary for nucleation and transient growth. Diffusional accommodation soon relaxes the driving stress to the point that further growth ceases. Void nucleation rates obtained from the SANS measurements were used to check the theory of void nucleation developed by RAJ and ASHBY [10] and others. In order to obtain nucleation rates from the theory which are comparable to the measured values, it is necessary to assume a tensile stress normal to the boundary of the magnitude of $10^{-2}E$, where E is Young's modulus. This value is comparable to the stress required for athermal fracture.

SANS studies of grain boundary cavitation in a ceramic containing a continuous glassy film on the grain boundaries (hot pressed SiC) gave considerably different results [2-4]. The creep conditions, SANS measurements and data analysis were similar to those in the sintered alumina experiments. Although few pores are

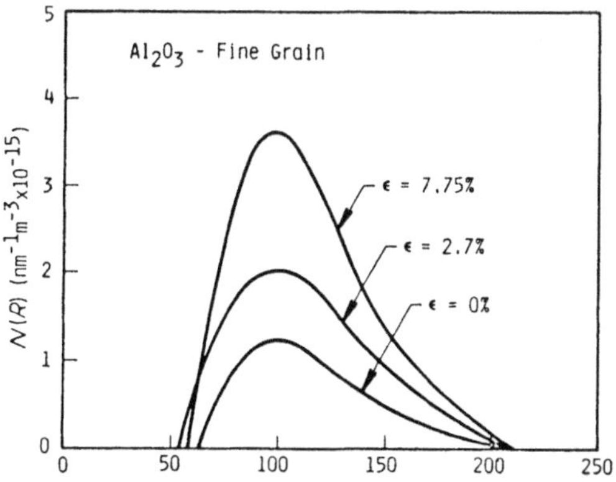

Fig. 1. Evolution of the cavity size distribution with creep strain ε in fine grained alumina [1].

present in the as-received material. it was necessary to subtract the scattering from an uncrept blank from that from the crept specimens, since the glassy phase as well as stable precipitates added to the scattering from the creep cavities. It was found that the number density of voids, Np/V, rises rapidly upon commencement of creep but soon stabilizes to a time-independent value. However the increase of $\Delta V/V$ is such that the individual voids must be growing at a continuing, though decelerating rate: $dv/dt \propto t^{-0.37}$, where t is time after void nucleation. Scattering isointensity contours on the two-dimensional detector show that the small angle scattering initially is isotropic, becomes strongly anisotropic as creep progresses (with the stronger scattering perpendicular to the creep axis), then eventually returns to a nearly isotropic behavior.

These observations can be used to model cavitation in a ceramic with a continuous glassy film on the grain boundaries. In the first place, the measured number density of voids is far too high to be consistent with the model of EVANS and RANA [5]. Most of the voids must be sited on two-grain faces, rather than solely at triple junctions. It is assumed by PAGE and co-workers [2,4,11] that the voids grow by viscous hole growth. The necessary tensile stress is provided by grain boundary sliding (Fig. 2). According to the model [11] the voids, which are initially equiaxed, become increasingly crack-like as they grow (Fig. 3). Depending upon the initial void spacing, the shape remains crack-like or returns to an equiaxed morphology as growth proceeds further. The observed behavior of the SANS isointensity contours is interpreted as evidence in support of the model.

In summary, the details of the creep cavitation process in two classes of ceramics have been revealed by SANS studies. It is unlikely that the differences between the cavitation behavior in the two materials would have been detected by other techniques, since in both types of ceramic the specimens end up with voids of about 100-150 nm diameter. Life in sintered alumina is controlled by the nucleation process, whereas failure in the ceramic with a glassy grain boundary film can be modelled as the slow growth to coalescence of a constant number of voids.

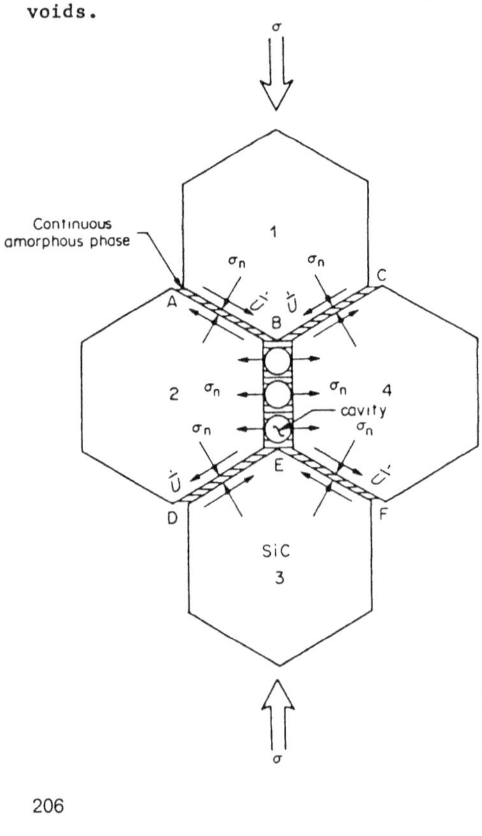

Fig. 2. Model of cavitating boundary, BE, showing origin of normal tensile stress [11].

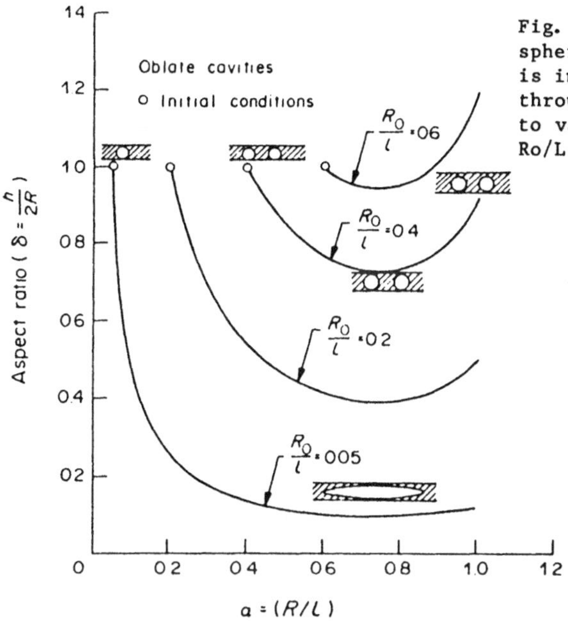

Fig. 3. Aspect ratio vs R/L for oblate spheroid cavities. R is cavity radius, L is initial void spacing, assumed constant throughout creep. The curves correspond to various initial values of the ratio, Ro/L [11].

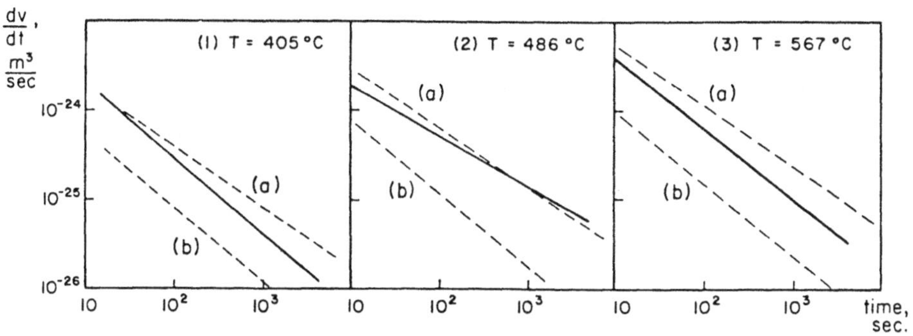

Fig. 4. Log of individual void growth rate vs. log time. Dashed curves are theoretical curves corresponding to (a) zero mean stress; (b) compressive mean stress equal in magnitude to the amplitude of the applied fatigue stress [18]. Solid curves are calculated from SANS dats. (1) T = 405°C, (2) T = 486°C, (3) T = 567°C. Time is measured from instant of void nucleation and is in seconds. From [14].

2. Cavitation in Single Phase Metals

Small angle neutron scattering has been used to study grain boundary cavitation in the metals Ni [12], Cu [13-15] and Cu-Al [16]. The SANS data were analysed by procedures similar to those described in the previous section. As in the case of ceramics, significant variations in cavitation behavior have been found. However in these studies it is the mode of deformation which is responsible for changes in the kinetics of cavitation. Extensive SANS studies of fatigue-induced cavitation in Cu and Cu-Al show that cyclic deformation with zero mean stress causes continuous nucleation of voids, but the growth rate dv/dt of individual voids is very transient and rapidly falls to zero after nucleation (Fig. 4). The evolution of void size distributions is similar to that seen in Fig. 1. The high nucleation rate in the fatigued metals ("high" as compared to the rate observed under time-independent applied stresses of similar magnitude) is not surprising. The high transient stress concentrations needed for void nucleation which are associated

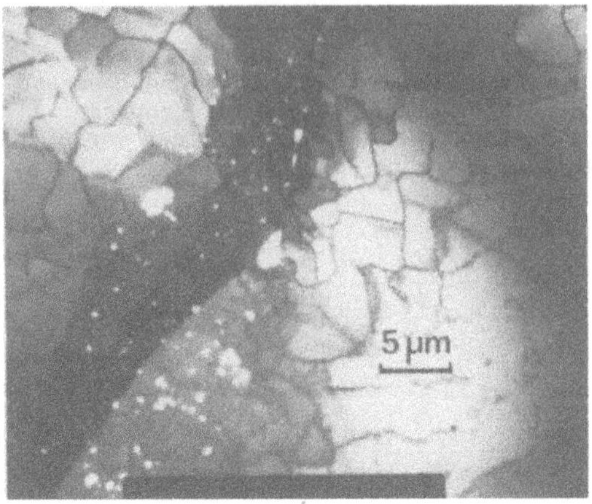

Fig. 5. Mobile grain boundary, cavities and dislocation substructure in copper bicrystal fatigued at 405°C [21].

with the onset of grain boundary sliding are quickly relaxed by diffusional accommodation. Continuous nucleation under creep conditions has been ascribed to the stochastic nature of grain boundary sliding [17]. In the fatigued copper specimens grain boundary sliding is continually being initiated by the cyclic nature of the driving stress. At least two factors contribute to the highly transient nature of individual void growth. Analysis [18] of void growth by vacancy diffusion under a completely reversed stress cycle predicts such behavior (Fig. 4). Furthermore,grain boundaries are subject to considerable migration during fatigue, as demonstrated by LANGDON and co-workers [19,20]. In the course of migration grain boundary voids are dumped into the matrix, where they no longer grow and may slowly shrink under surface tension forces. Figure 5 shows a large number of voids in the region over which a migrating grain boundary has passed in a copper bicrystal fatigued at 405°C [21]. (The voids in Fig. 5 were enlarged when they were perforated during preparation of the TEM foil.) Thus a substantial fraction of the voids which contribute to the small angle scattering from samples fatigued at high temperatures are no longer on grain boundaries and probably do not contribute to the failure process.

When specimens are subject to steady state stresses (creep conditions) the kinetics of cavitation change markedly [15,22]. At the same temperature (405°C, one half the melting temperature) and under comparable applied stresses, the void nucleation rate in copper drops by several orders of magnitude in going from cyclic to static loading. However the average individual void growth rate dv/dt in creep decreases only slowly with time after nucleation and is a few orders of magnitude greater than the fatigue growth rate. Void size distributions calculated from the SANS data show that creep voids are much larger than those produced by fatigue. Furthermore, unlike the case of fatigue, the creep size distributions shift to larger void sizes as deformation proceeds. The evolution of creep void size distributions, as measured by SANS, appears to be in excellent agreement with predictions based on nucleation rates measured by SANS combined with theories of void growth by vacancy diffusion [15].

3. Cavitation in Alloys Containing Second Phase Particles

In the absence of embrittling elements, large stress concentrations are required for the nucleation of grain boundary voids [17]. Hard grain boundary particles are excellent sources of stress concentrations and therefore are favored sites for the initiation of voids. The sort of detailed information on the kinetics of cavitation in engineering alloys which has been obtained for single-phase materials by SANS would be of great value as input into life prediction models. Several SANS investigations have been carried out of creep cavitation in alloys which contain hard particles in the form of carbides [23-25]. In the cases in which creep started with specimens in a solutionized condition [23,24], concurrent precipitation and cavitation make interpretation of the scattering data difficult. In a recent experiment PAGE [25] began with fully precipitated specimens of 304 stainless steel and alloy 800. The scattering intensity was observed to fall during the initial stages of creep but then increased with further deformation. It was concluded that the initial decrease was caused by carbide coarsening accelerated by deformation, the eventual upturn by the increasing contribution from cavities. A final short solutionizing treatment and quench has been used to eliminate the small carbides [23-25], leaving only cavities to contribute to the small angle scattering.

4. Vacancy Clusters in Fatigued Single Crystals

The production of point defects and vacancy clusters in crystals by cyclic deformation has been predicted by several models (e.g., 26). KETTUNEN, LEPISTO, KOSTORZ and GOLTZ [27,28] have used SANS to measure changes in copper single crystals fatigued at room temperature. Careful scattering measurements showed an extensive Guinier region which indicates the presence of small voids of approximately 2 nm diameter. Their volume fraction is $\sim 10^{-5}$. At smaller values of q a plot of $\ln d\Sigma/d\Omega$ vs $\ln q$ shows a slope of -3. Scattering in this region is dominated by dislocations. At even smaller q's the magnitude of the slope drops, which is interpreted as a manifestation of the presence of dislocation dipoles. The extensive TEM studies of fatigued Cu crystals in the literature confirm that a large fraction of the dislocations are arranged in dipoles or multipoles.

5. Bulk Microcracking in Brittle Materials

Experiments carried out by CASE and GLINKA [29] illustrate the type of information on bulk microcracking which can be obtained by SANS. (It should be noted that, in their case, the microcracking was produced by a phase transformation, not deformation.) The experiments were performed on the ceramic $YCrO_3$. This material undergoes a phase transformation at 1100°C which produces fine microcracks as it is cooled through the critical temperature. Prolonged annealing below 1100°C heals the cracks, so that the contribution by the cracks to the total small angle scattering can be determined by subtracting the scattering from a healed specimen from one which is cracked. That the scattering was not influenced by refraction effects was verified by measuring $d\Sigma/d\Omega$ as a function of q at several neutron wave lengths. Attempts were made to fit the radially averaged cross-sections to several theoretical curves. The best agreement between theory and data was found for the scattering predicted from an ensemble of randomly oriented thin disks. An average disk height of 24 nm was obtained from the curve-fitting procedure. An average crack radius and crack number density were obtained from a knowledge of the total surface area of the cracks (obtained from Porod's law) combined with the results of measurements of changes in Young's modulus and Poisson's ratio caused by the microcracking. (The moduli changes depend on crack density and average crack size.) An average crack radius of 5.7 μm was found. A model of microcracking caused by localized stresses predicts cracks of 1-2 grain diameters. The average grain size was 6 μm.

6. Accelerated Coarsening of Precipitates

While the accelerated coarsening of strengthening particles by high temperature creep or fatigue is not a deformation-induced defect in the same sense as is a dislocation, void or crack, such coarsening certainly changes mechanical behavior and can hasten failure. One example of accelerated coarsening was given in Section 3 [25]. SCHWAHN, KESTERNICH and SCHUSTER [30] have used SANS and TEM to study in detail the coarsening and "rafting" of γ' particles in a nickel base superalloy subjected to high temperature creep. Strong anisotropy in the two-dimensional scattering patterns developed during creep deformation, but dropped near the fracture region, probably because of the damage accumulated there in the form of cracks and voids. This observation suggests a potential application of SANS for nondestructive evaluation. SANS also has been used to study the influence of creep and cyclic deformation on the carbide population in a ferritic stainless steel [31].

7. Problems and Pitfalls in Interpretation of SANS Data

In order to calculate the volume fraction, size distribution and number density of an ensemble of scatterers from SANS data it is necessary to know the value of $d\Sigma/d\Omega$ over all values of q. The Guinier and Porod approximations must be used to extrapolate the data at the low q and high q ends of the measurements. In most of the scattering systems described in the previous sections, the defects were large enough that the SANS measurements reached well into the Porod region. However it is difficult to extend the measurements to sufficiently small values of q to obtain data in the Guinier region ($q R_{max} \lesssim 1$, where R_{max} is the radius of the largest scatterer in the ensemble which occurs with appreciable frequency). YOO et al. [12] have pointed out the effect which changing the Guinier extrapolation has on the calculated size distribution of the scattering ensemble. While global properties such as volume fraction and number density of the scatterers may be changed only by a factor of 2 or 3 (which usually is within the limits of uncertainty of the relevant models) the size distribution is seriously affected. These observations are illustrated by Fig. 6, which shows the changes in void size distribution, volume fraction and number density which result from extending the range of measured $d\Sigma/d\Omega$ from a minimum q of 0.039 down to 0.021 nm^{-1} [22]. (The scattering data were taken from specimens of copper fatigued at 405°C, 17 Hz, with a stress amplitude of 34 MPa.) Even in the case of the lower limit of data at $q=0.039\ nm^{-1}$, the extent and degree of linearity of the "straight line" portion of

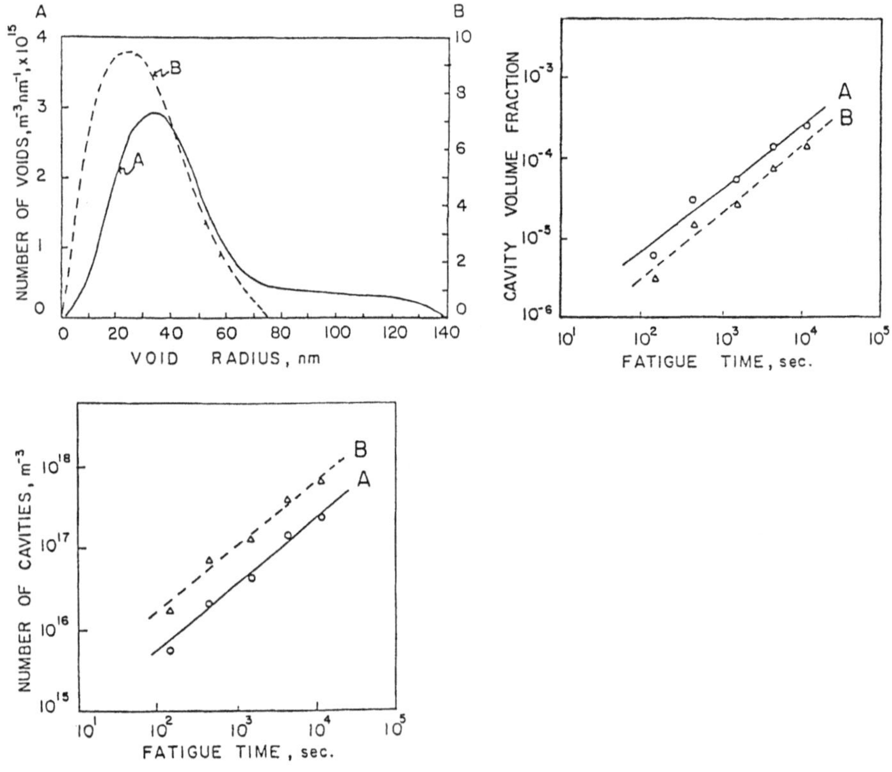

Fig. 6 (a) Comparison of void size distributions calculated from SANS data extending to a minimum q of 0.021 nm^{-1} (curve A), 0.039 nm^{-1} (curve B). Copper fatigued 7.5 x 10^4 cycles at 405°C, 17 Hz, stress amplitude of 34.0 MPa, (b) change in calculated void volume fraction and (c) void number density resulting from changing the lower limit of q. Copper fatigued for varying lengths of time under the conditions of (a). From [22].

the Guinier plots are comparable to those of the other cavitation and coarsening experiments described in the preceding sections. The values of R_p and R_G determined for voids in the specimen of Fig. 6a using the different minimum values of measured q are: R_p = 40.3 nm, R_G = 45.3 nm with q_{min} = 0.039 nm^{-1}, and R_p = 78.8 nm, R_G = 92.3 nm with q_{min} = 0.021 nm^{-1}. These large changes in characteristic radii with rather modest changes in the Guinier extrapolation signal the possibility of serious error in conclusions drawn about evolution of scatterer size based on changes in R_G or R_p. As the ensembles of scatterers in a series of specimens move the SANS data ever further from a valid Guinier approximation, spurious changes in R_G or R_p may be introduced.

It is almost impossible to assign error bars to a size distribution of scatters. Statistical or instrumental errors in the data, errors introduced by invalid Guinier or Porod extrapolations - all contribute in a complicated fashion to the final error in the size distibution. Several checks can be run on the SANS results. These include independent determination of scatterer volume fraction, e.g., by density measurements [2,32] or of size distributions by TEM; noting the agreement between measured $d\Sigma/d\Omega$ and values back calculated from the derived size distribution; comparison of R_G, R_p, $d\Sigma/d\Omega$ at q=0 determined directly from the appropriate extrapolations and integrations of the data with values calculated from the size distributions. These self-consistency checks do not necessarily guarantee that all is well. Figure 7 shows the excellent agreement between the measured values of $d\Sigma/d\Omega$(solid line) and those calculated back from size distribution A of Fig. 6a. However, the points calculated from curve B of Fig. 6a (not shown on Fig. 7) also agree well with the measured $d\Sigma/d\Omega$ curve, but only down to the minimum q of 0.039 nm^{-1}. Evidently self-consistency in the type of calculation show in Fig. 7 is a necessary but not sufficient condition for insuring accuracy of a size distribution.

Different sorts of problems arise when a specimen contains two (or more) types of scatterer: when both ensembles of scatterers change as the result of deformation; when the two types of scatterers are correlated. Both of these problems are illustrated by the studies of cavitation in stainless steel and alloy 800 [23-25]. Even if the carbides were entirely stable throughout deformation, the presence of three types of interface (matrix/carbide, void/carbide, matrix/void) complicates the interpretation of scattering data.

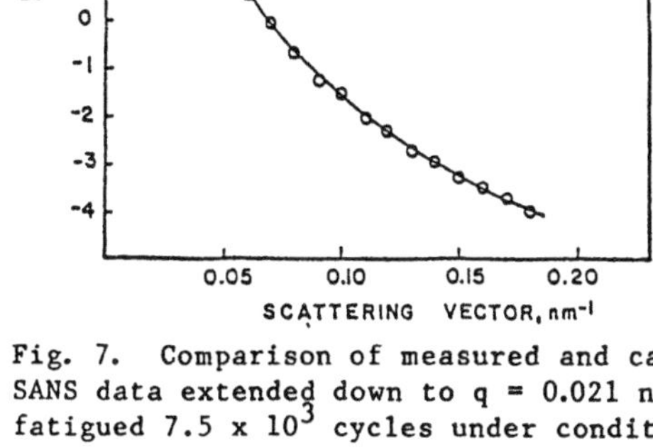

Fig. 7. Comparison of measured and calculated scattering cross-sections. The SANS data extended down to q = 0.021 nm^{-1}. Unit for ($d\Sigma/d\Omega$) is cm^{-1}. Copper fatigued 7.5 x 10^3 cycles under conditions of Fig. 6. From [22].

Acknowledgments

This work was supported by the National Science Foundation, Grant DMR-8320157 and the U.S. Department of Energy, Grant DE-AC02-81ER10960.

References

1. R.A. Page, J. Lankford and S. Spooner, J. Mats. Sci. 19, 3360 (1984.)
2. R.A. Page, J. Lankford and S. Spooner, Acta. Met. 32, 1275 (1984).
3. J. Lankford, K.S. Chan and R.A. Page, Frac. Mech. of Ceramics, in press.
4. R.A. Page and S. Spooner, J. Mats. Sci., in press.
5. A.G. Evans and A. Rana, Acta. Met. 28, 129 (1980).
6. Treatise on Materials Science and Engineering, Vol. 15, G. Kostorz, Ed. (Academic Press, 1975).
7. R. Raj, Met. Trans 6A, 1499 (1975).
8. I.S. Fedorova and P.W. Schmidt, J. Appl. Cryst. 11, 405 (1978).
9. O.L. Brill and P.W. Schmidt, J. Appl. Phys. 39, 2274 (1968).
10. R. Raj and M.F. Ashby, Acta Met. 23, 653, (1975).
11. K.S. Chan, J. Lankford and R.A. Page, Acta Met. 32, 1907 (1984).
12. M.H. Yoo, J.C. Ogle, B.S. Borie, E.H. Lee and R.W. Hendricks, Acta Met. 30, 1733 (1982).
13. T. Saegusa, J.R. Weertman, J.B. Cohen and M. Roth, J. Appl. Cryst. 11, 602 (1978).
14. R. Page, J.R. Weertman and M. Roth, Acta Met. 30, 1357 (1982).
15. M.S. Yang, J. R. Weertman and M. Roth, Scripta Met. 18, 543 (1984).
16. J.G. Cabañas-Moreno, M.S. Yang, M. Roth, Z.Y. Zhang, G.D. Wignall and W.C. Koehler, "Studies of Grain Boundary Cavitation by SANS", in ASTM STP 811, J. Lankford, D.L. Davidson, W.L. Morris and R.P. Wei, Eds., (American Soc. for Testing and Materials, 1983).
17. A.S. Argon, I.W. Chen and C.W. Lau, "Intergranular Cavitation in Creep: Theory and Experiment", in Creep-Fatigue-Environmental Interactions, R.M. Pelloux and N.S. Stoloff, Eds., (TMS-AIME, Warrendale, PA, 1980).
18. J.R. Weertman, Can. Metall. Q. 18, 73 (1979).
19. T.G. Langdon and R.C. Gifkins, Scripta Met. 13, 1191 (1979).
20. P. Yavari and T.G. Langdon, Scripta Met. 14, 551 (1980).
21. J.G. Cabañas-Moreno, Ph.D. Thesis, Northwestern University, Evanston, IL, June, 1984.
22. M.S. Yang, Ph.D. Thesis, Northwestern University, Evanston, IL, June, 1984.
23. G.G. Nilsson and M. Roth, Mat. Sci. Eng. 50, 101 (1981).
24. M.H. Yoo, J.C. Ogle, J.H. Schneibel and R.W. Swindeman, in Strength of Metals and Alloys, 2, R.L. Gifkins, Ed. (Pergamon Press, Oxford, 1983).
25. R.A. Page, Met Trans A, in press.
26. U. Essmann, U. Gösele and H. Mughrabi, Phil. Mag A, 44, 405 (1981).
27. P.O. Kettunen, T. Lepisto, G. Kostorz and G. Goltz, Acta Met. 29, 969 (1981).
28. G. Kostorz, Mats. Res. Soc. Symposium Proceedings, 41, 107 (Materials Research Society, Pittsburgh, PA, 1985).
29. E.D. Case and C.J. Glinka, J. Mats. Sci. 19, 2962 (1984).
30. D. Schwahn, W. Kesternich and H. Schuster, Met. Trans. A 12A, 155 (1981).
31. S. Kim, Ph.D. Thesis, Northwestern University, Evanston, IL, June 1985.
32. M.S. Yang, J.R. Weertman and M. Roth, "Use of SANS to Study Grain Boundary Cavitation", in Creep and Fracture of Engineering Materials and Structures, B. Wilshire and D.R.J. Owen, Eds., (Pineridge Press, Swansea, 1984).

Neutrons in Practical Metallurgy

R.N. Sinclair and C.G. Windsor

Materials Physics and Metallurgy Division Atomic Energy Research Establishment Harwell, Didcot, Oxon, U.K.

The virtues of slow neutrons as unique probes of the microscopic structure of materials are proven to the extent that industrial use is now made of neutron studies to aid alloy development. Since most metals have low absorption cross-sections for neutrons and scattering mean paths of the order of a centimetre, penetration of the neutron beam through most metals is easy and facilitates "in situ" studies of annealing effects, texture and residual stress. In many cases combinations of SANS and wide angle diffraction results are required to characterise the microscopic transformations. Studies of the development of the hardening phases in commercial alloys are described to illustrate this point. In each case a simple model for the kinetics of the process is found to be applicable. Radiation can cause pressure vessel steels to harden and embrittle if the copper content is too high. This has been known for many years but the inhomogeneities causing the effect are difficult to observe and characterise. A systematic investigation by the SANS technique of a large number of samples of a pressure vessel steel with differing concentrations of minority components is presented. A close correlation between SANS due to 30 A defects, copper content and radiation hardening is found. The nature of the defect is discussed. It would be both efficient and accurate to be able to observe SANS, wide angle diffraction, texture and macroscopic stress simultaneously. A multifacility based on a pulsed neutron source has been designed for this purpose.

1. Introduction

Neutrons penetrate distances of the order of centimetres in most metals and so allow measurements to be made on actual engineering components or large samples. The near transparency of container materials, in particular aluminium, permits the 'in situ' study of specimens within a furnace, cryostat, pressure vessel, stress cell or chemical reactor vessel. These are unique advantages of slow neutrons as a probe. As compared to X-ray or electron studies, complementary information is often gained because of the nonsystematic variation of the neutron scattering length with atomic number, which gives an alternative contrast to that of the other techniques in coherent scattering studies. In the neutron case this contrast can often be varied by application of a magnetic field to give another means of scattering centre identification.

The serious disadvantage of using neutrons for practical metallurgical studies arises from the non-portable nature of the source; either an expensive reactor or accelerator. This involves advance booking, conforming to programme constraints and travel to the facility. In the UK, two materials testing reactors DIDO and PLUTO and a pulsed electron accelerator source HELIOS are available to industry for neutron beam studies and almost immediate access is open to industry once a commercial contract has been placed [1]. In practice these formalities determine the minimum time required to get an investigation started.

Four techniques have, to date, proved to be most useful in metallurgical studies [2]:

i) Radiography gives an alternative view of a specimen to that obtained with X-rays because of the almost random cross-sections of the elements, as discussed earlier. Static measurements are routinely used to verify the integrity of cooling channels in complex components and to determine the distribution of neutron-absorbing elements in reactor components. The low cross-sections at long wavelengths, particularly for steel or aluminium, have allowed neutrons to penetrate through complete assemblies, such as auto or aero engines, to give dynamic imaging of fuels, lubricants or coolants;

ii) Small angle scattering provides information on precipitates, voids and other defects with sensitivity governed by the neutron scattering lengths. Additional information to that obtained with the electron microscope also comes from the ability to observe defects with sizes ranging down to almost interatomic dimensions;

iii)Diffraction studies detect phases and measure their volume fraction and texture. These can be carried out on bulk samples without the need for extraction methods, so enabling 'in situ' ageing experiments to be performed;

iv) High-resolution diffraction gives a unique non-destructive method of measuring the strain tensor over a defined volume within a bulk metal. It is thus possible to determine the stress state of the material non-invasively. Large engineering components, in particular weld regions, can be observed in this way.

This paper describes studies in the categories (ii) and (iii) above. Finally, a multifacility which would mimic and extend the options available with an electron microscope is discussed.

2. The Hardening Phase in Maraging Steel

An example of the use of neutron diffraction and small angle neutron scattering (SANS) combined to investigate the effect of heat treatment on alloys is a study of maraging steel [3] carried out in conjunction with electron microscopy studies at the UKAEA's Risley Nuclear Laboratories. Maraging steels are high alloy steels containing 18% nickel, 13% cobalt, 4% molybdenum, 1.5% titanium. Their exceptionally high strength is achieved by ageing conventionally for 3 hours at 480 C to produce a hardening precipitate (Ni(3)Mo). Overageing at temperatures greater than this is known to produce softening because of re-austenisation of the steel. The growth of the precipitate hardening phase was followed by SANS both as a function of ageing temperature over 10 hour periods and also as a function of ageing time at 480 C using an "in situ" specimen within a furnace. The mean particle diameter was found to grow from 2.4 nm (24 A) after 0.25 hours to 14.6 nm (146 A) after 43 hours following quite closely the 1/3 power law as a function of time predicted by LIFSHITZ and SLYOZOV [4]. The volume fraction of the hardening phase remains roughly constant at about 10%. Reduced plots of smoothed SANS data from the 'in situ' ageing experiment at 480 C are shown in Fig 1. Mean values of the precipitate diameters (D) are derived from the Q range above the interference maximum via the Guinier approximation. As the volume fraction is constant, the observed time dependence only arises through the value of D and the simple scaling used to produce Fig 1 is possible. The close agreement between the curves at QD values above the interference maximum confirms the applicability of the simple model.

The growth of the transformed austenite phase was followed as a function of ageing temperature by diffraction. The austenite volume fraction varied from 1.3% after 10 hours at 480 C (not observed by X-ray studies) to 25% after 10 hours at 600 C. Recent experiments have observed the 'insitu' development of the hardening phase at temperatures below 480 C. The aim is to establish the heat treatment needed to produce the required volume fraction and mean particle diameter under conditions such that the unwanted softening austenite phase is not formed.

3. The Microstructure of an Evaporated Aluminium Alloy

In the search for corrosion-resistant aluminium alloys with greater thermal stability and Young's modulus than current industrial alloys, vapour deposition methods of fabrication are being developed [5]. In particular aluminium chromium solid solutions have been condensed at temperatures between 85 C and 360 C.Ageing is observed on subsequent heat treatment (below 400 C) of alloys containing a uniform distribution of fine precipitates, and is a dominant contribution to the strength of the final material. In an experimental study commissioned by the Royal Aircraft Establishment, Farnborough [6], vapour-deposited samples of an aluminium alloy containing 7.5wt % Cr and 1% Fe were studied by both SANS and diffraction techniques. 'Insitu' furnace experiments were carried out at 300 C, 350 C and 400 C with the temperatures controlled to 3 C. The evolution of the SANS was followed for a period of 24 hours for three

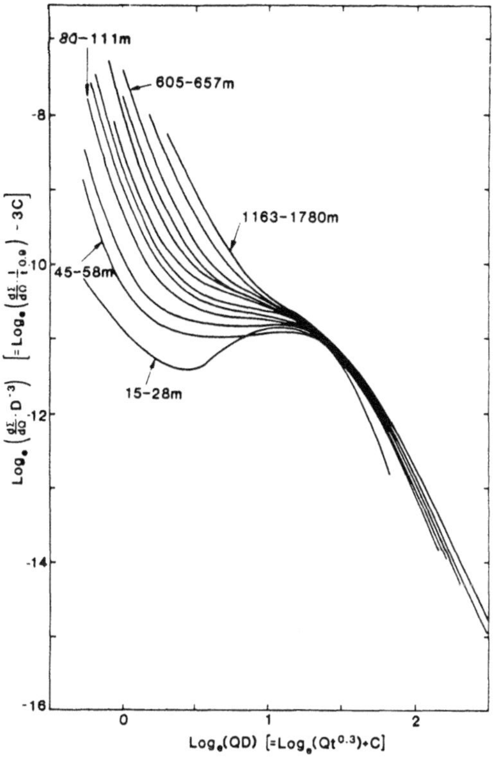

Figure 1. Reduced plot of smoothed SANS data from a maraging steel aged at 480 C [3].

samples at the three temperatures. Analysis gave the macroscopic differential cross section as a function of scattering vector (Q) and time t. It was found that the data for the three temperatures could be superimposed over a wide range of Q values by simple scaling of the time at each temperature. (t(400 C)=11t(350 C)=299t(300 C).

If an Arrhenius rate equation is applicable then the time taken to reach a given stage of the precipitation process t(y) at temperature T can be written

$$\ln t(y) = \text{const} - \ln A(A) + E(A)/k(B)T$$

where E(A) is the effective activation energy and A(A) is the effective multiplication factor. Figure 2(a) shows that such a model is correct in this instance and the precipitation is isokinetic. The derived activation energy of 43.7 Kcal/mol is closer to that for Fe diffusion in aluminium (46 Kcal/mol) than that for Cr (58 Kcal/mol). The scaling is confirmed by the agreement of the derived mean precipitate diameters (Fig 2(b)) when scaled in the same way.

Diffraction experiments are in hand to identify the fine precipitate responsible for ageing, and its relationship to the equilibrium CrAl(7) phase. The aluminium matrix has been shown to be both textured and heavily strained.

4. Microstructural Changes in Irradiated Pressure Vessel Steels

Neutron irradiation of ferritic steels such as those used for reactor pressure vessels can cause embrittlement. Residual elements in the steels can have large effects causing alloys of nominally the same composition to be radiation sensitive or insensitive according to the composition of these elements. Copper appears to have the most detrimental effect [7] and an increase in the ductile to brittle transition temperature as a function of copper content and fast neutron dose has been reported [8].

Attempts to characterise the radiation damage in irradiated samples by TEM have been unsuccessful and the features responsible for the hardening process have only

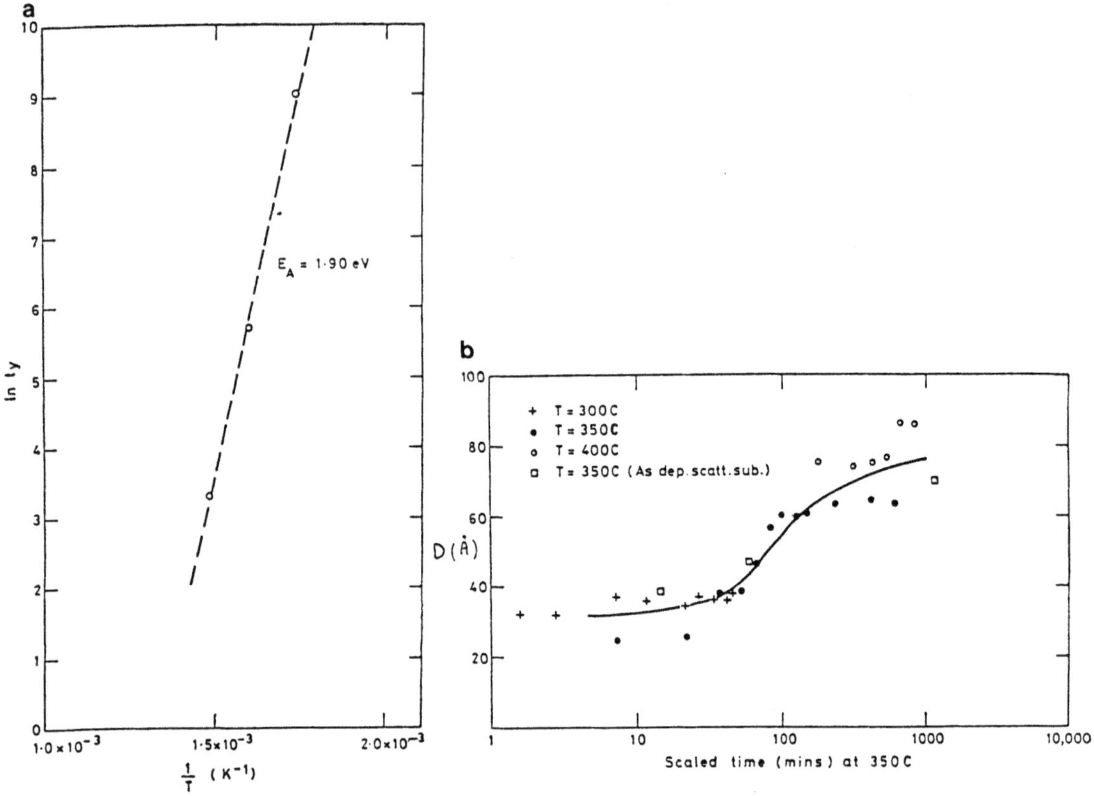

Figure 2. a) The Arrhenius plot for the evolution of CrAl7 precipitates in a vapour deposited aluminium-chromium alloy [6].
b) The change in mean precipitate diameter with equivalent ageing time at 350 C.

been observed through field ion microscope - atom probe studies [9] and SANS experiments [10][11].

In work for Berkeley Nuclear Laboratories of the Central Electricity Generating Board, UK, SANS has been used to investigate the influence of Cu, Ni and P on radiation hardening and embrittlement in nineteen samples of neutron - irradiated ferritic steels with differing minority element compositions [11][12]. In every case where significant copper is present, the subtracted intensity (irradiated-unirradiated) shows a significant increase in SANS which can be interpreted by the Guinier approximation to indicate defects with diameters less than 4 nm (40 A). The individual response of particular steels varies in a linear way with their copper content and also with the radiation conditions. Interestingly, an approximately linear relation exists between the induced SANS response and the measured hardening in all the steels (Fig 3). When comparing the SANS response for different steels, a strong microstructural effect is found, so as to make the response markedly different for bainitic and martensitic steels. The accentuating effects of both nickel and phosphorus are consistently reproduced and there is some evidence for a dose rate effect. The nature of the hardening features (which cannot be shown to be simple copper precipitates by magnetic contrast experiments), has been tentatively identified by correlating the SANS results with observed strength changes, using dispersed barrier hardening models [12]. The obstacles that have been identified in this way include copper precipitates (possibly containing iron or vacancies) and phosphides.

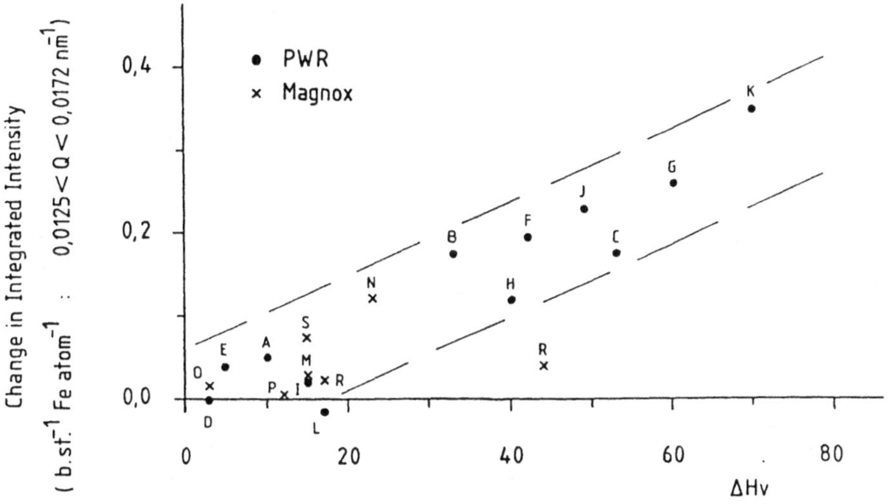

Figure 3. The change in SANS integrated intensity as a function of radiation-induced hardness change for a series of low alloy pressure vessel steel samples [12].

5. A Multi-Function Facility for Material Science

Experience with experiments from the Harwell applied neutron beam programme, such as those described above, has led to an appreciation of the advantage for this type of work of a facility where the most important neutron scattering experiments can be carried out simultaneously. This has resulted in a proposal [13] for such a multi-facility to be installed on the Spallation Neutron Source recently commissioned at the Rutherford Appleton Laboratory, UK. The thinking behind the instrument is illustrated by recent developments in electron microscopy. When a feature of interest is found by transmission microscopy, a battery of techniques can be applied to the same area of the specimen. Neutron scattering offers an even wider range of complementary techniques.

The proposed instrument (Fig 4) offers simultaneous measurement in eight different areas. These may be listed in an estimated priority order for an average problem. 1. Microstructure (SANS) 2. Phase Identification 3. Internal Stress 4. Defect Studies 5. Local Order 6. Texture Measurement 7. Diffusion Measurement 8. Vibrational Spectroscopy.

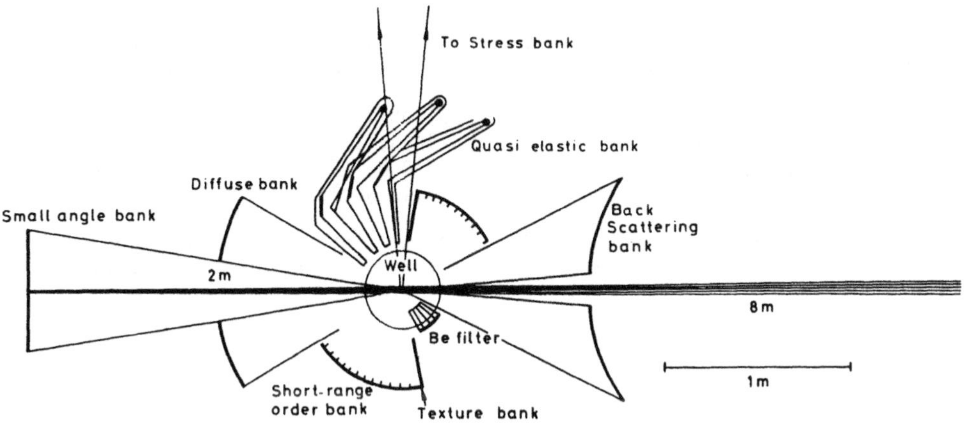

Figure 4. A schematic layout of a proposed materials multi-facility for installation on the SNS [13].

All these properties may be studied in a bulk specimen 'in situ' during a heat treatment or chemical reaction as a function of time. The possibility of using narrow incident and scattered neutron beams allows the simultaneous monitoring of these properties as a function of position across the sample. Only by using the 'white' neutron beam from a pulsed source can all these properties be measured at the same time and the same point on the specimen. Calculation of the performance of the facility for a series of key experiments in applied materials science shows that, with SNS at design intensity, a complete neutron examination of a defined volume in the sample can be made within minutes. The advantage in its use for kinetic experiments lies in the ability to relate together the many different processes which may take place at the same time. A secondary advantage is the neutron economy which results in a relatively low cost for a complete process study.

6. Acknowledgments

We would like to thank Berkeley Nuclear Laboratory, CEGB for permission to present data on pressure vessel steels and the Royal Aircraft Establishment, Farnborough for permission to present data on an evaporated aluminium alloy. We are also indebted to our colleagues in the applied neutron beam programme at Harwell and G Slattery at Risley Nuclear Laboratories.

References

1. C.G. Windsor, A.J. Allen, M.T. Hutchings, C.M. Sayers, R.N. Sinclair, P. Schofield and C.J. Wright: "Neutron Scattering in the 90's" (IAEA, Vienna, 1985)p.575.
2. G.F. Slattery and C.G. Windsor: The Metallurgist and Materials Technologist. Feb 67 (1983).
3. C.G. Windsor, R.N. Sinclair, S. Faulkner, V. Rainey and G.F. Slattery: "Microstructural Characterisation of Materials by Non-Microscopical Techniques" Ed. Anderson, Eldrup Hansen, Jensen, Leffers, Lilholt, Pedersen, Singh (Riso Nat.Lab. 1984)p.583.
4. I.M. Lifshitz and V.V. Slyozov: J.Phys.Chem.Solids 19, 35 (1961).
5. R.L. Bickerdike, D. Clark, J.N. Eastabrook, G. Hughes, W.N. Mair, P.G. Partridge and H.C. Ranson: Royal Aircraft Est., Technical Report TR84087 (1984).
6. A. Bowen (private communication).
7. L.E. Steele: IAEA Technical Report Series No.163, (IAEA, Vienna, 1975).
8. L.E. Steele: Proc of 2nd Int.Conf. in Structural Mechanics in Reactor Technology" G, paper G-5/2, (Publication Commission of the European Communities Luxembourg, 1974).
9. R.G. Lott, S.S. Brenner, M.K. Miller and A. Wolfenden: Trans.Amer.Nucl.Soc. 38, 303 (1981).
10. F. Frisius and D. Bunemann: Proc.Conf. on the Irradiation Behaviour of Metallic Materials for Fast Reactor Components,(French Atomic Energy Commission, 1979)p.247.
11. R.B. Jones and J.T. Buswell: Proc. BNES Conf. on Dimensional Stability and Mechanical Behaviour of Irradiated Metals and Alloys, 2, 105 (1985).
12. J.T. Buswell, E.A. Little, R.B. Jones and R.N. Sinclair: Proc. 2nd Int. Symp. on Environmental Degradation of Materials in Nuclear Power System - Water Reactors (1985).
13. C.G. Windsor and R.N. Sinclair: Proc. of the Int. Collaboration on Advanced Neutron Sources, ICANS VIII. RAL report (1985) - in press. Nucl.Inst and Meth.(1985)-1n press.

Point Defects in Metals in Thermal Equilibrium

W. Trost, K. Differt, K. Maier, and A. Seeger

Universität Stuttgart, Institut für Theoretische und Angewandte Physik,
and Max-Planck-Institut für Metallforschung, Institut für Physik,
Postfach 800665, D-7000 Stuttgart 80, F.R.G.

Abstract: A brief critical review is given of the status of the experimental determination of the equilibrium concentration of point defects in metals. It is argued that the question whether, in addition to vacancies, self-interstitials contribute significantly must be considered open and that accurate simultaneous measurements of the macroscopic and microscopic (unit-cell) thermal expansion might help in answering it. For high-accuracy determination of the differential thermal expansion at high temperatures (melting points of Cu, Ag, Au and above) a laser-interferometry–neutron diffraction experiment is being built up at the Institute Laue–Langevin. Preliminary experiments on Cu are reported, indicating that the desired accuracy may be achieved.

1. Introduction

From the point of view of ease and reliability of interpretation, measurements on point defects in crystals under thermal equilibrium conditions are extremely attractive. In contrast to techniques in which supersaturations of defects are introduced, e.g. irradiation, cold-work, or quenching from high temperatures, the uncertainties related to the generation of and the interaction between the defects are largely avoided. Because equilibrium measurements are necessarily performed at high temperatures, interactions with foreign atoms are much less disturbing than in low- or intermediate-temperature experiments involving defect supersaturations. Most important is the fact that in thermal equilibrium under fixed conditions of temperature T and pressure the defect concentrations are always the same, so that one can be sure that measurements by different techniques pertain to the same defect population.

In principle, any measuring technique that responds to the presence of point defects and that can be employed at high temperatures may be used for thermal–equilibrium studies of defects. In practice, however, one encounters substantial difficulties, which till today have not yet been solved in an entirely satisfactory manner.

A "background problem" arises because for most properties p the change Δp due to the thermal–equilibrium defects is small compared to the contribution p_0 of the "defect-free" crystal and because in the temperature range in which Δp makes itself felt p_0 cannot be determined by independent measurements. The problem is illustrated by Fig. 1 [1], which shows a stronger increase of the specific heat of gold with increasing T above about 1000 K than below that temperature.

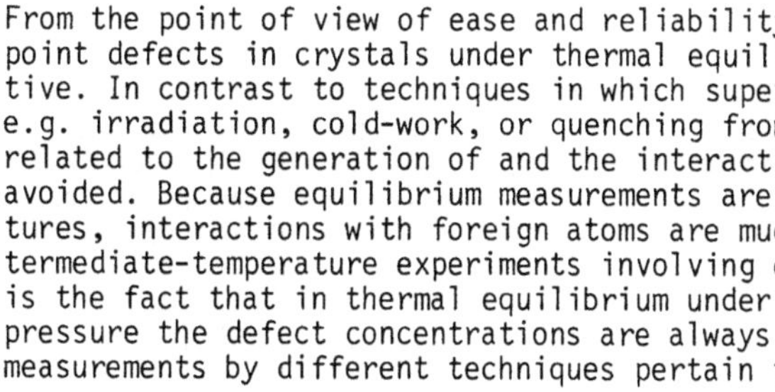

Fig. 1 Specific heat of gold at elevated temperatures (from [1])

Data such as Fig. 1, which is typical of many high-temperature measurements on metals, may be looked upon from two different viewpoints.

(i) $p_0(T)$ is determined by extrapolation from intermediate temperatures (dashed line), and Δp is identified with the difference between the high-temperature data and the extrapolation. In general, this procedure leads to rather high defect concentrations (at the melting temperature T_m of many metals of the order of magnitude 10^{-3} or even more).

(ii) $p_0(T)$ cannot be found by extrapolation (say, by means of power-law fits) from the temperature regime in which the defect concentration is negligible. It is argued that the anharmonic phonon contributions to p_0 (e.g., to the specific heat, to the thermal expansion coefficient, or to the electrical resistivity) may give rise to a T dependence that in practice cannot be separated from Δp, which hence might be responsible for only a small fraction of the high-temperature increase.

Viewpoint (ii) has received support from recent theoretical work [2] showing that certain pair potentials may give rise to an exponential temperature dependence of the high-temperature thermal expansion of metals. Viewpoint (i), which has been upheld for the last twenty years by Y.A. Kraftmakher [3], has been recently strengthened by the discovery of a relaxation effect in high-frequency specific heat measurements on tungsten [4]. Its attibution to the time-lag in defect generation or annihilation during small temperature changes implies quite high thermal-equilibrium concentrations near the melting point. However, on the basis of the information available at present we cannot exclude with certainty that the observed effect is partly or even entirely due to the experimental procedure and hence not intrinsic.

The background problem is substantially reduced if positron annihilation is used to study vacancy-type defects in thermal equilibrium [5]. Here another difficulty, the "prefactor problem", may arise. It is due to the fact that positron annihilation measurements give the product of the trap concentration times the trapping rate of the positrons per unit trap concentration, σ. Let us assume for the time being that it suffices to consider one type of defect with formation enthalpy H^F and formation enthalpy S^F only. We may write

$$\Delta p = p_{def} \exp(S^F/k_B) \exp(-H^F/k_BT), \tag{1}$$

where k_B denotes Boltzmann's constant and p_{def} the contribution of a unit concentration of defects to p. If σ is diffusion-controlled (this is in fact quite likely in metals at high temperatures), p_{def} will be temperature-dependent in a way which is difficult to determine independently [6]. This means that the quantity $-d(\ln\Delta p)/d(1/k_BT)$ as obtained from positron annihilation experiments should be interpreted as an effective formation enthalpy H^F_{eff} that may be different from the monovacancy formation enthalpy H^F. In close-packed metals the difference $H^F_{eff} - H^F$ may easily exceed the experimental uncertainty of H^F_{eff}. For the bcc Group-V transition metals it has been shown [7] that the high-temperature positron studies of vacancies are affected by detrapping which also leads to a temperature-dependent p_{def}. In the Group-VI transition metals the situation is even more open. It is feasible that detrapping from monovacancies is so pronounced that the published H^F_{eff} [7,8,9] pertain mainly to divacancies.

If the equilibrium defect concentrations are as high as favoured by Kraftmakher, a further difficulty may arise. As recently emphasized by Wautelet [10], the effects of the defects on the phonon spectrum can then not be entirely absorbed in the defect formation entropy S^F but will significantly change the spectrum of non-localized phonons as well. Then it is no longer possible to write $p = p_0 + \Delta p$ with Δp approximately given by (1).

2. The $\Delta l/l_0 - \Delta a/a_0$ method ("differential thermal expansion")

In view of the difficulties described in Sect. 1, it is fortunate that an experimental method exists that – in principle – is capable of overcoming most of them. The basic idea is that by comparing the macroscopic and microscopic (i.e. unit-cell) thermal expansion one may eliminate the unknown thermal expansion of the hypothetical "defect-free" crystal. For cubic crystals this leads to the following expression with the difference between the concentration of vacancies, C_V, and of self-interstitials, C_I, (see, e.g., [11]):

$$C_V - C_I = 3(\Delta l/l_0 - \Delta a/a_0)\ (1 + \text{terms in } \Delta l/l_0,\ \Delta a/a_0). \qquad (2)$$

Here $\Delta l/l_0$ denotes the relative change of the sample length and $\Delta a/a_0$ that of the mean interatomic distance, both with respect to a defect-free state (l_0, a_0).

In order for differential thermal expansion measurements to contribute to the solution of the above-mentioned problems and to furnish us with reliable information on H^F and S^F, rather stringent experimental conditions must be fulfilled. Between the defect-free temperature regions and T_m, $\Delta l/l_0$ and $\Delta a/a_0$ must both be determined with an accuracy of about 10^{-6}. This corresponds to a "precision measurement" and requires high mechanical stability of the set-up over a wide temperature range. Since the desired information is obtained as a small difference of two strongly temperature-dependent quantities, temperature gradients in the sample should not exceed 0.1 K.

Laser interferometry allows us to perform the $\Delta l/l_0$ measurement on samples that are only a few cm long. Since the cross-section of the available neutron beams are comparable with or larger than the specimen dimensions, $\Delta a/a_0$ measurement by neutron diffraction averages over unavoidably small temperature gradients in much the same way as the length measurements, in contrast to the X-ray measurements that were so far mainly employed. At elevated temperatures the strong absorption of X-rays by matter leads, in addition, to a "window problem": Openings in the heating and shielding elements allowing the passage of X-rays give rise to rather large temperature gradients. It is presumably this problem which has prevented the application of the X-ray technique to temperatures above the melting point of Cu, where $\Delta a/a_0$ was determined with an accuracy of about $7 \cdot 10^{-6}$ [12].

From what has been said it follows that the extension of the differential thermal-expansion technique to higher temperatures (i.e., to metals such as Fe, Ni, Co, Pd, Pt, or even Nb and Mo) requires the use of neutron diffraction. At present, this also offers the best chance to increase substantially the accuracy of the measurements on the Cu-Ag-Au group of metals that have already been investigated by means of X-rays [12-14] or to study the rather low equilibrium defect concentrations of Ge and Si.

The present paper is a progress report on laser-interferometry—neutron-diffraction experiments performed at the double crystal diffractometer S21 of the Institute Laue—Langevin (Grenoble). In order to keep the rocking curves sufficiently narrow, the temperature differences between sample and monochromator crystal has to be limited (for Cu/Cu to less than $\pm$ 100 K). Therefore, in the present measurement the temperature of the monochromator crystal is raised in steps of about 70 K. Nine different temperature settings of the monochromator crystal are used, covering the temperature interval $T_m/2 < T < T_m$. Since the thermal expansion coefficient α of metals obey Carnelley's rule $\alpha T_m \approx$ const ($\sim$ 0.02) the number of temperature settings required does not depend on the melting temperature and has not to be increased in measurements on high-melting point metals. The results reported in the following have preliminary character since the experimental conditions were not yet optimal.

3. Results and Discussion

The measurements were done on high-quality Cu and Au single crystals grown by W. Uelhoff and his associates at KfA Jülich. Fig. 2 shows the temperature dependence of the macroscopic thermal-expansion coefficient of Cu. Under the assumption of a defect volume of 0.8 atomic volumes an interpretation analogous to Fig. 1 gives for the defect concentration at the melting point $C(T_m) = 1.6 \cdot 10^{-3}$ in qualitative agreement with specific-heat measurements [3]. The temperature dependence of C_V-C_I

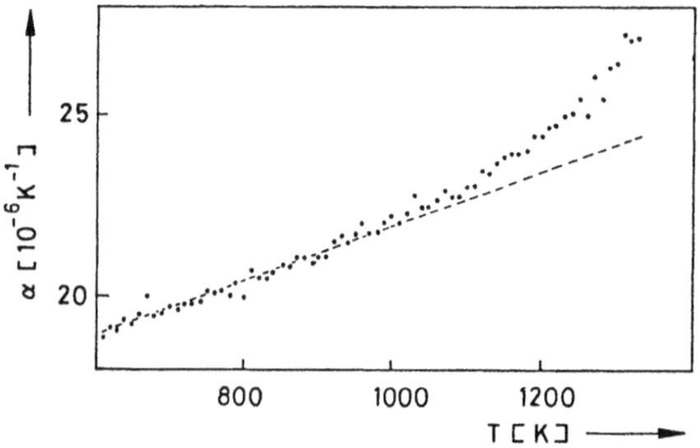

Fig. 2 Coefficient of macroscopic thermal expansion $\alpha=(1/l_0)\cdot dl/dT$ of Cu. The dashed line indicates the simplest assumption for the behaviour of the "defect-free" crystal

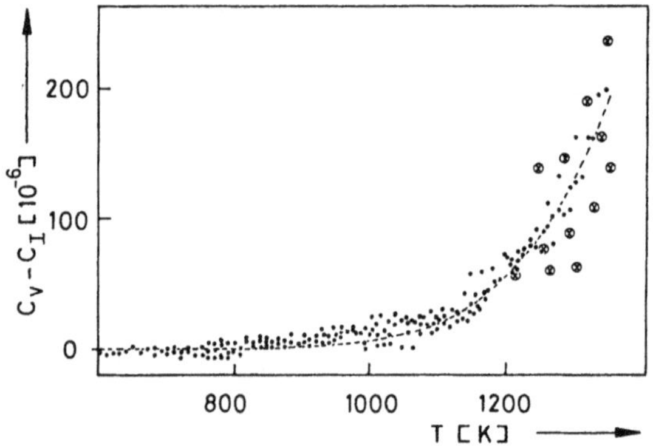

Fig. 3 Concentration difference C_V-C_I in Cu as a function of temperature. The dashed line corresponds to the fit $c=\exp(S/k_B-H/k_BT)$ with $S=1.6\ k_B$ and $H=1.18$ eV. Data from [12] are shown by circles.

following from an evaluation according to (2) is shown in Fig. 3. At the melting temperature one finds $C_V(T_m)-C_I(T_m) = (2.1 + 0.1) \cdot 10^{-4}$, in accordance with the Simmons–Balluffi value $(1.9 \pm 0.5) \cdot 10^{-4}$ [12].

In Au a negative contribution to the thermal expansion was observed immediately below the melting temperature (presumably due to evaporation or creep), which makes the extrapolation to T_m somewhat uncertain. Nevertheless, $C_V(T_m)-C_I(T_m)$ is in agreement with the Simmons–Balluffi value $(7.2+0.6) \cdot 10^{-4}$ [14]. Like in Cu, the $\Delta l/l$-data on Au are compatible with Kraftmakher's result $C(T_m) = 4 \cdot 10^{-3}$ from specific heat.

The agreement between our and Simmons' and Balluffi's $C_V(T_m)-C_I(T_m)$ on the one hand and between our very precise thermal-expansion measurements (statistical accuracy $\delta\Delta l/l_0 \approx 3 \cdot 10^{-6}$) and the specific heat data on the other hand suggests the following extension of viewpoint (i). It may be approximately correct to find $p_0(T)$ by extrapolation, but the difference $p-p_0$ obtained in this way should not be attributed to vacancy-type defects only (as has been done in most interpretations) but to both vacancy and self-interstitial contributions. The vacancy concentration would then be substantially higher than the C_V-C_I values found from (2), details depending on the properties (e.g. formation volumes) of interstitials and vacancies. While the defect concentration at the melting point is not very sensitive to slight modifications of the evaluation procedure this does not hold for a fit to determine the formation entropy and enthalpy according to (1).

For separate determination of S_{eff} and H_{eff} we have to apply a correction for the scattering geometry (in particular vertical divergence of the neutron beam and tilts of the samples). This leads to $S_{eff} = 1.6\ k_B$ and $H_{eff} = 1.18$ eV. The agreement with Simmons' and Balluffi's values [12] must be considered fortuitous, since on the one hand these authors assumed $S^F = 1.5\ k_B$ and our own values have to be considered with caution for the following reasons:
The correction for the scattering geometry is different for monochromatic or white

neutron beams. Since the neutron beam used lies somewhere between these extremes, we introduced a correction parameter K into the Bragg condition according to

$$(a^* + \Delta a)/a^* = k\sin\Theta^*/\sin(\Theta^* + \Delta\Theta), \quad \Delta a = a - a^*, \Delta\Theta = \Theta - \Theta^* \qquad (3)$$

where a^*, a and Θ^*, Θ are the atomic distances and Bragg angles of the monochromator and sample, respectively.

(3) results from a first-order approximation of possible corrections to the scattering angles. K is determined by the best fit to the experimental data in the defect-free temperature range and K typically differs from unity by $\sim 10^{-6}$. The validity of this procedure has yet to be tested by control experiments.

A further and even more critical problem arises in connection with the determination of the monochromator Bragg angle Θ^*. Not only is Θ^* at present not known with sufficient accuracy, a consideration of errors shows that it has to be known within a few seconds of arc, but it is expected to vary somewhat from setting to setting. To determine Θ^* from the measured data we had to use a model-dependent evaluation procedure. It leads to reasonable values of S_{eff} and H_{eff}, which have to be interpreted in terms of monovacancy and divacancy contributions [15]. A quantitative subdivision between these two contributions or a definite decision whether C_I is negligible or not is not possible at present.

A modification of S21 which permits a model-independent determination of Θ^* is urgently required and planned for the near future.

During our measurements on Cu and Au it turned out that the intensity of the neutron beam was subject to fluctuations presumably due to three other experiments in the beam in front of ours. We suspect that these fluctuations were accompanied by variations of the wavelength spectrum and hence affected the determination of the Bragg angle of the sample as a function of temperature in an ill-defined and not correctable manner.

The problem just mentioned will be solved by moving S21 to a location where the incoming neutron beam is no longer affected by other experiments. The new location provides enough space in order to shield the temperature-sensitive parts of equipment (e.g. analogue amplifiers and the support of mirrors of the interferometer) from fluctuations of the ambient temperatures exceeding about 0.5 K. In the experiments reported so far uncontrollable sudden variations of the environment temperature introduced additional scatter in the data (e.g., in Fig. 3 above 900 K).

It is expected that with the three improvements mentioned (constancy of the neutron beam, direct determination of the Bragg angle Θ^* of the reference crystal, constancy of the environment temperature within $\pm$ 0.5 K) an accuracy better than the achieved value of $2 \cdot 10^{-6}$ in $\Delta l/l_0 - \Delta a/a_0$ will be reached. Measurements on Si planned for the near future should provide an excellent test of whether this has been achieved, since in Si C_V-C_I is expected to be much smaller than in metals, so that quite high up in temperature $\Delta l/l_0$ and $\Delta a/a_0$ should coincide. We hope that subsequently we shall be able to investigate the high-melting point metals mentioned above (for Nb and Mo this will require a modification in the heating system) and to determine or put upper limits on the contributions of self-interstitials to the thermal-equilibrium defects in metals.

The main conclusions of the present paper are as follows:

(i) The difference $C_V(T_m)-C_I(T_m)$ of the vacancy and self-interstitial concentration at the melting points of Cu and Au found by Simmons and Balluffi [12,14] have been confirmed.

(ii) The question whether self-interstitials do or do not contribute substantially to the equilibrium concentration of point defects in metals must be considered as open.

(iii) The precision of the laser-interferometer–neutron diffraction equipment is $2 \cdot 10^{-6}$. With some further improvements the desired accuracy of $\Delta l/l_0 - \Delta a/a_0$ of about 10^{-6} of the total change of $\Delta l/l_0$ or $\Delta a/a_0$ can be achieved. This will allow us to answer a number of unsolved questions on defects in metals.

References

1 Y.A. Kraftmakher and P.G. Strelkow, Fiz. tverd. tela 8, 580 (1966)
2 R.C. Shukla and E.R. Cowley, Phys. Rev. B 31, 372 (1985)
3 Y.A. Kraftmakher, Vacancies and Interstitials in Metals, eds. A. Seeger et. al. (North-Holland, Amsterdam), 1970, p. 59.
4 Y.A. Kraftmakher, Fiz. tver. tela 27, 235 (1985)
5 H.E. Schaefer, in Positron Annihilation, eds. P.G. Coleman, S.C. Sherma and L.M. Diana (North-Holland, Amsterdam), 1982, p. 369.
6 A. Seeger, Proceedings of the 7th International Conference on Positron Annihilation New Delhi, India, to be published by World Scientific Publishers Co., Singapore.
7 K. Maier, M. Peo, B. Saile, H.E. Schaefer, and A. Seeger, Phil. Mag. A 40, 701 (1979)
8 J.L. Campbell and C.W. Schulte, Appl. Phys. 19, 149 (1979)
9 G.D. Loper, L.C. Smedskjaer, M.K. Chason and R.W. Siegel, Proceedings of the 7th International Conference on Positron Annihilation New Delhi, India, to be published by World Scientific Publishers Co., Singapore.
10 M. Wautelet, Phys. Lett. 108 A, 99 (1985)
11 A. Seeger, Crystal Lattice Defects 4, 221 (1973)
12 R.O. Simmons and R.W. Balluffi, Phys. Rev. 129, 1523 (1963)
13 R.O. Simmons and R.W. Balluffi, Phys. Rev. 119, 600 (1960)
14 R.O. Simmons and R.W. Balluffi, Phys. Rev. 125, 862 (1962)
15 H. Mehrer and A. Seeger, phys. stat. sol. 35, 313 (1969)

Saturation of Irradiation-Induced Precipitation

G. Martin

CECM-CNRS, 15, rue Georges Urbain, F-94407 Vitry sur Seine Cedex, France

1 INTRODUCTION

During the early study of irradiation-induced precipitation in undersaturated solid solutions [1], it was noticed that precipitation saturates before the matrix solute content decreases down to the solubility limit under irradiation [2]. Such a behaviour was explained by the change in defect balance which results from the freezing of mobile solute atoms to solute clusters : for a given irradiation flux and temperature, a decrease in the solute drift towards the precipitate results, the order of magnitude of which is compatible with the saturation observed in the Al Zn system under 1-MeV electron irradiation [2].

Recently, two sets of detailed data have been published concerning the saturation of precipitation in Cu Be dilute solid solutions. WAHI et al. [3] report a saturation with dose of the precipitate number density, as well as of the precipitate diameter in Cu 1.35 at.% Be under 300-keV Cu^+ ion irradiation at 600 K and at defect production rates ranging from 10^{-4} to 10^{-2} dpa s^{-1}. NAKAI et al. [4] report a similar behaviour in Cu 1.26 at.% Be under 1-MeV electron irradiation at various temperatures close to 600 K and at defect production rates ranging one order of magnitude around 10^{-3} dpa s^{-1}. While in the ion irradiation experiment [3] the precipitates seem to nucleate at displacement spikes, the nucleation looks homogeneous in the electron irradiation experiment.

In the following we propose a simple argument to account for the saturation of precipitate size and number density in the latter case. This argument is not linked to any recoil resolution process and yields interesting scaling laws with respect to the irradiation flux.

2 SIMPLE MODEL

Homogeneous irradiation-induced precipitation results from the irreversible recombination of vacancies trapped at solute clusters with mixed dumbbells arriving from the matrix [2]. Simple arguments result for the saturation of the precipitate number density on the one hand and of the precipitate size on the other.

2.1 SATURATION OF THE PRECIPITATE NUMBER DENSITY

To a first approxiation, we assume that precipitate nucleation will stop when the point defect elimination at the precipitate-matrix interface is a large fraction of the elimination process. In the case where vacancy-interstitial recombination is the dominant extra elimination mechanism, the above condition gives

$$4\pi\, NR\, D_i C_i = \alpha G, \qquad (1a)$$

$$\frac{4\pi\, R_c}{\Omega} (D_i + D_v)\, C_i C_v = (1-\alpha)\, G, \qquad (1b)$$

where N and R are respectively the precipitate number density and radius , D_i, C_i (D_v, C_v) the diffusion coefficient and concentration of interstitials (vacancies), r_c the v-i recombiation radius, Ω the atomic volume, G the defect production rate and α some saturation parameter (α<1). With the classical condition

$$D_iC_i = D_vC_v, \qquad (2)$$

equations (1) reduce to

$$NR = \left[\frac{G}{\langle D\rangle}\right]^{1/2} \sqrt{\frac{\alpha^2}{1-\alpha}\,\frac{r_c}{4\pi\Omega}}\,, \qquad (3a)$$

$$\langle D\rangle^{-1} = D_i^{-1} + D_v^{-1}\,. \qquad (3b)$$

As seen form (3a) the argument for the saturation of precipitate nucleation yields a condition on the product N X R.

2.2 SATURATION OF THE PRECIPITATE SIZE

In the case of irradiation-induced precipitation, solute is driven by point defect fluxes from the matrix, where the solute concentration is C_M, to the precipitate-matrix interface, where the concentration equals the classical solubility limit C_S. Let us notice that $C_S > C_M$. Precipitate growth ceases when the solute drift by point defect fluxes is balanced by the solute diffusion flux from the precipitate matrix interface ($C=C_S$) to the matrix ($C=C_M$). To a first approximation, the drift flux may be written

$$J_{drift} \propto C_M\,(\sigma_iD_iC_i + \sigma_vD_vC_v), \qquad (4a)$$

where σ_α (α=i,v) is a coupling coefficient between the defect flux ($D_\alpha C_\alpha$) and the solute flux.

Similarily, the back diffusion flux is

$$J_{diff} \propto (C_S-C_M)\,(f_iD_iC_i + f_vC_vD_v), \qquad (4b)$$

where f_α (α=i,v) is the proportionality constant between the solute diffusion coefficient by mechanism α and the corresponding atomic diffusion coefficient.

From (4a,4b) and (2), it is found that

$$\frac{C_S}{C_M} = 1 + \frac{f_i + f_v}{\alpha_i + \alpha_v}\,, \qquad (5)$$

where f_α and σ_α (α=i,v) are functions of the temperature but not of the irradiation flux. The matrix concentration C_M decreases as the precipitate volume fraction NR^3 increases :

$$C_M = \bar{C} - 4\pi\, NR\, C_P{}^3, \qquad (6)$$

where $\bar{C}$ is the nominal concentration and C_P the solute concentration in the precipitate.

Equation (6) together with (5) indicates that the precipitate volume fraction is independent of the defect production rate

$$NR^3 \propto G_o. \qquad (7)$$

3 SCALING LAWS AND DISCUSSION

From §§2 the arguments for the saturation of nucleation and growth yield respectively

$$NR \propto G^{1/2}, \quad \text{and } NR^3 \propto G^0. \qquad (8)$$

Simple algebra results in

$$N \propto G^{3/4}, \quad \text{and } R \propto G^{1/4}. \qquad (9)$$

As shown by Fig.1, (9) are consistent with the lower-flux electron irradiation data of [4]. The agreement is lost at higher irradiation flux or for ion irradiation [3]. In the latter case, the precipitate number density is found to be almost independent of the irradiation flux, and the saturation of the precipitate size could be accounted for by recoil resolution processes [3].

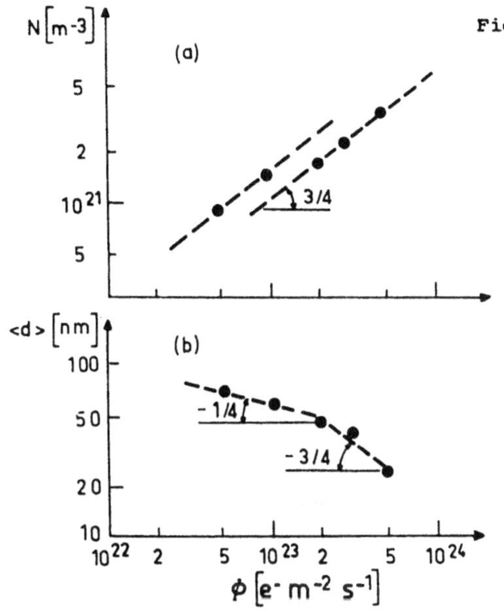

Fig. 1 : Fit of (9) to Nakai's et al. data [4]

As a summary, in the case where irradiation-induced precipitates nucleate homogeneously (e^- irradiation) simple mechanisms of solute-defect coupling are proposed to account for the saturation of the precipitate size and number density, even in the absence of recoil resolution processes. Simple scaling laws with the irradiation flux are expected (9) and found to be in fair agreement with available lower-flux irradiation data.

ACKNOWLEDGEMENTS

Useful discussions with Pr Wollenberger, and Drs Abromeit and Wahi are gratefully ackowledged.

REFERENCES

1 For a Review : F.V. NOLFI (editor) : Phase transformations during irradiation. Applied Science, London (1983) ; K.C. RUSSEL : Prog. Mat. Sci., 28, (1985), 229
2 R. CAUVIN and G. MARTIN : Phys. Rev., B23, (1981), 3322 ; Phys. Rev., B23, (1981), 3333 ; Phys. Rev., B25, (1982), 3385
3 R.P. WAHI, R. KOCH, C. ABROMEIT, H. WOLLENBERGER : J. Nucl. Mater., 127, (1985), 175
4 K. NAKAI, C. KINOSHITA, S. KITAJEIMA, T.E. MITCHELL : J. Nucl. Mater., 133-134, (1985), 284

Analysis of Defect Microstructures in Irradiated Ferritic Alloys

P.A. Beaven, F. Frisius, R. Kampmann, and R. Wagner

GKSS-Forschungszentrum, Institut für Physik, D-2054 Geesthacht, F.R.G.

The defect microstructures in a series of neutron-irradiated Fe-Cu binary alloys (0 - 0.86 at% Cu) and reactor pressure vessel steels (> 0.1 wt% Cu) have been investigated using magnetic and nuclear small angle neutron scattering (SANS) and transmission electron microscopy (TEM). Specimens were irradiated in the FRG-2 reactor in Geesthacht to fluences > 10^{23} n/m^2 at 290 °C. Under these conditions, the defect microstructures in the irradiated binary alloys could be analyzed using both SANS and TEM, and consist of distributions of Cu precipitates, voids, and interstitial loops; in the steels, for which only SANS could be applied, the scattering data and the post-irradiation annealing behaviour appear to be consistent with a defect microstructure consisting of Cu-rich precipitates and voids, at least in steels containing > 0.1 wt% Cu.

1. Introduction

The influence of alloying elements and residuals, particularly Cu, Ni and P, on the radiation-induced embrittlement of reactor pressure vessel steels is empirically well established [1]. However, it has so far not been possible to understand the changes in the radiation-induced defect structure, which in the presence of these elements, give rise to the observed increases in irradiation hardening and ductile-brittle transition temperature. The failure of TEM to resolve the nature of the damage in such steels suggests that the defects responsible for the irradiation hardening are extremely small (< 2 nm diameter), and this has led to the application of other techniques such as SANS and atom probe field ion microscopy to the problem, e.g. [2, 3]. In order to avoid some of the complexities associated with steels, and in order to isolate the effects of specific alloying elements e.g. Cu, some previous workers in this area have used binary model alloys [4, 5] and this is the approach that we have adopted. Using simple model alloys and relatively high neutron fluences, some opportunities exist for using TEM to assist in the interpretation of SANS data.

The work to be reported here forms part of a programme concerned with the effect of alloying elements on the radiation damage process in iron-based alloys and effects on mechanical properties. Major emphasis has so far been placed on attempting to identify and quantify the irradiation-induced damage. Preliminary results indicating that irradiation of Fe-Cu alloys at 290 °C leads to the radiation enhanced precipitation of Cu and the formation of interstitial loops and voids have been published elsewhere [6, 7]. In this paper we present quantitative data obtained primarily from SANS experiments on Fe-Cu alloys and Cu-containing steels irradiated to two different fluences, together with information obtained from post-irradiation annealing experiments. One of the aims of these experiments is to assess to what extent the analysis of the defect microstructures of the model alloys can be helpful in interpreting the SANS data from the complex steels.

2. Experimental Details

Binary Fe-Cu alloys containing 0, 0.09, 0.21, 0.64 and 0.86 at% Cu with interstitial impurity levels of the order of 120 ppm C, 20 ppm N, 36 ppm P and 415 ppm O, were homogenized at 840 °C and quenched prior to irradiation. Analyses of the Cu-containing pressure vessel steels are given in Table I. Specimens for SANS,

Table I: Chemical compositions of the steels (wt%)

Steel	C	Mn	Mo	Ni	Si	Cr	Cu	P	S
20MnMoNi55 [KS12,Plate]	0.23	1.44	0.64	0.63	0.31	0.20	0.17	0.014	0.009
22NiMoCr37 [KS07, Plate]	0.3	0.62	0.99	0.73	0.29	0.46	0.26	0.023	0.023
22NiMoCr37 [Weld]	0.08	1.75	0.64	1.01	0.18	0.4	0.25	0.013	0.008

TEM, and compression testing were irradiated together with Charpy bars in temperature-controlled capsules in the FRG-2 reactor in Geesthacht to target neutron fluences of 2×10^{23} n/m^2 and 7×10^{23} n/m^2 at 290 °C (neutron flux ~5×10^{16} n/m^2/s). The binary Fe-Cu alloys actually received neutron doses somewhat lower than the target fluences - see Table III.

Results of the mechanical testing on the Fe-Cu alloys have been reported previously [6], as have details of the radiation-induced mechanical property changes observed in the steels [8, 9]; these will not be discussed further here.

3. SANS Analysis

The SANS intensity is measured as a function of scattering angle Θ, by means of position-sensitive detectors parallel (‖) and perpendicular (⊥) to the magnetic field applied to the specimen. The ratio of the corresponding scattering cross-sections is then:

$$A = (d\Sigma/d\Omega)\perp/(d\Sigma/d\Omega)\| = (\Delta\eta^2_{nuc} + \Delta\eta^2_{mag})/\Delta\eta^2_{nuc} \qquad (1)$$

which gives A = 11.0 for Cu-particles and A = 1.4 for voids. Quantitative evaluation of the measured SANS curves allows determination of the particle radius, and if the nature of the particle is known, the number density and volume fraction. If the nature of the scattering centers is not known (e.g. via other techniques), then the additional information obtained from magnetic scattering experiments can be useful in helping to identify their nature. For Fe-Cu alloys the magnetic scattering from Cu precipitates is relatively strong ('A' values approaching the theoretical ratio) and can be measured over an extended range of κ. It is therefore possible to determine not only a Guinier radius for the particles, but also a precipitate size distribution. Good agreement between calculated and measured scattering curves is obtained using a distribution of the logarithmic normal type. Measurements of the absolute scattering intensity then allow the volume fraction to be determined. A similar procedure, using the nuclear scattering curves, can be used to analyze for voids.

Under circumstances where both kinds of defect produce scattered intensity, the 'A' ratio is sensitive to the volume fractions and particle sizes of the defects present. Figs. 1, 2 and 3 show typical measured SANS cross-sections of irradiated Fe, Fe-0.3% Cu and the 0.17% Cu steel (KS12). In Fig. 1, the ratio of the SANS cross-sections approaches 1.4 and is virtually independent of scattering vector κ indicating the presence of voids. Fig. 2 illustrates the high value of 'A' associated with Cu precipitates; in the κ range corresponding to smaller particle radii the 'A' ratio decreases due to the contribution from voids. For the steel, the SANS cross-sections in Fig. 3 show 'A' has an average value of ~ 2.9. This could either indicate the presence of a single type of defect with scattering properties that do not correspond to simple Cu precipitates or voids, or, that some combination of voids and precipitates with similar sizes give rise to the SANS.

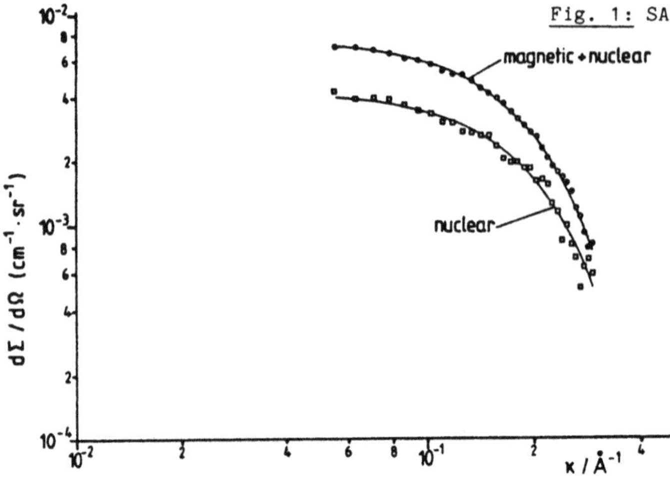

Fig. 1: SANS curves for irradiated Fe

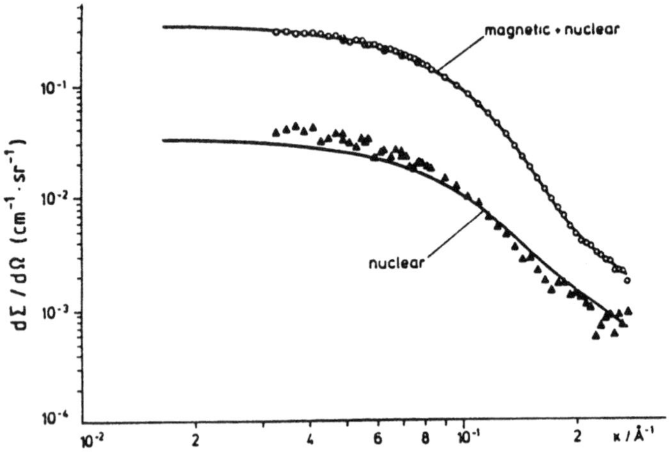

Fig. 2: SANS curves for Fe-0.3 at% Cu

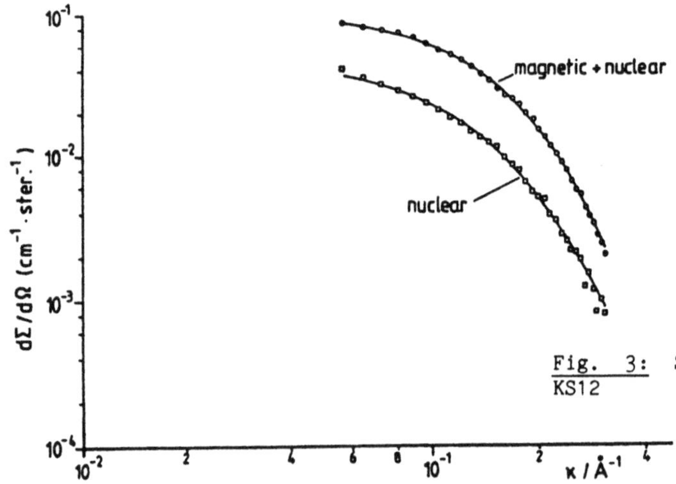

Fig. 3: SANS curves for irradiated KS12

The SANS data for the model alloys could be analyzed quantitatively in terms of Cu precipitates and voids, with some degree of overlap in the dilute alloys. For the steels we have adopted an approach whereby the nuclear and magnetic scattering are analyzed separately, the former being associated predominantly with voids, and the latter with "precipitates" or "magnetic holes" (data given for precipitates include also the void contribution). We have then used post-irradiation annealing experiments to see to what extent this approach is valid before attempting a more iterative analysis.

4. Results

Table II shows SANS data for iron specimens irradiated to 2 x 10^{23} and 5 x 10^{23} n/m^2 at 290 °C. The void distribution parameters have been derived from nuclear scattering, since data obtained from magnetic scattering overestimate the voidage by producing values for the void diameter which are ~ 10 % larger. This suggests that the size of the magnetic disturbance at such a void is of somewhat larger spatial extent than the actual void diameter.

Table II: SANS-data for irradiated Fe

Fluence $[n.m^{-2}]$	N_V $[m^{-3}x10^{22}]$	$\bar{D}_V$ $[nm]$	f_V $[\%]$
$2\cdot10^{23}$	5·1	1·5	0.008
$5\cdot10^{23}$	4·8	1·8	0.015

Table III shows the SANS data for two irradiated Fe-Cu alloys. Data for the voids are limited to their volume fractions; number densities and sizes cannot be established with certainty for these low void fractions. Sizes are, however, roughly comparable with those determined for iron.

Consideration of the SANS data in Tables II and III allows the following general observations to be made:

1. At the lower fluence, precipitation of Cu is largely complete, with a number density (~ $10^{23}/m^3$) which is independent of the Cu supersaturation before irradiation; the particle diameter is a strong function of Cu content.
2. At the higher fluence, the precipitate distribution in the high Cu alloy remains virtually unchanged; in the low Cu alloy, significant coarsening has taken place.
3. The voidage in Fe-Cu alloys is generally lower than in Fe.

TEM gives confirmation of the presence of voids in Fe, and of Cu precipitates in Fe-Cu alloys containing > 0.3 at% Cu. In addition, the interstitial component

Table III: SANS-data for irradiated Fe-Cu alloys

Alloy	Fluence $[n.m^{-2}\cdot10^{23}]$	$N_p\cdot10^{23}$ $[m^{-3}]$	$\bar{D}_p$ $[nm]$	f_p $[\%]$	f_V $[\%]$
0.86 % Cu	1.8	1.1	5.2	0.83	0.01
	5	1.1	5.2	0.83	0.01
0.16 % Cu	1.7	1.15	2.9	0.15	0.004
	4.3	0.6	3.6	0.14	0.005

of the damage has been identified as consisting of distributions of loops, chiefly with Burgers vectors of the a <100> type, with number densities in the range of 10^{21} - 10^{22} m^{-3} and diameters which range from ~ 15 nm in iron down to ~ 6 - 8 nm in the high Cu alloys.

Specimens irradiated to the higher fluence (~ 5 x 10^{23} n/m^2) have been subjected to post-irradiation annealing treatments at 400 °C for various times up to ~ 100 hrs. For iron specimens SANS shows that the voidage steadily decreases with time, such that after 100 hours the volume fraction of voids is < 0.007 % with the mean void diameter remaining roughly constant (~ 1.8 nm). In the Fe-Cu alloys a similar void annealing process occurs at 400 °C such that after ~ 100 hours the void volume fraction is reduced to ~ 0.0015 %. Changes in the Cu precipitate distribution during annealing have also been monitored. For the Fe-0.21 at% Cu alloy, for example, post irradiation annealing for ~ 100 hours is accompanied by some reduction in number density 0.7 x 10^{23} m^{-3} → 0.4 x 10^{23} m^{-3}, and volume fraction 0.21 % → 0.18 %, with a slight increase in mean diameter 3.8 → 4.3 nm.

Table IV: SANS data for the irradiated steels

Alloy	Fluence $[n.m^{-2} \cdot 10^{23}]$	$N_v \cdot 10^{-23}$ $[m^{-3}]$	$\bar{D}_v$ [nm]	f_v [%]	$N_p \cdot 10^{-23}$ $[m^{-3}]$	$\bar{D}_p$ [nm]	f_p [%]
Fe-0.24 %Cu	1.6	--	--	0.005	1.1	3.3	0.21
22NiMoCr37 Weld 0.25%Cu	1.8	0.56	2.7	0.06	3.6	2.4	0.25
KS12 Plate 0.17% Cu	2	8.8	1.1	0.05	6.8	1.6	0.15
	7	1.3	1.9	0.05	3.8	2.2	0.23
KS07 Plate 0.26% Cu	2	0.46	1.8	0.015	8.1	1.6	0.17
	7	1.7	1.7	0.04	4.1	2.3	0.27

Table IV shows SANS data for the steels investigated together with data for the Fe-0.24 wt% Cu alloy for purposes of comparison. Since TEM examination of the steels failed to provide additional information on the nature of the defect microstructures, the analysis procedure described in the SANS section was adopted. This means that the data for voids must be regarded as upper limits, since precipitates also produce a relatively weak nuclear scattering contribution; the "precipitate" data derived from the net magnetic scattering include all "magnetic holes" in the specimens (i.e. also voids).

Inspection of the defect diameter data in the tables indicates the presence of scattering centers whose nuclear and magnetic spatial extensions are similar; thus 'A' ratios characteristic of voids or Cu precipitates cannot be expected. There is, however, evidence that 'A' increases with Cu content, which would be consistent with scattering from a void component together with an increasing contribution from a Cu precipitate component. Such defects, with sizes < 2.5 nm, would be difficult to detect by TEM.

Analysis of the nuclear scattering data from the plate steels gives void diameters (~ 1.8 nm) which are comparable with those found in Fe. However, the levels of voidage are significantly higher and result from a factor of ten increase in number density compared to Fe. For the weld material the void distribution is significantly coarser.

Analysis of the magnetic scattering data gives distributions of "magnetic holes" with number densities in the $10^{23}/m^3$ range, defect diameters in the range 1.5 - 2.4 nm, and volume fractions between 0.15 and 0.27 %. These figures reflect a finer particle distribution than obtained in the binary alloy with a comparable Cu content. Interpretation of the magnetic scattering data in terms of precipitates would require that a further subtraction be made in order to remove the void contribution allowing comparison with the available Cu supersaturations. Such subtractions could lead to volume fractions consistent with complete precipitation of Cu at the higher fluence.

In order to test this interpretation of the SANS data, some of the steels have been subjected to post-irradiation annealing at 400 and 450 °C. Fig. 4 shows the annealing time dependencies of the volume fractions of voids and "magnetic holes" at 400 °C and 450 °C in steel KS07 (0.26 wt% Cu) after the high fluence irradiation. At 400 °C the slight reductions in volume fractions measured by both nuclear and magnetic scattering may be associated with some void annealing (as occurs in the binary alloys). Defect sizes remain essentially constant, the decreases in volume fraction being associated with a reduction in number density. At 450 °C, the nuclear SANS decreases rapidly giving void volume fractions ~ 0.01 % after annealing times of less than 1 hour. This void annealing is accompanied by decreases in the volume fraction of "magnetic holes", but significantly, by substantial increases in the value of the SANS cross-sections' ratio 'A' to ~10, indicating the presence of Cu precipitates with a volume fraction of ~ 0.1% after annealing for 48 hours. Steel KS12 shows similar behaviour during annealing at 450 °C.

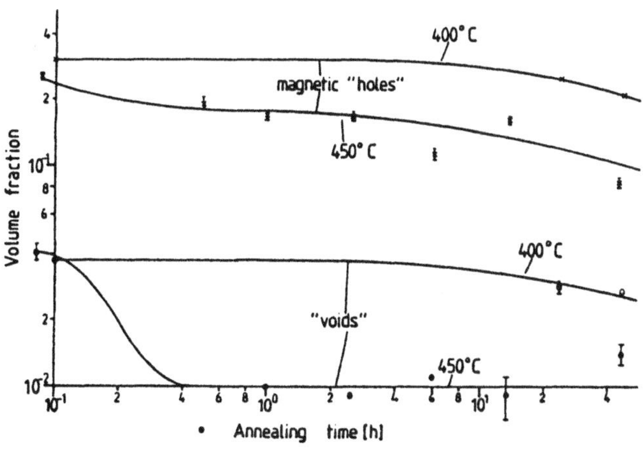

Fig. 4: Annealing behaviour of irradiated KS07 at 400 and 450 °C

5. Summary

Radiation hardening in the low sink density Fe-Cu alloys is associated with a radiation damage component together with a Cu precipitation hardening component. In the Cu containing pressure vessel steels the SANS data suggest that the defect microstructures consist of voids and Cu-rich precipitates, although we cannot rule out the possibility that other types of precipitates or segregations are present. Further work is in progress to understand the kinetics of the radiation-enhanced precipitation of Cu and its relationship to radiation hardening and embrittlement [10].

References

1. G.R. Odette and G.E. Lucas: Proc. Specialists Meeting "Radiation Embrittlement and Surveillance of Reactor Pressure Vessel Steels", ASTM, to be published (1985).

2. F. Frisius and D. Bünemann: Proc. Conf. on "Irradiation Behaviour of Metallic Materials for Fast Reactor Core Components", French Atomic Energy Commission, 247 (1979)
3. M.K. Miller and S.S. Brenner: Res Mechanica, 10, 161 (1984)
4. F.A. Smidt and J.A. Sprague: "Effects of Radiation on Substructure and Mechanical Properties of Metals and Alloys", ASTM STP 529, 78 (1973)
5. S.S. Brenner, R. Wagner, and J.A. Spitznagel: Met. Trans. A, 9, 1761 (1978)
6. F. Frisius, R. Kampmann, P.A. Beaven, and R. Wagner: Proc. BNES Conf. "Dimensional Stability and Mechanical Behaviour of Irradiated Metals and Alloys", 1, 171 (1983)
7. R. Wagner, F. Frisius, R. Kampmann, and P. A. Beaven: Proc. 5th ASTM-Euratom Symposium on Reactor Dosimetry, Commission of the European Communities, 549 (1984)
8. J. Ahlf, D. Bellmann, J. Föhl, H.D. Hebenbrock, F.J. Schmidt, and W. Spalthoff: GKSS-Report (to be published)
9. J. Ahlf and F.J. Schmitt: J. Nucl. Mat., 105, 48 (1982).
10. P. A. Beaven, F. Frisius, R. Kampmann, and R. Wagner: Proc. 2nd Int. Symposium on Environmental Degradation of Materials in Nuclear Power Systems-Water Reactors, A.N.S., in the press (1985)

Influence of Electron Irradiation on Decomposition Kinetics and Morphology in CuNiFe Alloys

W. Wagner, R. Poerschke, and H. Wollenberger*

Hahn-Meitner-Institut für Kernforschung Berlin, Glienicker Str. 100, D-1000 Berlin 39

1 Introduction

The influence of particle irradiation on phase transformations in solid solutions is of great interest in theory and nuclear technology. One of the most fundamental questions regards the effects caused by radiation-enhanced interdiffusion as compared to those due to thermal diffusion. The present work focuses attention to the irradiation influence on decomposition kinetics and morphology within the miscibility gap of the comparatively simple ternary system Cu-Ni-Fe near the binary Cu-Ni boundary system (Fe content 4 at.%). Electron irradiation has been applied in order to reduce cascade effects with respect to simple diffusional processes. Small Angle Neutron Scattering (SANS) on null-matrix samples (enriched in ^{62}Ni) has been chosen as method of detection because of its high sensitivity to spatial concentration fluctuations. Irradiations were performed at temperatures between 423 and 723 K.

The paper presents the structure functions after irradiation in comparison to those obtained after thermal annealing at comparable temperatures. Scaling analysis is applied for the SANS data evaluation. The kinetics of the resulting scaling parameters is discussed.

2 Experimental

For the specimens isotopically enriched Ni-material (97 % ^{62}Ni) was used to obtain optimal contrast for order-disorder scattering and simultaneously reduced fundamental scattering. The nominal composition of the alloy is Cu-46 at.% Ni-4 at.% Fe.

After levitation melting in an induction furnace the 50 % cold worked ingots were homogenized by a 20 h annealing at 1323 K, water quenched to room temperature, and rolled to foils of ~ 150 µm thickness. The foils were solution treated at 1073 K for 15 h, cut into pieces of 10 x 10 mm^2 and annealed for 2 h at 973 K to obtain a uniform reference state of short-range clustering prior to irradiation.

The irradiation was performed with a 2 MeV electron beam from a Van-de-Graaff accelerator. For homogeneous irradiation the beam with a natural spot size of about 4 mm (full width at half maximum of the Gaussian intensity profile) was scanned over the specimen area. Thus, inhomogeneities across the specimen were kept below ± 5%. The irradiation temperature was adjusted by a preheated He gas flow and controlled by two thermocouples spot-welded to the edges of the specimens. The beam heating was in the order of 50 K for the applied beam current of 50 µA/cm^2, resp. 3.1 x 10^{14} electrons/cm^{2}s. Assuming a displacement cross-section of 25 x 10^{-24} cm^2 determined for binary CuNi [1], the corresponding displacement rate is 8 x 10-9 dpa (displacement per atom) per second. The irradiation times extended from 1 to 200 h.

*now: DPG Bad Honnef, D-5340 Bad Honnef, Germany

The neutron diffraction measurements were done at room temperature, using the SANS facility at RISØ National Laboratory, Denmark, and the diffuse diffractometer T2 at the BER II, Berlin. The neutron wavelength at RISØ was chosen as 0.53 nm^{-1} and 1.8 nm^{-1}. The wavelength at the T2 is fixed at 0.24 nm^{-1}. The covered range in momentum transfer κ (defined in the usual way as $\kappa = 4\pi/\lambda \cdot \sin\Theta$, with λ, the neutron wavelength and 2θ, the scattering angle) extended from 7×10^{-2} nm^{-1} up to 10 nm^{-1} with a comfortable overlap of both instruments around 1 nm^{-1}. The intensity data, corrected for background, transmission and detector acceptance, and radially averaged for the RISØ area sensitive detector, were computed to absolute values of differential scattering cross-section $d\sigma/d\Omega$ by referring to the incoherent scattering of simultaneously measured vanadium.

3 Results

Typical examples of scattering curves measured after irradiation are shown in Fig. 1a, along with the scattering curve from the solution annealed reference specimen. We note that the critical decomposition temperature of the 4 at.% Fe alloy is located around 790 K [2]. Figure 1b gives for direct comparison scattering curves of nonirradiated specimens which were thermally annealed at comparable temperatures inside the miscibility gap.

The scattering curves shown in Fig. 2 were previously scaled following a procedure suggested by Fratzl et al. [3]. For this procedure, the double logarithmic plotted scattering curves like the ones in Fig. 1 are superposed by shifts of the plots parallel to abscissa and ordinate. All so superposed scattering curves were scaled to a Gaussian (Guinier approximation) in the range $2 \leq \kappa \cdot R \leq 4$ and retransferred to the linear, normalized coordinates of Fig. 2. The scaling parameters which result from this procedure, i.e. the relative shifts of abscissa and ordinate, are equivalent to the parameters "peak position" κ_m and "peak intensity" I_m. In Fig. 3 the kinetics of these parameters is shown, again comparing thermal ageing with ageing under irradiation. The abscissa in Fig. 3 is the real ageing respectively irradiation time.

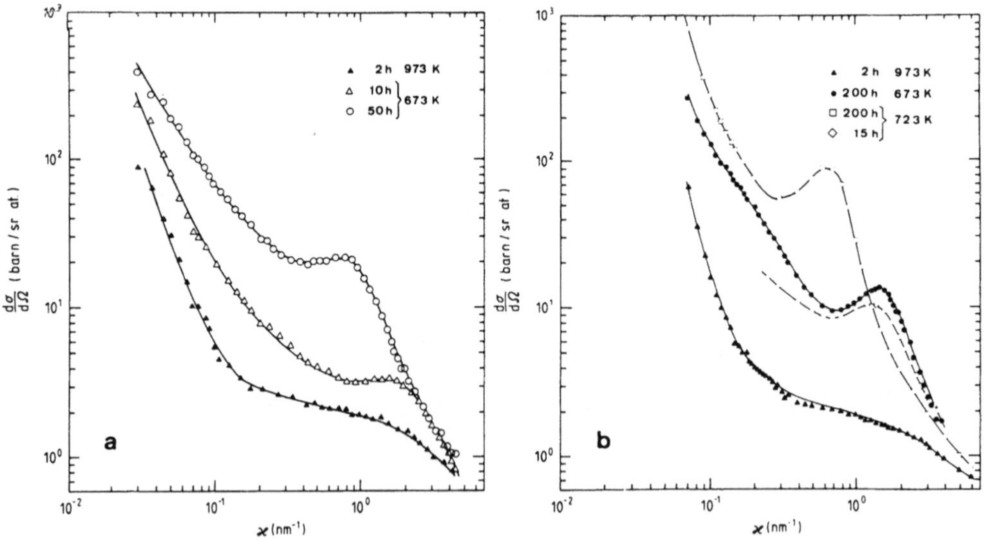

Fig. 1
Scattering cross-section vs. momentum transfer of Cu-46 at.% Ni-4 at.% Fe after electron irradiation (a, open symbols) and thermal ageing (b), along with an annealed reference specimen.

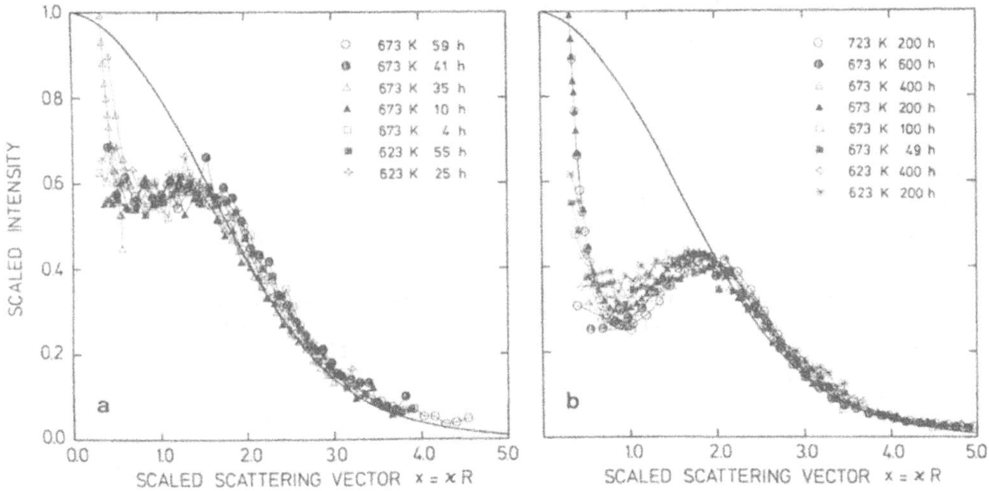

Fig. 2

Scaled scattering curves after electron irradiation (a) and after thermal annealing (b). The solid line is a Gaussian, representing the Guinier approximation.

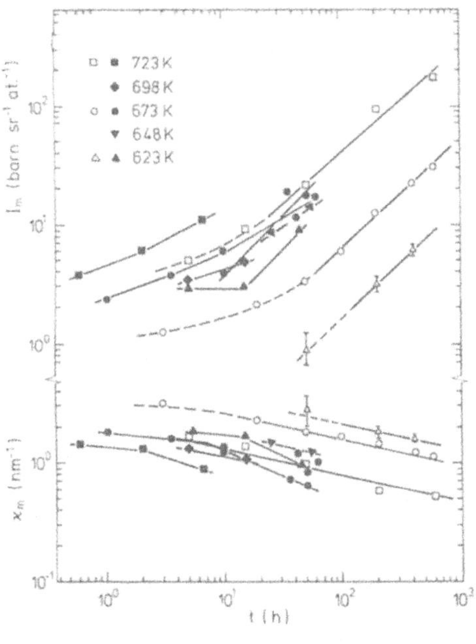

Fig. 3

Kinetics of the scaling parameters (peak intensity I_m and peak position κ_m) for thermal annealing (open symbols) and irradiation (filled symbols).

4 Discussion

In view of the kinetics parameters shown in Fig. 3 the time dependence of κ_m and I_m seems to be unchanged by the irradiation within the straggle of the data points: After a transient period at the beginning, the parameters follow the same power laws quoted earlier for the thermal decomposition [4]. However, a significant acceleration of the diffusion is obvious from the kinetics curves under irradiation, which are shifted to shorter times for equal temperatures. Further, the kinetics curves for irradiations at different temperatures have moved closer on the time scale compared to those of the thermal decomposition. This means that the effective activation enthalpy for the diffusion under irradiation is smaller than the thermal activation enthalpy. Indeed, preparing an Arrhenius-plot of the time shift necessary to superpose the kinetics curves of different temperatures, an activation enthalpy of 0.3 to 0.5 eV is obtained for the decomposition under irradiation as against 2.2 to 2.4 eV thermally. Both effects, acceleration and reduced effective activation enthalpy, are well-known phenomena for diffusion processes under irradiation, so called "Radiation-Enhanced Diffusion" (RED). They are explained by the defect balance considering the production of defect pairs, vacancies and interstitials, in excess to the thermally existing defects, and the irreversible annihilation at sinks and/or by mutual recombination [5].

If the influence of irradiation was just the diffusion enhancement, or more general, the modification of diffusion velocity, one would not expect any changes in the decomposition morphology nor in the general kinetics behaviour under irradiation. The kinetics, indeed, seems to be unchanged, however, the neutron scattering reveals a change in morphology: Comparing Fig. 1a with Fig. 1b there is an obvious difference in the shape of the structure function in the range of the intensity maximum around $\kappa = 1\ nm^{-1}$. After irradiation, the intensity maximum is less pronounced or even reduced to a steplike plateau. Further, at comparable heights of the maximum resp. the plateau, the step is located at lower κ after irradiation than the maximum after pure thermal treatment.

In the scaled scattering curves shown in Fig. 2, the difference between irradiation and non-irradiation is also obvious. Those scattering curves measured after irradiation could not be scaled to those after thermal ageing, although the structure functions for all irradiated samples do scale with each other as also do those for all nonirradiated ones. Comparing the scaled structure functions with the Gaussian, we see that after irradiation the structure functions follow the Gaussian for an extended range. The modified structure function can only be

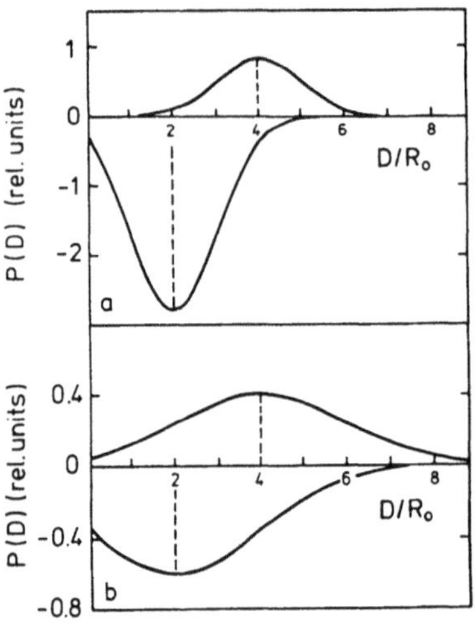

Fig. 4

Correlation function for the distance distributions between adjacent two phase regions, for unlike phases at $D = 2R_0$ (R_0 is the average precipitate radius) and for the next like phases at $D = 4R_0$, representing the arrangement establishing thermally (a) and under electron irradiation (b)

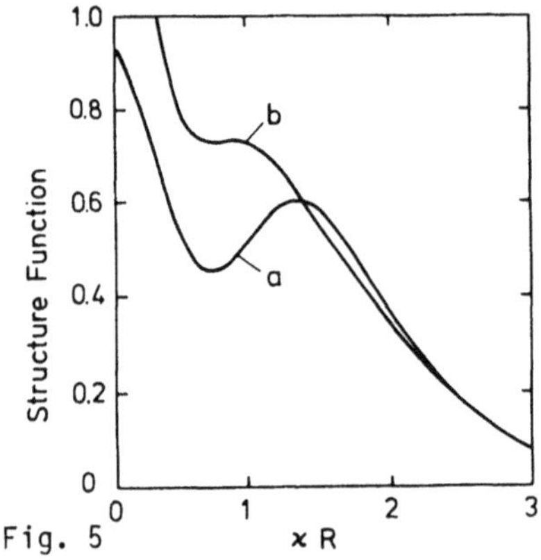

Fig. 5

Structure function, calculated for a dispersion of spheres with the spatial correlations functions of Fig. 4a (curve a) and Fig. 4b (curve b), representing the structure functions measured after thermal ageing (a) and ageing under irradiation (b).

explained by a radiation-induced modification of the decomposition morphology. The extended Gaussian shape as well as the observed "smearing" of the intensity maximum to a plateau indicate that under irradiation the regularity of the thermally establishing two-phase structure [6] is disturbed. This disturbance can be expressed as a broader distribution of the distances between adjacent two-phase regions.

The influence of a broadened distance distribution to the structure function is demonstrated by the model calculations shown in Figs. 4 and 5. For both cases a) and b), the pair correlation function P(D) (Fig. 4) consists of two Gaussians, one with a negative amplitude for adjacent unlike clusters, and the second one with a positive amplitude for the next like clusters at double distance. Just increasing the width of these distributions by a factor of 2 (with unchanged normalisation) modifies the resulting structure function from curve a) to curve b) in Fig. 5. Curve a) shows the characteristic shape of structure functions observed after thermal ageing, and curve b) after ageing under irradiation.

The number of jumps per atom required for transforming the structure characterized by Fig. 4a to that by Fig. 4b is estimated to be more than 10^2. This order of magnitude can by far not be provided by mechanisms like replacement collision sequences or inverse Kirkendall effect [7] with the total applied fluence of 10^{-2} dpa. The effect can only be caused by a difference in the diffusional atom transport mechanism. As the interstitials only can introduce a difference in the transport mechanisms for thermal and irradiation treatments, we attribute the observed effect to interstitialcy mass transport. The efficiency of this transport for the different alloy constituents must strongly depend on the local alloy composition.

Acknowledgement

We would like to thank Dr. Dorte Juul Jensen for her assistance during the SANS-measurements at RISØ.

References

1 R. Poerschke and H. Wollenberger: J. Phys. F: Metal Phys. 6, 27 (1976)
2 R. Poerschke, W. Wagner and H. Wollenberger: J. Phys. F. (1985), in press
3 P. Fratzl, J.L. Lebowitz, J. Marro and M.H. Kalos: Acta Metall. 31, 1849 (1983)
4 W. Wagner, R. Poerschke and H. Wollenberger: Decomposition in Alloys: The Early Stages, Proc. 2nd Acta-Scripta Met. Conf., Sonnenberg, Pergamon Press, New York, pp. 170ff
5 R. Sizmann: J. Nucl. Mater. 69 & 70, 386 (1968)
6 J. Piller, W. Wagner, H. Wollenberger and P. Mertens: Decomposition in Alloys: The Early Stages, Proc. 2nd Acta-Scripta Met. Conf., Sonnenberg, Pergamon Press, New York, pp. 156ff
7 P.R. Okamoto and L.G. Rehn: J. Nucl. Mater. 83, 2 (1979)

Index of Contributors

Beaven, P.A. 228
Beck, H. 155
Bellissent, R. 32
Bigot, J. 32
Binder, K. 12

Cadeville, M.C. 28,78
Calvayrac, Y. 32
Chieux, P. 42

Dahmani, C.E. 78
Degischer, H.P. 89,94
Desre, P. 42
Differt, K. 219
Dosch, H. 106
Driesen, G. 126

Eckerlebe, H. 66
Eckert, J. 170

Faux, D.A. 116
Flottmann, T. 134
Frisius, F. 228

George, B. 37
Gerold, V. 17
Goldstone, J.A. 170

Haubold, H.-G. 22
Hauck, J. 176
Heidemann, A. 134
Hein, W. 89
Hempelmann, R. 121,126, 170

Janot, C. 37
Jensen, D.J. 89

Kampmann, R. 66,73,228
Kern, J. 17
Kesternich, W. 197
Klamt, A. 165
Kostorz, G. 1

Lamparter, P. 49
Lefebvre, S. 32
Leroux, C. 78
Lyon, O. 99

Magerl, A. 161
Maier, K. 219
Martin, G. 225
McKergow, M.W. 116

Neumaier, K. 161

Parlinski, K. 176
Peisl, J. 106,181
Petry, W. 134
Pierron-Bohnes, V. 28,78
Piller, J. 94
Poerschke, R. 235

Quéré, Y. 188
Quivy, A. 32

Richter, D. 121,126,170
Ross, D.K. 116
Rowe, J.M. 170
Ruebenbauer, K. 139
Rush, J.J. 170

Schelten, J. 197
Schober, H.R. 149

Schroeder, H. 197
Schroeder, K. 143
Schwahn, D. 83,197
Schweika, W. 22
Schwerin, A.,von 106
Seeger, A. 219
Sepioł, B. 139
Simon, J.P. 99
Sinclair, R.N. 213
Sinha, S.K. 116
Solt, G. 155
Spooner, S. 197
Stewart, R.J. 56
Steyrer, G. 181
Stoneham, A.M. 149
Strecker, H. 89

Tonks, D. 170
Trost, W. 219

Ullmaier, H. 197

Vogl, G. 134

Wagner, R. 66,73,228
Wagner, W. 89,94,235
Wahi, R.P. 89
Weertman, J.R. 204
Wilson, D.L.T. 116
Windsor, C.G. 213
Wipf, H. 161
Wollenberger, H. 89, 94,235

Yavari, R. 42
Yoo, M.H. 83

B. Dorner

Coherent Inelastic Neutron Scattering in Lattice Dynamics

1982. 47 figures. VIII, 96 pages. (Springer Tracts in Modern Physics, Volume 93). ISBN 3-540-11049-6

The book gives an overview on the possibilities and the limitations of inelastic neutron scattering in lattice dynamics. Within aspects of experimental technique particular attention is payed to questions of resolution, intensity and focussing. Symmetry operations contained in the space groups of the samples play an important role for the performance of the experiment. These influences are explained on experimental grounds with examples avoiding heavy mathematics. The background of theoretical models to interprete the experimental results is given. The discussion of completed investigations which follows starts with the information one can obtain from a precise determination of phonon frequencies (peak position), continues with a qualitative intensity analysis of phonon peaks and leads to an extended description of the information extracted from a quantitative analysis of phonon intensities, i. e. the determination of phonon eigenvectors. The knowledge of eigenvectors gives a more microscopic insight in lattice dynamics than only the dispersion curves. An entire chapter is devoted to the investigation of anharmonic effects.
The discussion begins with a line broadening due to a reduced life time of the phonons. Soft mode and central peak phenomena as observed near structural phase transformation follow. Finally frequency-dependent damping producing a double peak for a one phonon response at He temperature is presented in detail.
All along the discussions are given on the basis of examples of investigated substances. The reader will benefit from a general insight in lattice dynamics and at the same time from a detailed interpretation of selected substances.

G. Leibfried, N. Breuer

Point Defects in Metals I

Introduction to the Theory

1978. 138 figures, 22 tables. XIV, 342 pages. (Springer Tracts in Modern Physics, Volume 81). ISBN 3-540-08375-8

GPSR Compliance
The European Union's (EU) General Product Safety Regulation (GPSR) is a set of rules that requires consumer products to be safe and our obligations to ensure this.

If you have any concerns about our products, you can contact us on

ProductSafety@springernature.com

In case Publisher is established outside the EU, the EU authorized representative is:

Springer Nature Customer Service Center GmbH
Europaplatz 3
69115 Heidelberg, Germany

www.ingramcontent.com/pod-product-compliance
Lightning Source LLC
Chambersburg PA
CBHW061205240926
51846CB00006B/1122
* 9 7 8 3 6 4 2 7 1 0 0 9 4 *